Regulation of Synthetic Biology

ELGAR STUDIES IN LAW AND REGULATION

Series Editor: Roger Brownsword, *Professor of Law, King's College London, UK*

Regulation is a ubiquitous concept in today's world. The proliferation of new technologies necessitates continual re-appraisal of the rules that govern them, whilst globalization increases the complexity of interaction between governance systems on a national, regional and international level. As a field of study, regulation continues to grow and evolve, both in consideration of specific sectors, and at a conceptual level, as the rationale and motivation for regulation are scrutinized.

This new and exciting series is an important forum for original works of scholarship that explore regulation and its interaction with the law. It favours work that has a critical, innovative or analytical perspective and, whilst having a primary focus on legal writing, welcomes approaches that draw on other disciplines. The scope of the series encompasses a wide range of regulatory fields, from biotechnology and ICT to machine learning and AI, from security, food, and environment to health, leisure, and employment. At the same time the series plays host to broader discussions on the nature of risk, governance models, the role of democracy, regulatory theory and many others.

The primary mission of the series is to stimulate the development of original thinking across law and regulation and to foster the best theoretical and empirical scholarship in the field.

Titles in the series include:

Gambling Regulation and Vulnerability
Malgorzata Anna Carran

Regulation of Synthetic Biology
BioBricks, Biopunks and Bioentrepreneurs
Alison McLennan

Regulation of Synthetic Biology

BioBricks, Biopunks and Bioentrepreneurs

Alison McLennan

Assistant Professor, School of Law and Justice, Faculty of Business, Government and Law, University of Canberra, Australia

ELGAR STUDIES IN LAW AND REGULATION

Edward Elgar
PUBLISHING

Cheltenham, UK • Northampton, MA, USA

Published by
Edward Elgar Publishing Limited
The Lypiatts
15 Lansdown Road
Cheltenham
Glos GL50 2JA
UK

Edward Elgar Publishing, Inc.
William Pratt House
9 Dewey Court
Northampton
Massachusetts 01060
USA

A catalogue record for this book
is available from the British Library

Library of Congress Control Number: 2017955176

This book is available electronically in the **Elgar**online
Law subject collection
DOI 10.4337/9781785369445

FSC www.fsc.org MIX Paper from responsible sources FSC® C013056

ISBN 978 1 78536 943 8 (cased)
ISBN 978 1 78536 944 5 (eBook)

Typeset by Columns Design XML Ltd, Reading
Printed and bound in Great Britain by TJ International Ltd, Padstow

Contents

Acknowledgements

Much of the research in this book was undertaken at the ANU College of Law, Australian National University. I am grateful to Timothy Bonyhady for his guidance during this project. I have learned a great deal from him and have been given many opportunities to develop as a researcher and teacher. I am also grateful for the guidance of Barbara van Leeuwen and Anton Wasson. Many members of the ANU College of Law offered advice and comments on my work, including Don Anton, Peter Cane, Tom Faunce, Judy Jones, Hitoshi Nasu, Matthew Rimmer and Daniel Stewart. I am grateful for their generosity and their helpful feedback.

I also wish to thank the two Deans of the ANU College of Law during my time at the College, Stephen Bottomley and Michael Coper, for creating a wonderful culture in which to work. I also wish to thank colleagues Leighton McDonald and Mark Nolan for their support for this project. I am grateful for the friendship and encouragement of colleagues including Elinor Jean, Brad Jessup, Rebecca Monson, Mary Spiers-Williams, Fanny Thornton and Ruth Townsend. I also received helpful feedback from Janet Hope during my time at the ANU. I am also grateful to Beth Beckmann and Inger Mewburn for providing excellent training courses for ANU researchers.

Work on this book has also been undertaken at the Faculty of Business, Government and Law, University of Canberra, Australia. I would like to thank the Dean, Lawrence Pratchett, the two Heads of School during this time, Lorana Bartels and Maree Sainsbury, and the Associate Dean for Research, David Carter, for their guidance, encouragement, and support for this project. I have received helpful advice from colleagues including Ben Authors, Patricia Easteal, Susan Priest, Ben Sheehy and Fanny Thornton. Research assistance by Ivana Damjanovic and Zebo Nasirova has been valuable. I have also benefited from the University of Canberra's excellent researcher development program, led by Melanie Haines.

I am grateful to the scholars who allowed me to visit their research centres during this project, including Jane Calvert, Jeremy de Beer, Michael Geist, Ysolde Gendreau, Hank T Greely, Yann Joly, Ian Kerr, Bartha Maria Knoppers, Graeme Laurie, Stephen M Maurer, Aurora

Plomer and Kenneth A Oye. I have benefited from discussions with many scientists during this research, particularly Robert Carlson, Drew Endy, Michael Fero and Nathan Hillson. I greatly appreciate Drew Endy's openness to engaging with early career researchers and scholars from disciplines outside the life sciences.

In developing this book, I have benefited from insightful feedback from Roger Brownsword, LeRoy C Paddock, and anonymous reviewers. I have greatly appreciated editorial guidance and assistance provided by Laura Mann.

I am deeply appreciative of the support and assistance which my parents, Margaret McCulloch and Dennis McLennan, have provided throughout this process. Lastly thank you to my wonderful partner Sean Flynn for his love, good humour and encouragement.

Abbreviations

Asilomar Conference	Asilomar Conference on Recombinant DNA Molecules
BBSRC	Biotechnology and Biosciences Research Council
The BIOFAB	International Open Facility Advancing Biotechnology
bp	base pair
BPA	BioBrick Public Agreement
cDNA	Complementary DNA
The Code of Conduct	the *IASB Code of Conduct for Best Practices in Gene Synthesis*
The Commission	the Presidential Commission for the Study of Bioethical Issues (or, the Presidential Commission)
Coordinated Framework	US Coordinated Framework for the Regulation of Biotechnology
CRISPR	CRISPR-Cas 9. CRISPR stands for 'clustered, regularly interspaced, short palindromic repeat'
DARPA	Defense Advanced Research Projects Agency
DNA	Deoxyribonucleic acid
DOE	Department of Energy
EBRC	Engineering Biology Research Consortium
ELSI	Ethical, legal and social issues
EPA	Environmental Protection Agency
EPSRC	Engineering and Physical Sciences Research Council
FBI	Federal Bureau of Investigation
Gates Foundation	Bill and Melinda Gates Foundation
GPL	The General Public License

the Guidance	*Screening Framework Guidance for Providers of Synthetic Double-Stranded DNA*
the House Committee	United States House of Representatives Committee on Energy and Commerce
IASB	International Association Synthetic Biology
iGEM	International Genetically Engineered Machine Competition
IGSC	International Gene Synthesis Consortium
JBEI	Joint Bioenergy Institute
kb	Kilobase
MIT	Massachusetts Institute of Technology
NEST	New and Emerging Science and Technology
NIH	National Institutes of Health
NIH Guidelines	NIH Guidelines for Research Involving Recombinant or Synthetic Nucleic Acid Molecules
NSABB	National Science Advisory Board for Biosecurity
NSF	National Science Foundation
PCR	polymerase chain reaction
the Presidential Commission	The Presidential Commission for the Study of Bioethical Issues (or, the Commission)
the Protocol	Harmonized Screening Protocol: Gene Sequence and Customer Screening to Promote Biosecurity
the Registry	the Registry of Standard Biological Parts
RNA	ribonucleic acid
SynBERC	Synthetic Biology Engineering Research Centre
USDA	United States Department of Agriculture
US FDA	United States Food and Drug Administration
USPTO	United States Patents and Trademark Office
Venter Institute	J Craig Venter Institute
WHO	World Health Organization
Wyss Institute	Wyss Institute for Biologically Inspired Engineering

PART I

The new world of synthetic biology: possibilities, challenges and debates

1. Introduction

1.1 INTRODUCTION

On 20 May 2010, scientists in the United States announced that they had made the world's 'first self-replicating synthetic bacterial cell'.[1] This achievement is the most famous outcome of the emerging field of synthetic biology.

Synthetic biology is a burgeoning field which draws on molecular biotechnology, biochemical engineering, genomics and information technology to create new tools and techniques. It offers to transform how we tackle our most difficult global problems, through applications such as greener energy and synthetic vaccines. However, it also poses risks such as damage to the environment, bioterrorism, and harm arising from 'do-it-yourself biology'. As leading biologist George Church has pointed out, 'For all the benefits it promises, synthetic biology is potentially more dangerous than chemical or nuclear weaponry, since the organisms can self-replicate, spread rapidly throughout the world, and mutate or evolve on their own.'[2] Vast amounts of public and private funding are being invested in synthetic biology research, with this field being identified by *Technology Review* as one of the top ten breakthrough technologies most likely to change the world.[3]

The accelerating pace of scientific research and uncertainties about the risks and benefits create complex challenges for regulation. What should the law try to achieve when new technologies emerge? Does a new technology require a new regulatory approach? How can the law possibly keep up with the science? How should scientists be involved in developing regulation and policy? Who should be able to access the technology?

---

1 Daniel G Gibson et al, 'Creation of a Bacterial Cell Controlled by a Chemically Synthesized Genome' (2010) **329**(5987) *Science* 52.

2 George Church and Ed Regis, *Regenesis: How Synthetic Biology will Reinvent Nature and Ourselves* (Basic Books, 2013), 231.

3 Katherine Bourzac, 'Synthetic Cells' (2011) **114**(3) *Technology Review*, 56.

How can regulation be used to promote innovation that benefits us all, while protecting us from harm?

This book explores the interplay between law and emerging technologies by analysing these questions in the synthetic biology context. As synthetic biology promises transformative benefits, but also poses risks of harm, these are critical issues to address. They are questions that matter not only to scientists and lawyers. They are also being grappled with by governments around the world, the biotechnology industry, environmentalists, bioethicists, philosophers, sociologists, and the public as end-users such as patients.

Synthetic biology aims to create novel organisms and biological systems which do not exist in nature, and to redesign natural organisms in substantial ways.[4] This field is revolutionary in the new tools and techniques it offers and in the scale at which it allows us to experiment with biology. We are now able to modify organisms to a much greater extent than was possible with traditional genetic engineering. Scientists have moved from modifying one or several genes to being able to make large-scale changes to genomes, or even design and build completely new synthetic genomes. Synthetic biology is part of a new industrial revolution, 'the information-genomics revolution'.[5]

Synthetic biologists aim to make biology 'easy to engineer', to make organisms more like machines.[6] 'Synthetic biology' is really a collection of approaches to this goal: building biological machines from bricks of deoxyribonucleic acid (DNA), building human-made genomes and cells from scratch, and using large-scale genetic engineering to make new products. These approaches are supported by the emergence of technologies that allow us to write and build DNA to our own designs.

During the short history of synthetic biology, there have been two moral panics arising from new developments in the field. Both developments related to research of the J Craig Venter Institute (Venter Institute). In 2006, the Venter Institute identified the smallest, most streamlined set of genes required for a particular bacterium to survive in a laboratory:

---

4 See, e.g., Synthetic Biology Research Community, *FAQ – What is Synthetic Biology?* <http://syntheticbiology.org/FAQ.html>.

5 Church and Regis (n 2) 152, 210.

6 See Drew Endy, 'Synthetic Biology: Can We Make Biology Easy to Engineer?' (2008) Winter *Industrial Biotechnology* 340; ibid., 181–2.

'the minimal bacterial genome'.[7] This was followed by a patent application, and a related application for 'synthetic genomes'. A panic ensued regarding whether such things could rightly be owned, and whether these patents might hinder scientific research.[8]

The second panic occurred following the 2010 announcement that a 'self-replicating synthetic bacterial cell' had been created at the Venter Institute. This was publicised prominently and followed by explosive media coverage that triggered a wave of policy discussion in the US and around the world.[9] The announcement alerted people to what could become possible with synthetic biology. Many would not have heard of synthetic biology prior to 2010, but the synthetic cell announcement brought the field and its implications sharply into focus.

This announcement was game-changing for debates not only about synthetic biology, but also about regulation of the biological sciences generally. In the United States, President Barack Obama joined the ensuing debate. His Presidential Commission for the Study of Bioethical Issues (the Commission) was tasked with investigating 'the implications of this scientific milestone, as well as other advances that may lie ahead in this field of research'.[10] It was directed to consider 'the potential medical, environmental, security, and other benefits of this field of research', in addition to considering its risks.[11] As Obama stated, 'it is

---

[7] John I Glass et al, 'Essential Genes of a Minimal Bacterium' (2006) **103**(2) *Proceedings of the National Academy of Sciences of the United States of America* 425.

[8] See, e.g., ETC Group, 'Extreme Genetic Engineering: An Introduction to Synthetic Biology' (ETC Group, January 2007) <http://www.etcgroup.org/upload/publication/602/01/synbioreportweb.pdf>, 51; ETC Group, 'Venter Institute Builds Longest Sequence of Synthetic DNA (that Doesn't Work)' (News Release, 24 January 2008); Jocelyn Kaiser, 'Attempt to Patent Artificial Organism Draws a Protest' (2007) **316**(15 June 2007) *Science* 1557.

[9] See, e.g., Ewen Callaway and Andy Coghlan, 'The Meaning of Artificial Life' (2010) **206**(2762) *New Scientist* 6; Staff Writer, 'And Man Made Life: Artificial Life, the Stuff of Dreams and Nightmares, Has Arrived' (2010) **395**(8638) *The Economist* 11.

[10] The Presidential Commission for the Study of Bioethical Issues, 'New Directions: The Ethics of Synthetic Biology and Emerging Technologies' (The Presidential Commission for the Study of Bioethical Issues, December 2010) <http://bioethics.gov/sites/default/files/PCSBI-Synthetic-Biology-Report-12.16.10_0.pdf>, vi.

[11] Ibid.

vital that we as a society consider, in a thoughtful manner, the significance of this kind of scientific development'.[12]

These two episodes have highlighted the issues to be explored in this book: the controversy over the risks and benefits of synthetic biology and the regulatory challenges created by this technology as it emerges. By understanding these challenges, we can move towards developing law that protects us from synthetic biology's risks, while capturing its potential to improve our world.

1.2 THIS BOOK'S PROJECT

This book lays the groundwork for bespoke regulation of synthetic biology. It considers to what extent the technology and its social context are distinctive from those of earlier technologies. It examines the regulatory issues that arise from the emergence of this new field, with a view to identifying the key regulatory challenges and exploring the extent to which these are distinctive. It analyses whether synthetic biology can be regulated within existing regulatory structures or whether new mechanisms are needed.

While new technologies emerge into an environment where there are existing regimes developed to regulate earlier technologies, these regimes should not merely be 'mechanically transplanted' to apply to a new technology.[13] In synthetic biology's case, the existing regimes address genetic engineering and other previous biotechnologies. Of course, we should recognise 'common regulatory challenges', which arise for emerging technologies generally.[14] Nonetheless, when a new technology like synthetic biology emerges, it will be necessary to consider the application of existing regulation and whether this is sufficient to address the new technology, including its impact on health, safety and the environment.[15] While not wishing to 'reinvent the regulatory wheel', it is important to

---

[12] Ibid.

[13] Roger Brownsword, 'So What Does the World Need Now? Reflections on Regulating Technologies' in Roger Brownsword and Karen Yeung (eds), *Regulating Technologies: Legal Futures, Regulatory Frames, and Technological Fixes* (2008) 23, 32.

[14] Ibid., 30–32.

[15] Roger Brownsword and Morag Goodwin, *Law and the Technologies of the Twenty-First Century: Text and Materials* (Cambridge University Press, 2012) 301, 312–16.

'understand whether particular technologies present their own distinctive problems'.[16] This involves asking whether new or specific regulation is required.[17]

The main criteria to consider are legitimacy of the regulatory regime, its effectiveness, its 'prudence' or response to risk, its 'connection' or ability to keep up with the science, and its 'cosmopolitanism' or ability to deal with differing values and approaches internationally.[18] These criteria can be used to identify the regulatory 'hot spots' or 'pressure points on regulators'.[19] I will draw on them to assess the regulatory challenges posed by synthetic biology, the adequacy of existing regulatory regimes, and potential solutions to the challenges. This discussion will focus significantly on regulatory effectiveness and connection. These are the main areas of regulatory challenge for synthetic biology at this stage in its development. The pursuit of procedural legitimacy through public consultation and discussion is also a key consideration in this work.

This discussion will draw primarily on legal analysis, but also on science policy, sociology of science, and the science itself. Considering the regulatory challenges posed by a technology also requires considering whether the technology *itself* is distinctive in some way.[20] Accordingly, this book critically examines the complex and shifting science of synthetic biology, the emergence of the field, its key drivers and its potential impacts. An understanding of how synthetic biology departs from other scientific fields will contribute to developing a picture of the regulatory challenges it poses.

The US has been a focal point of both scientific research efforts and policy discussions and is the focus of much of the discussion in this book. A number of large reports into synthetic biology have been released in various jurisdictions including the US, the UK and the EU.[21]

---

16 Roger Brownsword, *Rights, Regulation and the Technological Revolution* (Oxford University Press, 2008), 27; Brownsword (n 13) 30–32.

17 Brownsword (n 13) 31.

18 Ibid., 31–2.

19 Ibid.; Brownsword and Goodwin (n 15) 109–66, 316; Roger Brownsword and Han Somsen, 'Law, Innovation and Technology: Before We Fast Forward – A Forum for Debate' (2009) **1**(1) *Law, Innovation and Technology* 11–12, 31–7.

20 Brownsword (n 13) 31–2.

21 For example, Andrew Balmer and Paul Martin, 'Synthetic Biology: Social and Ethical Challenges, An Independent Review Commissioned by the Biotechnology and Biological Sciences Research Council' (Biotechnology and Biological Sciences Research Council, 2008) <http://www.bbsrc.ac.uk/organisation/policies/reviews/scientific_areas/0806_synthetic_biology.pdf>; European Group on Ethics in Science and New Technologies to the European Commission, 'Ethical

The Presidential Commission's report is the most thorough and detailed of these considerations and I will analyse and critique its recommendations in depth.

Two key scientific figures appear repeatedly throughout the story of synthetic biology told in this book; those of J Craig Venter and Drew Endy. They are both leading scientists in the field of synthetic biology. Venter has been a visionary and divisive figure in biotechnology for several decades. He is in favour of the aggressive use of the patent system and light-handed governmental intervention in synthetic biology. Drew Endy has been at the helm of the engineering approach to biology since the early 2000s and has founded key synthetic biology institutions to be discussed in this book. Endy has taken a leading role in community development and engagement with the legal and ethical issues raised by synthetic biology. He is a critic of intellectual property and has strived to introduce arrangements for sharing in synthetic biology. These two figures are important in the story of synthetic biology because of the differing approaches to regulation and intellectual property they represent, and their key role in shaping not only the science of synthetic biology but also the regulatory debate surrounding it.

This book forms part of a wider effort to investigate, understand and respond to the issues raised by emerging technologies. In addition to legal issues, there are also issues of politics, justice and public consultation. There are issues regarding promotion of innovation and technology transfer. There are also ethical issues in relation to manipulating and patenting living things, and philosophical issues regarding whether synthetic, human-made organisms are actually 'life', and how they will impact upon our understanding of ourselves. There are issues regarding the impact of synthetic biology-based industry on traditional methods of production, and impact on land use and livelihoods in the developing world.[22]

Aspects of Synthetic Biology: Proceedings of the Roundtable Debate' (European Commission, 19 May 2009) <http://ec.europa.eu/bepa/european-group-ethics/docs/publications/round_table_ethical_aspects_of_synthetic_biology.pdf>; The Royal Society and Organisation for Economic Co-operation and Development, 'Symposium on Opportunities and Challenges in the Emerging Field of Synthetic Biology: Synthesis Report' (The Royal Society, Organisation for Economic Co-operation and Development, 2010) <http://www.oecd.org/science/biotech/45144066.pdf>.

[22] See ETC Group, 'The New Biomassters: Synthetic Biology and the Next Assault on Biodiversity and Livelihoods' (ETC Group, 2010) <http://www.etcgroup.org/upload/publication/pdf_file/biomassters_27feb2011.pdf>; ETC Group

Some existing scholarship in relation to synthetic biology has focused on these ethical, sociological and political perspectives, with less discussion of legal issues. Legal discussion has tended to consider the individual regulatory issues in isolation. This book extends the boundaries of discussion of synthetic biology to include greater consideration of regulatory issues and explore the extent to which the regulatory challenges posed by synthetic biology are distinctive.

1.3 STARTING POINTS – THE ROLE OF REGULATION OF EMERGING TECHNOLOGIES

Answering the questions posed in this book requires, as a starting point, considering what the role of regulation should be in relation to emerging technologies. I have begun from the position that regulators and regulation properly have a dual role: both in maximising the benefit to the public from emerging technology and minimising its potential harms.[23] That is, the role of regulation encompasses managing of issues of risk associated with a new technology, but also facilitating societally beneficial innovation.[24] This means that 'It falls to politicians and regulators, and ultimately to the law, to set the limits of technological innovation, to co-ordinate the assessment and management of risk, to design procedures for public participation, and to set the terms of compensatory responsibility.'[25]

Regulation thus has a role in both constraining and enabling the development and use of the technology. In relation to regulation's facilitative role, it 'falls to regulators and the law to establish a governance environment that is supportive of desirable technological innovation and that ensures the benefits are fairly shared'.[26] The goal is for 'law and technology … to work together to improve the basic conditions of human social existence'.[27] To do this, we need a regulatory environment that

et al, 'The Principles for the Oversight of Synthetic Biology' (ETC Group, 13 March 2012) <http://libcloud.s3.amazonaws.com/93/ae/9/2287/1/Principles_for_the_oversight_of_synthetic_biology.pdf>; Rachel Wellhausen and Gautam Mukunda, 'Aspects of the Political Economy of Development and Synthetic Biology' (2009) **3** *Systems and Synthetic Biology* 115.

[23] Following Brownsword and Somsen (n 19), 2–3.

[24] Ibid.

[25] Ibid., 2.

[26] Ibid.

[27] Ibid., 3.

promotes beneficial innovation, the deployment of such technologies and benefit-sharing, while also allowing us to manage risks.[28]

This conception of the role of regulation in relation to emerging technologies is drawn from the work of regulatory scholars Roger Brownsword, Han Somsen and Morag Goodwin. It is based upon the idea that, in democracies, the development of science and technology occurs with a social licence – a sort of permission from society – which is subject to constraints imposed by protection of human rights and human dignity.[29] The Commission also pointed out the responsibility of scientists to society:

> Society as a whole has a stake in what scientists and engineers do. In turn, scientists and engineers should recognize the potential impact of their research on those who will experience both its benefits and burdens and their responsibility to those who provide the means, directly or indirectly, for their research.[30]

In considering regulating for societal benefit through technological innovation, I have also taken inspiration from the work of regulatory scholar Hailemichael Demissie. Demissie has proposed that for nanotechnology regulation, the focus on risk management needs to be supplemented with a distinct discourse and regulatory jurisdiction for *benefit* management.[31] Discussion of emerging technologies can tend to focus on risk, to the detriment of consideration of benefit issues.[32] Taking 'regulating for the positives'[33] as part of the role of regulation in relation to emerging science is preferable because it acknowledges that emerging technologies present potentially transformative *benefits* as well as risks.

For these reasons, the role of regulation of synthetic biology should be considered more broadly than as management of risk and setting of limits on scientific development. The discussion in this book is aimed at promoting the beneficial development of the synthetic biology at the

---

28 Ibid.; see also Brownsword and Goodwin (n 15) 46.

29 Brownsword and Somsen (n 19) 2.

30 The Presidential Commission for the Study of Bioethical Issues (n 10) 141.

31 See, e.g., Hailemichael Teshome Demissie, 'Is Beneficent Regulation the New Better Regulation? Nano-Regulation in the Wake of the "New Better Regulation" Movement' (2010) **2**(1) *Law, Innovation and Technology* 115, 137–42.

32 See, e.g., ibid., 137–41.

33 Brownsword and Somsen (n 19) 49.

research stage, by looking into the issues of intellectual property, ownership, and sharing of information.

Reflecting the dual goals of regulation, I will divide the discussion of particular regulatory challenges posed by synthetic biology into regulation for the risks (Part II) and regulation for the benefits (Part III).[34]

As a further starting point, it is necessary to consider the role and aims of science in a liberal democratic society. That is, why do we 'do' science? A key role of science is to improve the public's well-being and happiness[35] through the pursuit and generation of knowledge of our natural world, development of beneficial technologies, and economic stimulation. While recognising that not *all* scientific developments turn out to have publicly beneficial uses, and that some can be put to dangerous use or misused in potentially dangerous ways, my starting point is that the *overarching* role and objective of science in a democracy is the improvement of collective well-being and happiness. This is a societal view of the benefit to be obtained from scientific innovation. Indeed, 'the regulation of biotechnology should be guided by a concern for the common good'.[36]

In the specific case of synthetic biology, generation of beneficial applications is a clear, stated objective. The BioBricks Foundation, a leading organisation of scientists within this field, states that: 'When people have the tools and infrastructure to work with one another, we can meet global needs for food, medicines, shelter, clean water and air. To achieve this better future, BioBricks Foundation supports the open and ethical development of biotechnology that benefits all peoples and the planet.'[37]

Of course, there will not always be agreement as to what is desirable in terms of science and technological innovation. The notions of 'public

---

34 In a similar fashion to Brownsword and Somsen, ibid.

35 The use of the term 'happiness' is inspired by Demissie's discussion of 'beneficence' as a regulatory virtue. Drawing on the philosophy of Imanuelle Kant, Demissie explains that beneficence is a broader concept than 'charity' or 'humanity' because it encompasses providing for the 'happiness' of others as well as their basic 'wellbeing'. See Hailemichael Teshome Demissie, 'Justice or Beneficence: What Regulatory Virtue for Nano-Governance?' (2011) **2**(3) *European Journal of Law and Technology* 1, 8.

36 D J Galligan, 'Citizen's Rights and Participation in the Regulation of Biotechnology' in Francesco Francioni (ed), *Biotechnologies and International Human Rights* (Hart Publishing Ltd, 2007) 342, 344.

37 The BioBricks Foundation, *The BioBricks Foundation* (2016) <http://www.biobricks.org/index.php>.

interest' and 'public benefit' are highly contested. Democratic decision-making and wide public engagement in decision-making are important in such debates.[38]

1.4 ASSESSING SYNTHETIC BIOLOGY'S REGULATORY ENVIRONMENT

This section will explain the criteria that can be used to assess the adequacy of synthetic biology's existing regulatory environment: regulatory legitimacy, effectiveness, prudence (which can be seen as an aspect of effectiveness), connection and cosmopolitanism.

Initially, it is necessary to consider the meaning of terms relating to 'regulation'. In this book, I will use 'regulation' to refer to interventions that are put in place by regulators to 'control and channel conduct in the desired way'.[39] 'Regulators' are government agencies authorised to engage in such regulating activities.[40] Interventions which are not put in place by agencies of government, but by industry or scientists, will be described as 'self-regulation'. I will use the term 'regulatory environment' as a shorthand to mean both of these sets of interventions, including the applicable legislative regime, government regulatory institutions and voluntary mechanisms.

I will use 'social and scientific context' as a shorthand to indicate all the surrounding circumstances into which the regulation is introduced, including – in the case of an emerging technology – the relevant science, the regulatees and other interested parties, the nature of the industry, the relevant debates and ethical or other disagreements, uncertainties, risks and potential benefits.

1.4.1 Regulatory Legitimacy

What does it mean to create regulation that is 'legitimate'? The concept of regulatory legitimacy has two limbs. The first is legitimacy of regulatory purpose; that is, whether regulators are doing 'the right kind of thing'.[41] This is a question of the ethical appropriateness or justifiability

---

[38] As emphasised by Galligan in relation to the importance of elected decision-makers, broad consultation and public participation in biotechnology regulation generally: Galligan (n 36) 342–3.

[39] Brownsword (n 16) 7.

[40] See the sense in which 'regulators' is used in ibid.

[41] Brownsword and Somsen (n 19) 11.

of the regulatory action.[42] Secondly, legitimacy involves considering the 'regulatory means'; are the regulators 'going about their business in the right kind of way?'[43] For example, a regulatory intervention that creates concerns about autonomy or privacy may not be legitimate.[44]

Regulatory legitimacy also has a procedural element.[45] Regulators should be able to show that they are trying in good faith to implement regulatory regimes that accord with the community's moral commitments.[46] Procedural legitimacy requires regulators to be transparent, accountable and inclusive.[47] Decision-making should be done by elected representatives, and must also be well-informed and take account of all perspectives.[48]

Public participation is an essential part of procedural legitimacy.[49] There should be opportunities for those who are interested to express their views and decision-making should be responsive to those views.[50] Indeed, intellectual property and emerging technology law scholar Gregory N Mandel has pointed out the importance of public trust and confidence in regulation of emerging technologies generally.[51] He has emphasised the crucial role of broad stakeholder engagement in developing effective regulation in this context.

In the story of synthetic biology, important questions have arisen as to who should be involved in debates about the limits of synthetic biology research, its regulation and oversight. Further, to what extent should ethical and philosophical perspectives be part of the regulatory debate, along with scientific information? 'External accountability' is critical in this discussion.[52] That is, synthetic biology research efforts need to be

---

[42] Brownsword and Goodwin (n 15) 51.
[43] Brownsword and Somsen (n 19) 11.
[44] Brownsword and Goodwin (n 15) 60.
[45] Roger Brownsword, *Rights, Regulation and the Technological Revolution* (n 16) 179.
[46] Ibid., 127.
[47] Brownsword and Somsen (n 19) 17–19.
[48] Galligan (n 36) 342–3.
[49] Brownsword and Goodwin (n 15) 48.
[50] Brownsword, *Rights, Regulation and the Technological Revolution* (n 16) 128.
[51] Gregory N Mandel, 'Regulating Emerging Technologies' (2009) **1** *Law, Innovation & Technology* 75, 90.
[52] Joy Y Zhang et al, 'The Transnational Governance of Synthetic Biology: Scientific Uncertainty, Cross-borderness and the "Art" of Governance' (BIOS Working Paper No: 4, BIOS (Centre for the Study of Bioscience, Biomedicine,

accountable across all the various disciplines which make up synthetic biology, and outside science to 'all those who may be affected'.[53]

This notion of external accountability raises particularly difficult questions where those doing the research are not 'traditional' scientists, but 'backyard' or 'do-it-yourself' biologists. This book will investigate the risks, benefits and oversight issues that arise from the emergence of such unique communities.

This book considers regulation of a technology at a relatively early stage in its development, when few regulatory interventions have been introduced specifically for synthetic biology. Thus procedural legitimacy is the major form of legitimacy discussed. Issues of controversy about the risks and benefits of synthetic biology, who is consulted about its regulation and who is able to influence its development are key themes in this discussion. Some see procedure as important in managing a plurality of views, for example the Chair of the Commission, Amy Gutmann, espoused an approach of democratic deliberation for bioethical issues.[54] This book will explore the application of this approach in synthetic biology regulation.

1.4.2 Regulatory Effectiveness

Assessing regulatory effectiveness involves asking 'whether the regulation is making a difference'; that is, whether it is 'achieving the intended regulatory effects'.[55] This is most helpfully seen as a matter of degree, rather than categorically defining approaches as effective or ineffective.[56] Effectiveness is an assessment that is made when regulation is already in place; that is, the relevant standards have been set and implemented.[57]

Assessing regulatory effectiveness raises issues such as whether enough action has been taken to address risks, whether the regulation has

---

Biotechnology and Society), May 2011) <http://www2.lse.ac.uk/BIOS/publications/Working%20Papers/WP4TransnationalGovernance.pdf>, 4.

[53] Ibid.

[54] Amy Gutmann, 'The Ethics of Synthetic Biology: Guiding Principles for Emerging Technologies' (2011) **41**(4) *Hastings Center Report* 17, 19–21; Brownsword and Goodwin (n 15) 58–9. See also Amy Gutmann and Dennis Thompson, 'Deliberating about Bioethics' (1997) **27**(3) *The Hastings Center Report* 38.

[55] Brownsword and Goodwin (n 15) 277.

[56] Brownsword, *Rights, Regulation and the Technological Revolution* (n 16) 135.

[57] Brownsword and Somsen (n 19) 11.

clarity, whether the regulation is being complied with, whether there is resistance from regulatees, whether regulatory overreach has occurred, and whether problematic unintended consequences of the regulation have developed.[58] Ineffective regulation can be due to problems relating to regulators, problems relating to regulatees, and disruptive externalities.[59]

(a) Regulator failure

The potential failures by regulators include regulatory corruption, regulatory capture and regulatory competence.[60] Insufficient regulatory competence is the main area of regulator failure discussed in this book. This can occur where there has been inaction in addressing a new technology.[61] For example, in relation to nanotechnology, it has been argued that the existing regulations were not sufficient due to the way materials were defined, and the regime failed to adequately take into account the different properties of materials on the nano-scale.[62] Thus, there were significant gaps in regulation in relation to nanotechnologies.[63] The regulators did not take sufficient proactive measures to manage the risks posed by nanomaterials.[64]

Similar kinds of regulatory effectiveness are considered in this book in relation to synthetic biology. I have investigated whether existing regulatory mechanisms, designed for regulating earlier technologies, are rendered ineffective by their application to synthetic biology. That is, I have queried whether synthetic biology's emergence creates gaps and deficiencies in the existing regulation. Where there are deficiencies, I have considered the questions of how these gaps ought to be filled, and whether something new and specific to synthetic biology is needed.[65] Where I identify that new intervention is required, my aim is to identify

---

58 Ibid., 24–5; Brownsword, *Rights, Regulation and the Technological Revolution* (n 16) 134–45, 148–9; Brownsword and Goodwin (n 15) 290–95, 301, 312–16.

59 Brownsword and Somsen (n 19) 23.

60 Regulatory 'corruption' occurs when there is a lack of integrity and problems arise such as bribery, ignoring non-compliance or issuing improper permissions. The related problem of 'regulatory capture' arises where regulatees have 'undue influence' in relation to regulatory standard-setting, or monitoring and compliance: Brownsword and Goodwin (n 15) 297–300.

61 Ibid., 296, 301.

62 Ibid., 301, 313.

63 Ibid., 301, 315.

64 Ibid., 296, 301, 313–16.

65 Ibid., 315.

the reasons why synthetic biology necessitates a distinctive approach, and move towards a response to the unique regulatory challenge.

(b) Unintended consequences of regulation

Regulatory effectiveness can also be diminished by unintended consequences of the regulation.[66] The unintended consequences discussed in this book arise from intellectual property law's interaction with emerging technologies. The intellectual property system, in particular the patent system, is designed to incentivise innovation. In theory, the grant of monopoly property rights in an invention provides a reward for the inventive effort and capital that has been expended, in turn creating an incentive to innovate.[67] In return for the legal monopoly, the patent holder cannot keep their invention secret; it must be fully and publicly disclosed (though the patent holder can charge fees for its use).[68] Thus the invention becomes part of general human knowledge.[69] By creating a reward for innovating in response to market demands, the patent system is intended to form a 'self-correcting innovation policy'.[70] In this way, the patent system is underpinned by the goal of promoting the progress of science and technology.[71]

Patent law is thus designed to have a facilitative role in fields such as genetics, where the 'working assumption' is that the possibility of being rewarded for innovation with a patent right creates an incentive to invest in research.[72] However, in some areas of technology it has been argued that the patent system can actually impede innovation, contrary to this facilitative role.[73] This is an unintended negative consequence of the

---

[66] Ibid.

[67] James Boyle, *The Public Domain: Enclosing the Commons of the Mind* (Yale University Press, 2008), 5–7; Hector MacQueen et al, *Contemporary Intellectual Property: Law and Policy* (Oxford University Press, 2008), 10–11.

[68] Boyle (n 67) 6; MacQueen et al (n 67) 366.

[69] Boyle, ibid.

[70] Ibid., 5–8.

[71] Indeed, the legislative goal for enactment of the provisions of the patent system in the US is found in the Constitution, which grants Congress power to 'promote the progress of science and useful arts, by securing for limited times to authors and inventors the exclusive right to their respective writings and discoveries': *Constitution of the United States* Art I § 8.

[72] Brownsword, *Rights, Regulation and the Technological Revolution* (n 16) 20.

[73] See, e.g., in relation to synthetic biology: Arti Rai and James Boyle, 'Synthetic Biology: Caught Between Property Rights, the Public Domain and the Commons' (2007) **5**(3) *PLoS Biology* 0389.

regulatory intervention.[74] In the context of gene patents, concern about the impact of such negative consequences on both researchers and patients has led to high-stakes litigation. In light of these issues, regulators should take an interest in how the patent system affects new technologies, both in principle and in practice.

That is, patent law should be designed, interpreted and applied in a way which facilitates rather than impedes scientific innovation. If the intellectual property system was not working in this way, but was instead hindering innovation, there would be regulatory ineffectiveness. The design or application of the patent system would need to be changed, in order to achieve its goals. These are big-ticket questions in the synthetic biology context, where huge amounts of money are being poured into research and revolutionary beneficial applications are being promised. The ownership and control of these applications will influence the trajectory of development of the field and its ability to make a positive difference in the world.

An alternative to seeking ownership of technological innovations is to share these freely with others. In the synthetic biology world, some scientists have created their own 'commons' for sharing of data, tools and ideas.[75] These scientists value openness of scientific research, collaboration and community. The tension between 'open' and 'closed' research is pervasive in debates around synthetic biology, and is a recurring theme throughout this book. In addition to exploring the operations of this synthetic biology commons, this book will engage with the disagreement about the extent to which synthetic biology research should be open to all, particularly in the context of biosafety and biosecurity risks.

(c) Regulatee resistance

Regulatory failure can also arise where regulatees resist the regulatory intervention.[76] This may be for economic reasons – such as the cost of compliance compared to the potential costs of being found in breach of obligations – or there may be cultural reasons.[77] Matters of legitimacy and effectiveness can be linked.[78] For example, regulatees may not

---

74 See Brownsword and Goodwin (n 15) 290–95.

75 See Boyle (n 67) 176–8; Sapna Kumar and Arti Rai, 'Synthetic Biology: The Intellectual Property Puzzle' (2007) **85** *Texas Law Review* 1745, 1763–7; Rai and Boyle (n 73) 0391–2.

76 Brownsword and Somsen (n 19) 24–5.

77 Brownsword, *Rights, Regulation and the Technological Revolution* (n 16) 137; Brownsword and Goodwin (n 15) 318–19.

78 Brownsword and Somsen (n 19) 11.

comply with regulation because they do not view it as legitimate.[79] Commentary on information technology regulation has proposed that regulation is more likely to work if regulators seek to generate consensus with stakeholders (including regulatees, such as those who might have the ability to resist regulation).[80]

In a related vein, the influential 'responsive regulation' theory of legal academic and economist Ian Ayres and sociologist of law John Braithwaite proposes industry self-regulation as a starting point, with increasingly interventionist measures if this is not implemented.[81] Under this model, regulators should seek a cooperative model in which they try to 'persuade' regulatees, in order to encourage effective self-regulation and compliance.[82] The threat of more intrusive government action and escalating sanctions if regulated firms 'defect' is said to promote compliance.[83]

Similarly, the theory of 'smart regulation' – developed by regulatory scholar Neil Gunningham, political scientist and regulatory scholar Peter Grabosky and environmental scientist and regulation specialist Darren Sinclair – proposes that 'less interventionist' measures should be used initially.[84] The regulatory action can be escalated as needed to achieve the policy goals.[85] In smart regulation, escalation can occur across multiple instruments and these instruments may be implemented by government, regulatees or third parties.[86]

However, there are objections to self-regulatory approaches in terms of effectiveness and legitimacy, as regulatees may seek self-interested regulation.[87] While smart regulation has worked well in environmental protection, some of its principles have 'very limited application' to

---

79 Ibid.

80 Stuart Biegel, *Beyond Our Control?: Confronting the Limits of Our Legal System in the Age of Cyberspace* (The MIT Press, 2001), 221–5, 357–61.

81 Ian Ayres and John Braithwaite, *Responsive Regulation: Transcending the Deregulation Debate* (Oxford University Press, 1992), 35–41.

82 Ibid., 35–51.

83 Ibid., 35–41.

84 See Neil Gunningham et al, *Smart Regulation: Designing Environmental Policy* (Clarendon Press, 1998) 377, 391–408.

85 Ibid.

86 See ibid., 399–400.

87 Brownsword, *Rights, Regulation and the Technological Revolution* (n 16) 153–5.

biotechnology.[88] Starting with less interventionist approaches would be less appropriate here, where the consequences of regulatory failure could potentially be catastrophic.[89] Also, smart regulation assumes that there is a level of agreement on the regulatory goals, but there are often fundamental disagreements about this in the biotechnology sphere.[90] The story of synthetic biology regulation, including attempts at self-regulation, highlights such disagreements and related challenges.

(d) Regulatory prudence

Emerging technologies create difficulties for regulation because there is uncertainty both about the risks they pose and their potential benefits.[91] 'Regulatory prudence' refers to 'prudential assessment of risk and benefit'.[92] It involves making sure risks are at an 'acceptable level', which can be seen as an aspect of regulatory effectiveness.[93]

Decisions made on the basis of an assessment of risk and benefit should be 'provisional' so that regulatory instruments can be modified as the technology, and public response to it, develops.[94] Further, prudent regulators should engage with the public before taking regulatory action.[95] Regulators need to assess public views and preferences regarding risks associated with new technologies, but this can be challenging.[96] The considerations of democratic deliberation are therefore important here.[97]

What approaches are then open to regulators? Regulatory strategies for emerging technologies are 'permissive, promotional, precautionary or preventive', depending upon their impact upon the technology.[98] 'Promotional' policies increase the spread of the technology; those which slow

---

[88] Neil Gunningham, 'Regulating Biotechnology: Lessons from Environmental Policy' in Hans Somsen (ed.), *The Regulatory Challenge of Biotechnology: Human Genetics, Food and Patents* (Edward Elgar Publishing, 2007) 3, 6–8.

[89] Ibid., 7.

[90] Ibid., 7–8.

[91] Brownsword and Goodwin (n 15) 113.

[92] Ibid., 114.

[93] Ibid., 316.

[94] Ibid., 136.

[95] Ibid., 166.

[96] Ibid., 120, 136.

[97] See Gutmann (n 54) 19–21; ibid., 58–9.

[98] Jennifer Kuzma and Todd Tanji, 'Unpackaging Synthetic Biology: Identification of Oversight Policy Problems and Options' (2010) **4**(1) *Regulation & Governance* 92, 106.

down the spread are 'precautionary', those which halt the development of the technology are 'preventive' and those which are neutral in relation to spread of the technology are 'permissive'.[99]

To choose an approach, regulators need information about the likelihood that a technology could cause harm or be abused, and possible options for preventing this.[100] Generally, the more information there is available, the more likelihood there is of devising the 'right kind' of regulatory approach.[101] However, in the case of relatively 'immature' technologies, such as synthetic biology, the information available is not adequate.[102] Where there is uncertainty, such as scientific uncertainty about the likelihood of harm, the type of harm or its severity, these analyses are more complex. Thus managing scientific uncertainty is a key aspect of the challenge of responding to risk.[103]

In relation to synthetic biology, there are novel risks and a high degree of uncertainty. Because synthetic biology falls within a category of unpredictable and still emerging technologies of which novelty is a key feature, the uncertainty surrounding the risks and benefits of the technology is pervasive.[104] The nature and extent of potential harms, as well as their likelihood, are often unclear. As the field is progressing at a very rapid pace, the directions in which it will develop and the time frames for developments to occur are also uncertain.

Tools such as the precautionary principle have been proposed to manage scientific uncertainty. The precautionary principle can be defined as 'a principle of public decision-making that requires decision-makers in cases where there are "threats" of environmental or health harm not to use "lack of full scientific certainty" as a reason for not taking measures to prevent such harm'.[105] There are criticisms of the precautionary principle, some of which are valid. However, because of the substantial uncertainty surrounding synthetic biology, the principle has utility. It can be applied in ways which need not halt the development of this new field.

---

[99] Ibid.

[100] Brownsword and Somsen (n 19) 39–40.

[101] Ibid.

[102] Ibid.

[103] Roger Brownsword, 'Responsible Regulation: Prudence, Precaution and Stewardship' (2011) **62**(5) *Northern Ireland Legal Quarterly* 573, 582, 588–91.

[104] Brownsword and Goodwin (n 15) 113, 129.

[105] Elizabeth Fisher et al, 'Implementing the Precautionary Principle: Perspectives and Prospects' in Elizabeth Fisher, Judith Jones and Rene von Schomberg (eds), *Implementing the Precautionary Principle: Perspectives and Prospects* (Edward Elgar Publishing, 2006) 1, 6.

This book will explore precaution and other tools for managing uncertainty such as 'adaptive management' and 'planned adaptation'.[106]

1.4.3 Regulatory Connection

Because modern technological development moves faster than regulatory development and uses of technology change over time, regulation can become 'disconnected' from the technology.[107] Regulators need to be proactive about understanding the risks posed by a technology.[108] They also need to foster debate regarding the ethical and social issues that arise.[109]

While regulators may try to incorporate flexibility to allow the regime to keep pace with technology, the goal of flexibility will always be in tension with the need for clear and consistent regulation.[110] If regulatees become unsure of how regulatory instruments will apply to them, the regime becomes ineffective.[111]

Mandel has proposed that connection requires governance to follow an adaptive and iterative model.[112] This can occur through a 'continuing cycle' of regulation, data gathering on the risks associated with the technology, evaluation and modification of the regime.[113] In this model, Mandel recommends that regulatory agencies be proactive in addressing the gaps that new technologies can illuminate or exacerbate. In this book, I consider approaches such as planned adaptation, which can be used to promote regulatory connection in relation to the fast-changing science of synthetic biology.[114]

---

106 See, e.g., Judith Jones, 'Regulatory Design for Scientific Uncertainty: Acknowledging the Diversity of Approaches in Environmental Regulation and Public Administration' (2007) **19**(3) *Journal of Environmental Law* 347; Lawrence C McCray et al, 'Planned Adaptation in Risk Regulation: An Initial Survey of US Environmental, Health, and Safety Regulation' (2010) **77**(6) *Technological Forecasting & Social Change* 951.

107 Brownsword and Goodwin (n 15) 372.

108 Ibid.

109 Ibid.

110 Brownsword and Somsen (n 19) 29.

111 Ibid; Brownsword, *Rights, Regulation and the Technological Revolution* (n 16) 160.

112 Mandel (n 35) 83.

113 Ibid., 89.

114 See McCray et al (n 106); Gautam Mukunda et al, 'What Rough Beast? Synthetic Biology, Uncertainty, and the Future of Biosecurity' (2009) **28**(2) *Politics and the Life Sciences* 2.

1.4.4 Regulatory Cosmopolitanism

'Regulatory cosmopolitanism' refers to the challenge of developing transnational regulation in relation to emerging technologies, which may evoke differing attitudes across jurisdictions.[115] Cosmopolitanism involves seeking to find common values across nations, but also to respecting differences.[116] The challenge of addressing a plurality of perspectives and ethical positions across national jurisdictions is largely beyond the scope of this book. However, it is important to bear in mind the international nature of the field of synthetic biology research, the synthetic biology industry, and the regulatory issues posed. These considerations will be touched on in relation to the major risks posed by synthetic biology, particularly the risk of environmental harm and loss of biodiversity. The development of transnational regulation is a significant challenge confronting synthetic biology, and a key avenue for further investigation.

1.5 OVERVIEW OF THIS BOOK

The remaining three chapters in Part I will set the scene for the in-depth analysis of specific regulatory issues to follow. To start with, the discussion of the regulation of synthetic biology must be grounded in an understanding of the science. Chapter 2 introduces the science of synthetic biology and explores the development of the field, the leading scientists, the location and funding of research and its key applications in biofuels, medicine, agriculture and environmental remediation.

Chapter 3 analyses key policy positions in the debate over regulation of synthetic biology, represented by Venter, Endy, sociologists Joy Y Zhang, Claire Marris and Nikolas Rose, and the anti-technology civil society organisation ETC Group.

Chapter 4 considers the role and recommendations of the Presidential Commission's inquiry into synthetic biology. The recommendations, based on the principles of 'public beneficence', 'democratic deliberation', 'intellectual freedom and responsibility', 'responsible stewardship' and 'justice and fairness', are critiqued.

In Part II, Chapters 5–7 consider issues of risk and uncertainty. These are pervasive in debates about synthetic biology. Chapter 5 focuses on the

---

[115] Brownsword and Somsen (n 19) 11; Brownsword, *Rights, Regulation and the Technological Revolution* (n 16) 185–211.

[116] Brownsword and Somsen (n 19) 31–2; see also Brownsword, ibid.

potential for synthetic biology to cause environmental damage. It considers strategies for managing risks to the environment in light of legal and policy approaches to regulating where there is scientific uncertainty. I critique the Commission's approach to issues of risk and uncertainty, which is based on the principle of 'responsible stewardship' and the concept of 'prudent vigilance'.

Chapter 6 examines the potential biosafety issues arising from synthetic biology. It focuses on the emerging community of do-it-yourself biologists, their culture and motivations to elucidate the regulatory challenges posed.

In Chapter 7, I examine regulatory approaches to synthetic biology's security risks with reference to new scientific capabilities, the existing regulatory environment and self-regulation. I consider the debate that has been occurring in relation to regulation, and oversight efforts including the existing legislation, voluntary guidelines for oversight of gene synthesis, self-regulation by the synthetic biology community and voluntary standards developed by gene synthesis companies.

In Part III, Chapters 8 and 9 explore regulation for the benefits of synthetic biology, both within and outside the patent system. Chapter 8 investigates the interaction between synthetic biology and the patent system, in the context of the longstanding debate over patenting and the life sciences. The interpretation and expansion of patentable subject matter in relation to biotechnology is one of the key areas of controversy in this debate.[117] The chapter considers the likelihood of patents having an unintended negative impact on innovation in synthetic biology.

Chapter 9 investigates the ability of a unique synthetic biology 'commons', built by the BioBricks Foundation and its collaborators, to promote innovation in the field.[118] This commons involves sharing of research information and data, as well as shared norms and goals. Chapter 9 considers why this approach has been adopted by some synthetic biologists, how it is governed and how successful this project is likely to be in achieving its goals.

Together, these chapters will demonstrate that synthetic biology involves a revolutionary set of technologies that present regulatory challenges in the areas of environmental risk, biosafety, biosecurity and

---

[117] See Graham Dutfield, 'Patent Law, Biotechnologies and Language in Subject Matter Expansionism' in Matthew Rimmer and Alison McLennan (eds), *Intellectual Property and Emerging Technologies: The New Biology* (Edward Elgar Publishing, 2012) 117.

[118] See Boyle (n 67) 176–8; Kumar and Rai (n 75) 1763–7; Rai and Boyle (n 73) 0391–2.

intellectual property. They will highlight that the regulatory challenges posed by synthetic biology are distinctive from those posed by earlier technologies. The challenges arise from the distinctiveness of the science – including the use of engineering principles and the development of new tools – the novel communities of scientists and amateur scientists, and the uncertainty surrounding the risks. In Chapter 10, I will bring together the key regulatory changes needed and highlight critical issues for future consideration. I will consider the creation of a new institution for enhanced regulation of synthetic biology, to manage risk and promote desirable innovation.

2. Making biology easy to engineer: the science of synthetic biology, the emergence of the field and its major applications*

2.1 INTRODUCTION

Synthetic biologist Bonnie Bassler explains that

> Synthetic biologists promise to take the foundational knowledge acquired by traditional biologists and, using principles from engineering, refine, excel and apply it at scales that could achieve unprecedented good for the public. We now have the hope of efficiently applying biological solutions to some of the grandest problems facing the world.[1]

Synthetic biology refers to 'the design and fabrication of biological components and systems that do not already exist in the natural world', and 'the re-design and fabrication of existing biological systems'.[2] Synthetic biology 'aims to apply standardized engineering techniques to biology and thereby create organisms or biological systems with novel or specialized functions to address countless needs'.[3]

Synthetic biology combines several existing scientific disciplines including molecular biotechnology, chemical engineering, genomics (the

* The phrase 'Making biology easy to engineer' is taken from the following article: Drew Endy, 'Synthetic Biology: Can We Make Biology Easy to Engineer?' (2008) Winter *Industrial Biotechnology* 340.

1 Bonnie Bassler, 'Evidence to The Presidential Commission for the Study of Bioethical Issues, Synthetic Biology' (8 July 2010) Washington DC.

2 Synthetic Biology Research Community, *FAQ – What is Synthetic Biology?* <http://syntheticbiology.org/FAQ.html>.

3 The Presidential Commission for the Study of Bioethical Issues, 'New Directions: The Ethics of Synthetic Biology and Emerging Technologies' (The Presidential Commission for the Study of Bioethical Issues, December 2010) <http://bioethics.gov/sites/default/files/PCSBI-Synthetic-Biology-Report-12.16.10_0.pdf>, 2.

study of whole genomes) and information technology. Leading synthetic biologist Drew Endy has suggested that synthetic biology research is driven by four groups: biologists, chemists, 're-writers' (those who aim to 're-write' natural genes in order to optimise their useful functions) and engineers.[4] Synthetic biology research also includes bioinformatics – design and use of computer software programs – in addition to biological research.

The previous genetic technology, classical recombinant DNA technology, aimed to manipulate the genes of various organisms, for example by adding a gene to confer a new property on an organism. Synthetic biology attempts to build new organisms with functions that might not exist in nature.[5] While classical recombinant DNA technology can be seen as 'softening unpleasant edges' of existing organisms, 'synthetic biology creates life forms which are meant not to have any edges from the start', representing a shift from *manipulating* organisms to actually *creating* new organisms.[6] In this way, synthetic biologists aim to 'create pre-programmed cellular robots that can be used as biosensors, or self-sufficient factories in which to create new biofuels, drugs, and chemicals'.[7] Synthetic biology pioneer Venter suggests that one of the key advantages offered over classical biotechnology is 'that there is no need to have access to a physical supply of a particular DNA sequence … fragments are simply created de novo by chemical synthesis and assembled into entire chromosomes and organisms'.[8]

By contrast, anti-technology organisation the ETC Group describes synthetic biology as 'extreme genetic engineering' and is concerned about the dangers it presents.[9] They warn that 'Today, scientists aren't

---

[4] Drew Endy, 'Foundations for Engineering Biology' (2005) **438**(7067) *Nature* 449, 449.

[5] Joachim Boldt and Oliver Müller, 'Newtons of the Leaves of Grass: Certain Ethical Implications of Synthetic Biology Research Go Beyond Those of Genetic Engineering' (2008) **26**(4) *Nature Biotechnology* 387, 387.

[6] Ibid., 387–8 (emphasis added).

[7] Vid S Mohan-Ram and J M Waxman, 'Synthetic Biology Patent Applications Expected to Present New Challenges' (2008) **2**(12) *Life Sciences Law & Industry Report* 1, 1.

[8] Prepared Statement of J Craig Venter, 'Evidence to United States House of Representatives Committee on Energy and Commerce, Effects of Developments in Synthetic Genomics and Implications for Energy and Health' (27 May 2010) Washington DC.

[9] ETC Group, 'Extreme Genetic Engineering: An Introduction to Synthetic Biology' (ETC Group, January 2007) <http://www.etcgroup.org/upload/publication/602/01/synbioreportweb.pdf>.

just mapping genomes and manipulating genes, they're building life from scratch – and they're doing it in the absence of societal debate and regulatory oversight.'[10]

This chapter explains that three types of biological engineering come within the term 'synthetic biology', and explores how synthetic biology has developed as a field. It examines the applications of synthetic biology as well as mapping where synthetic biology research is occurring and how it is being funded. This discussion aims to give sufficient information about the field to provide a useful backdrop for the discussion in later chapters of debates about synthetic biology and the regulatory issues it raises.

2.2 TYPES OF SYNTHETIC BIOLOGY

Leading synthetic biologist George Church suggests that while the term synthetic biology had been used as early as 1912, it has been used since 2004 'to provide an intentional new feel to the enterprise, to distance ourselves from previous fields like genetic engineering, and to encourage a new rigor around the dozen or so fundamental engineering principles'.[11] This desire of the field of synthetic biology to differentiate itself from genetic engineering – possibly for the reasons of wishing to minimise 'public controversy' and external regulation – has also been suggested in science and technology studies literature.[12]

The term 'synthetic biology' can mean a number of different things, and does not have an agreed meaning even among relevant scientists. Often synthetic biology conferences and workshops include a discussion of what the term 'synthetic biology' actually means.

Scientists and commentators have suggested that there are several types of synthetic biology, and various typologies have been proffered. For example, patent examiner Berthold Rutz has argued that there are three approaches to synthetic biology in the literature:

- altering existing organisms to achieve useful functions – such as producing useful chemicals – using engineering practices like standardisation,

---

[10] Ibid., 1.

[11] George Church and Ed Regis, *Regenesis: How Synthetic Biology will Reinvent Nature and Ourselves* (Basic Books, 2013), 163.

[12] Maureen A O'Malley et al, 'Knowledge-making Distinctions in Synthetic Biology' (2008) **30**(1) *BioEssays* 57, 62.

- using small 'chemical building blocks' to synthesise genes or whole genomes, which can be put into cells to create a 'synthetic organism', and
- reducing 'the complexity of naturally existing organisms by deleting "superfluous" genes', in order to produce a 'streamlined microorganism' which has a 'minimal genome' and is highly efficient in performing a particular function.[13]

Church has suggested that since the intentional adoption of the term 'synthetic biology', the field has diverged into different camps.[14] These include 'genome engineering', 'recombinant DNA applied to metabolic engineering', 'engineering-inspired' biological research, and study of evolution.[15] Philosophy of science expert Maureen O'Malley and colleagues have identified three categories of synthetic biology: construction of devices from DNA components; engineering of natural organisms at the whole genome level; and 'protocell creation'.[16] 'Protocell creation' aims to build synthetic or semi-synthetic cells.[17]

Thus it is clear that synthetic biology refers to a collection of approaches within biological research. This is in contrast to earlier biotechnologies such as human genetic tests or embryonic stem cells, where only one type of technology is encompassed. It is important to understand the multifaceted nature of synthetic biology because this impacts upon the regulatory challenges that arise. For example, the question of how the law on patentable subject matter applies to synthetic biology is a more complex question than for a single technology like an embryonic stem cell line. Answering the question requires breaking synthetic biology down into its component approaches. Through these different approaches, greater changes to natural organisms are made possible than with previous technologies. This also impacts upon the nature of the regulatory challenges raised by synthetic biology, particularly in the areas of environmental risk and biosecurity risk.

I use the term 'synthetic biology' as an umbrella term for an emerging field comprising several different but interlinked approaches to biological

---

[13] Berthold Rutz, 'Synthetic Biology and Patents: A European Perspective' (2009) **10**(Special Issue) *EMBO reports* S14, SS14–15.

[14] Church and Regis (n 11) 163.

[15] Ibid.

[16] O'Malley et al (n 12) 57–9.

[17] Ibid., 59.

engineering. I call these the 'construction approach', the 'parts agenda',[18] and 'metabolic engineering'. The construction approach involves attempting to build human-made genomes and cells, as well as minimal bacterial genomes and cells. A minimal genome is one that has been pared back to the basic genes required for survival of the organism in the laboratory.[19] The parts agenda refers to creating 'standard biological parts' from small pieces of DNA, which can be combined to create new biological devices and systems. Metabolic engineering refers to altering whole biological processes within organisms. These approaches are interconnected, and underpinned by common enabling technologies, including DNA sequencing, synthesis and assembly techniques, DNA 'editing' tools for making changes to genetic information, and software for designing constructs of DNA and modelling biological systems.

Rather than being in competition to be seen as the 'real synthetic biology', these different approaches and their leading scientists interact in a mutually accepting and cooperative way within the broader field of 'synthetic biology'. There is collaboration between some of the key scientists, both in academic research and in establishment of biotechnology companies. There is some overlap between approaches, for example it is possible that metabolic engineering in synthetic biology could make use of standardised biological components or standard biological parts.

2.2.1 Construction Approach

In the construction branch of synthetic biology, scientists are trying to build entirely synthetic organisms. Most famous among such projects are the endeavours of the Venter Institute to build a synthetic bacterial cell. Venter has described the goals of this approach as to explore the questions: 'What is life? Can we pare it down to its most basic components? Can we digitize it?'[20]

The construction approach includes efforts to construct large pieces of DNA and whole genomes. The construction approach also includes

---

[18] This term has been used by Joachim Henkel and Stephen M Maurer, see, e.g., Joachim Henkel and Stephen M Maurer, 'Parts, Property and Sharing' (2009) **27**(12) *Nature Biotechnology* 1095.

[19] See John I Glass et al, 'Essential Genes of a Minimal Bacterium' (2006) **103**(2) *Proceedings of the National Academy of Sciences of the United States of America* 425.

[20] J Craig Venter, 'Designing and Synthesising Life' (Speech delivered at Ted Talks, 2008) http://www.youtube.com/watch?v=nKZ-GjSaqgo.

efforts to generate a minimal organism for use as a 'basic platform' to which genes to express new functions or traits can be added.[21] It also includes designing and assembling synthetic, human-made DNA in the laboratory. Scientists at the Venter Institute also aim to use these techniques to develop methods of biofuel production.

2.2.2 Parts Agenda

In synthetic biology's parts agenda, traditional engineering principles are being combined with molecular and cell biology in order to reduce the complexity of biological engineering and increase its predictability and reliability.[22] These engineering principles include standardisation, abstraction and decoupling. Endy argued in 2005 that there was a need to develop the foundational technologies of synthetic biology in order to promote progress in the field, using these engineering principles.[23]

In relation to standardisation, Endy noted that this has occurred in other fields of biology such as in measuring the performance of genetic sequences, as well as in science and engineering more broadly.[24] However he suggested that the lack of standards in some basic aspects of biology creates large costs for researchers.[25]

In the parts agenda, standardisation is achieved in two ways: through the development, definition and distribution of standard biological parts themselves, and through the development of technical standards for use in parts research. A 'standard biological' is a genetic unit (usually of DNA) with well-characterised properties that acts as a 'functional unit'.[26] It encodes a particular biological function.[27] These parts are intended to operate much like LEGO^{R} bricks. As described by synthetic biologist Christina Smolke, these 'very simple functions' can be added together to form larger devices.[28] She explains that the aim is to have a 'very large

---

21 Boldt and Muller (n 5) 387.

22 See Endy (n 4).

23 See ibid.

24 Ibid., 450–51.

25 Ibid.

26 Registry of Standard Biological Parts, *Help: About the Registry* <http://partsregistry.org/Help:About_the_Registry>.

27 Synthetic Biology Engineering Research Centre, *Parts* (2013) SynBERC <http://synberc.org/content/parts>.

28 Christina Smolke, 'Evidence to The United States National Academies, The Organisation for Economic Co-operation and Development, and The Royal Society, Opportunities and Challenges in the Emerging Field of Synthetic Biology' (9 July 2009) Washington DC, 138–9.

library of refined parts that you can plug and play into the device platforms that you build'.[29] Although there are many challenges associated with standardisation of biological parts, standards are seen as very important by synthetic biologists working in the parts agenda.[30]

Systematic design in synthetic biology uses the concepts of biological parts, 'devices' and 'systems'.[31] These are objects of increasing complexity, with a standard biological part being the most basic unit used in synthetic biology design.[32] A combination of one or more standard biological parts can form a 'device' with more complex function than a part.[33] Because of standardisation, such devices would not need to be created anew every time, but could be assembled from standard parts as in other areas of engineering.[34] Parts and devices are designed to be modular, so that they can be assembled to create synthetic biological 'systems' which perform specific tasks.[35]

'Decoupling' is 'the idea that it is useful to separate a complicated problem into many simpler problems that can be worked on independently, such that the resulting work can eventually be combined to produce a functioning whole'.[36] For example, in a building project different tasks would be separated including architecture, construction and project management.[37] In synthetic biology, the design of biological systems is being decoupled from their construction.[38] This is due to increasing capacity to synthesise DNA.[39] This means that construction of DNA can be a task conducted separately from the researcher, by DNA synthesis companies.

'Abstraction' is the principle of creating separate levels of complexity, so that expertise across all levels is not required to work at a particular

---

29 Ibid., 142.

30 See Adam Arkin, 'Setting the Standard in Synthetic Biology' (2008) **26**(7) *Nature Biotechnology* 771.

31 Geoff Baldwin et al, *Synthetic Biology – A Primer* (Imperial College Press, 2012), 20.

32 Synthetic Biology Engineering Research Centre, *Parts* (n 27).

33 Ibid; Synthetic Biology Engineering Research Center, *Devices* (2013) SynBERC <http://synberc.org/content/devices>; Registry of Standard Biological Parts, *Parts* Registry of Standard Biological Parts <http://partsregistry.org/Help:Parts>.

34 Baldwin et al (n 31) 19–20.

35 Ibid., 20, 51–2.

36 Endy (n 4) 451.

37 Ibid.

38 Ibid.

39 Ibid.

level.[40] The hierarchy of parts, devices and systems is an example of an abstraction hierarchy.[41] In practice it means, for example, that experts on the design of biological systems need not be experts on the function and characteristics of biological parts.[42]

Church gives a comprehensive list of 'fundamental engineering principles' as they apply to the engineering of biology. These principles include: 'the use of interoperable parts' such as standard biological parts, 'standards (e.g., protocols for connecting one biological part to the next)', 'hierarchical functional designs', 'computer-aided design' (e.g., DNA design software), 'cost-effectiveness (e.g., choosing organisms that replicate quickly and do not require lots of expensive nutrients, land, energy, or other scarce resources)', 'physical isolation' (e.g., laboratory biosafety procedures), 'redundant systems' (e.g., multiple biosafety mechanisms), 'ruggedness and safety testing' (e.g., testing before environmental release), 'user licensing' (e.g., registration of laboratory workers), 'user surveillance' (e.g., reporting requirements), 'functional integration', and 'evolution'.[43]

Synthetic biologists expressly aim to make biological engineering techniques more accessible, reducing the requirement for large amounts of 'tacit knowledge' and skill gained through years of experience in laboratories.[44] This is the goal of 'de-skilling' biology.[45] One of the goals of the parts agenda is to make standardised biological parts that can be easily assembled to make small machines that work in predictable ways. Synthetic biologists are trying to achieve this de-skilling by applying the principles of engineering, such as standardisation and modularity, to biological systems.[46]

A challenge for this field of synthetic biology is to move towards greater use of synthetic DNA circuits in practical applications.[47]

---

40 Baldwin et al (n 31) 51–2; Endy (n 4) 451–2.

41 Baldwin et al ibid.

42 Ibid.

43 Church and Regis (n 11) 161–2.

44 Gautam Mukunda et al, 'What Rough Beast? Synthetic Biology, Uncertainty, and the Future of Biosecurity' (2009) **28**(2) *Politics and the Life Sciences* 2, 14.

45 Ibid.

46 Ibid.

47 D Ewen Cameron et al, 'A Brief History of Synthetic Biology' (2014) **12**(5) *Nature Reviews Microbiology* 381, 388–99.

2.2.3 Metabolic Engineering

Metabolic engineering is the 'redirection of metabolism inside cells to produce a specific chemical or to degrade an environmental contaminant'.[48] This approach to the engineering of biology involves manipulating the 'metabolic pathways' of microorganisms. A metabolic pathway is the process by which a cell produces a particular substance from another substance. Such a pathway is 'composed of a set of these enzymes working together to sequentially convert one compound in the cell, through a series of selective chemical transformations, to a different compound'.[49] These metabolic processes within microbial cells can be used to convert a 'feedstock' such as plant biomass or sugar into a useful product such as a liquid biofuel.

These scientists are using a 'top down' approach to look at whole metabolic processes in microorganisms and identify ways in which these could be usefully altered with the techniques of synthetic biology. Through manipulating these metabolic processes, scientists can make them more efficient or able to produce different useful products from those produced by natural processes.[50] Alternatively, manipulation of a metabolic process might be used to create an organism which produces a product with more desirable characteristics than that produced through the natural process.[51]

This type of synthetic biology also involves searching for bacterial metabolic systems useful for biofuels production by using 'comparative genomics':

> You can look through algae and cyanobacteria for those that make trace amounts … of the alkanes by taking fatty acids, reducing and decarbonating the aldehydes. To find those genes, you can look at the genomes that produced and those that didn't produce these trace amounts and then you can identify the gene and overproduce them.[52]

Leading synthetic biologist Jay Keasling has given this take on the development of synthetic biology through metabolic engineering:

---

[48] Keasling Lab, *Metabolic Engineering* Keasling Lab, Lawrence Berkeley National Laboratory <http://keaslinglab.lbl.gov/research_areas/index.html>.

[49] LS9, *Technology Overview* (2011) LS9 <http://ls9.com/technology/technology-overview>.

[50] Baldwin et al (n 31) 111–13.

[51] Ibid., 113.

[52] George Church, 'Evidence to The Presidential Commission for the Study of Bioethical Issues, Synthetic Biology' (8 July 2010) Washington DC.

> When I started my career in 1992 … biological components were relatively crude, making the engineering of microorganisms time-consuming and costly. But my colleagues and I had the idea that one could engineer microorganisms into chemical factories that produce nearly any important chemical from sugar. Unfortunately, there were few tools available to us at the time. So we began by developing tools to accurately produce a chemical of interest by controlling the expression of genes that had been transferred into cells. At that time, there was no name for what we were doing, but now it is known as synthetic biology.[53]

At the University of California, Berkeley, Keasling and colleagues have used metabolic engineering to engineer a strain of yeast to be able to synthesise a precursor to the anti-malarial drug artemisinin.[54] This is usually extracted from a plant, but it is not readily available or affordable.[55] The researchers used a metabolic pathway from yeast, but constructed it synthetically within the bacterium *E. coli* (commonly used in laboratory work), using biological components.[56] They adapted the metabolic pathway to this different environment of *E. coli*.[57] They then engineered the pathway for optimal function within *E. coli*, before transferring it back into yeast.[58] The yeast containing the optimised pathway was then able to produce the precursor to artemisinin, artemisinic acid, in large quantities in a cost effective way.[59] The technology was licensed to the drug company Sanofi for production of 150 million doses.[60] This is a leading example of success in developing useful applications with synthetic biology.

---

[53] Jay D Keasling, 'Evidence to United States House of Representatives Committee on Energy and Commerce, Developments in Synthetic Genomics and Implications for Health and Energy', 27 May 2010.

[54] Amyris, *Artemesinin* (2013) Amyris <http://www.amyris.com/Products/176/Artemisinin>; Jay D Keasling, 'Synthetic Biology for Synthetic Chemistry' (2008) **3**(1) *ACS Chemical Biology* 64; Ahmad S Khalil and James J Collins, 'Synthetic Biology: Applications Come of Age' (2010) **11**(5) *Nature Reviews Genetics* 367, 372; C J Paddon et al, 'High-level Semi-synthetic Production of the Potent Antimalarial Artemisinin' (2013) **496**(7446) *Nature* 528.

[55] Keasling (n 54) 70; Priscilla E M Purnick and Ron Weiss, 'The Second Wave of Synthetic Biology: From Modules to Systems' (2009) **10** *Nature Reviews Molecular Cell Biology* 410, 415.

[56] Paddon et al (n 54); Keasling (n 54) 70–72; Khalil and Collins (n 54) 372.

[57] Purnick and Weiss (n 55).

[58] Khalil and Collins (n 54) 372.

[59] Ibid.

[60] Amyris (n 54).

Microorganisms such as *E. coli* and yeast continue to be the key organisms used as chassis, but research into the use of mammalian cells as chassis is also occurring.

2.3 THE DEVELOPMENT OF SYNTHETIC BIOLOGY

The emergence of synthetic biology has been catalysed by key scientists, and by development of a range of enabling technologies. The enabling technologies include DNA sequencing, DNA synthesis, DNA assembly techniques and standard methods for assembling DNA, and openly accessible online databases of DNA sequences.[61] Tools for genome editing are also becoming key enabling technologies.

2.3.1 Key Scientists in Synthetic Biology

I have identified key figures in the emergence and development of synthetic biology as including molecular biologists Venter and Church, biological engineer Endy and biochemist Keasling. They have been influential in the development of synthetic biology research, and Venter, Endy and Church in particular have been involved in discussion of its regulatory challenges. I explore aspects of their approaches to these challenges in later chapters.

Venter is a prominent figure in the development of biotechnology generally. He has developed new techniques in molecular biology and genetic sequencing.[62] After a career in biochemistry, Venter moved into human genetics and researched at the National Institutes of Health.[63] After leaving the public sector, he established genomics research institutes using private funding, and led the privately funded effort to sequence the human genome.[64] Venter then established the Venter Institute. He runs the Institute and the company Synthetic Genomics Inc., which commercialises its research.

---

61 Baldwin et al (n 31) 42–8.

62 See e.g., development of expressed sequence tags for genetic sequencing: J Craig Venter, *A Life Decoded, My Genome: My Life* (Viking, 2007), 121–7.

63 See ibid., 130–38, 171–2.

64 Ibid., 153–88, 255–332; see also Bruce Alberts and Sir Aaron Klug, 'The Human Genome Itself Must be Freely Available to all Humankind' (2000) **404**(23 March) *Nature* 324; Robert Cook-Deegan, *The Gene Wars: Science, Politics and the Human Genome* (Norton, 1994).

Venter has significant economic clout in terms of the large amounts of public funding and private investment he commands through his research institute and companies. For example, the Venter Institute has received funding from the Department of Energy (DOE) and the National Institutes of Health (NIH). Synthetic Genomics Inc. sources funding from the private sector and has partnerships with large energy companies such as British Petroleum. Thus research at the Venter Institute feeds into a business strategy for the generation of commercial products.

Endy is a visionary engineer who – along with developing a new avenue in biological parts research – has a strong interest in intellectual property management and community-building in this new field and is engaged in discussions about ethical and regulatory issues. After originally training in civil and environmental engineering, Endy moved into biotechnology and biochemical engineering for his PhD. Endy and a colleague at the Massachusetts Institute of Technology (MIT), Adam Arkin, developed an early project proposal for 'a standard parts list for biological circuitry'.[65] At MIT, this work was developed with scientists including computer scientists Tom Knight and Randy Rettberg and biologist Pamela Silver.

Endy has been involved in establishing key institutions in synthetic biology. He co-founded the BioBricks Foundation and Registry of Standard Biological Parts (the Registry), discussed below. He also co-taught the early synthetic biology courses which developed into the annual International Genetically Engineered Machine Competition (iGEM), a synthetic biology competition for undergraduate students.

Endy is now at Stanford University, where he co-directs the International Open Facility Advancing Biotechnology (known as the BIOFAB). Endy was also involved in synthetic biology start-up company Codon Devices, a now-closed gene synthesis company. He co-founded the synthetic biology and gene synthesis company Gen9.[66] He has been involved with the multi-institution Synthetic Biology Engineering Research Centre (SynBERC), as Director of Practices (later known as 'Policy and Practices'). Policy and Practices referred to the research arm which studied matters of biosafety and biosecurity, sought to foster dialogue with regulators, policy makers and scientists, considered risks associated with the use of organisms engineered with synthetic biology outside the lab,

---

[65] Drew Endy and Adam Arkin, 'A Standard Parts List for Biological Circuitry' (7 October, 1999) *DARPA White Paper* <http://dspace.mit.edu/bitstream/handle/1721.1/29794/Arkin.Endy.DARPA.pdf?sequence=1>.

[66] See Gen9, *Powering the Synthetic Biology Revolution* (2013) Gen9 <http://gen9bio.com/>.

looked into ownership and sharing issues, and fostered leadership and community building in synthetic biology.[67] SynBERC's funding ended in 2016, and a new Engineering Biology Research Consortium (EBRC) was then established.

Church is a leading figure in the development of both DNA sequencing and synthesis technologies. He was one of the founders of the Human Genome Project. He is also one of the pioneers of synthetic biology and did early research into the development of a minimal cell. He is a member of the core faculty at the Wyss Institute at Harvard University and Professor of Genetics at Harvard Medical School. He has also been involved in SynBERC with Keasling and is part of the new EBRC.

Church has been involved in founding several companies including genome sequencing, genomic diagnostics, synthetic biology and biofuels companies.[68] The synthetic biology and biofuels companies include Joule Unlimited (solar fuels), Gen9 (gene synthesis and synthetic biology), and LS9 (biofuels and renewable petroleum). LS9 has now been sold to the Renewable Energy Group. He also co-founded Codon Devices, the now closed synthetic biology company, along with Endy and Keasling. He does advisory work for numerous other companies.[69]

Like Venter, Church also commands significant amounts of funding, including from the United States DOE, NIH, Defense Advanced Research Projects Agency (DARPA) and the National Science Foundation (NSF). Thus in his numerous roles, Church is involved in both academic and corporate models of research in synthetic biology.

Keasling is a Professor in the Departments of Chemical and Biomolecular Engineering and Bioengineering at the University of California, Berkeley. His research focuses on the metabolic engineering of microbes. He co-founded the biotechnology company Amyris, a significant player in the commercial landscape of synthetic biology. He has been involved in SynBERC along with Church and Endy and Chairs the Board of Directors of the EBRC. Like Venter and Church, Keasling also commands large amounts of funding including public, private and philanthropic funds. His University of California, Berkeley laboratory has received funding from the NSF, NIH, DOE, and the Office of Naval Research.

---

[67] Synthetic Biology Engineering Research Centre, *Practices* (2013) Synthetic Biology Engineering Research Centre <http://www.synberc.org/practices>.

[68] George Church, *George M Church's Tech Transfer, Advisory Roles, and Funding Sources* (2012) Harvard Molecular Technology Group & Lipper Center for Computational Genetics <http://arep.med.harvard.edu/gmc/tech.html>.

[69] Ibid.

Other important figures in the synthetic biology field include Randy Rettberg, Pamela Silver and Ron Weiss. Rettberg is a computer architect who began working with Drew Endy on synthetic biology at MIT.[70] He now runs the iGEM competition.[71]

Silver is a molecular biologist who worked with Endy, Rettberg, and Knight at MIT early in the development of synthetic biology.[72] She is now a Professor of Biochemistry and Systems Biology at Harvard Medical School. Her lab is involved in energy research among other areas.

Weiss is Professor of Biological Engineering at MIT. His lab is investigating some medical applications of synthetic biology using biosensors, along with other projects.

2.3.2 Development of the Construction Approach

Understanding key enabling technologies, such as DNA synthesis and assembly of large DNA molecules, is an important aspect of understanding the regulatory challenges presented by synthetic biology. These new capabilities impact upon what can be achieved with synthetic biology and the nature of the risks it presents. Understanding the key developments in the construction approach to synthetic biology is important because patent applications on some of these developments – and the potential problems they may pose for innovation in synthetic biology – are analysed in later chapters.

(a) Traditional recombinant DNA technology

This section explains the basics of recombinant DNA technology, which will assist in understanding synthetic biology. The primary structure of DNA is a chain of chemical units called nucleotides. There are four nucleotides in DNA, characterised by the purine or pyrimidine bases that they carry. The two purines are adenine and guanine, and the two

---

[70] Randy Rettberg, 'Evidence to The United States National Academies, The Organisation for Economic Co-operation and Development, and The Royal Society, Opportunities and Challenges in the Emerging Field of Synthetic Biology' (10 July 2009) Washington DC, 25.

[71] iGEM Foundation, *iGEM: Synthetic Biology Based on Standard Parts* iGEM <http://ung.igem.org/Main_Page>.

[72] Pamela Silver, 'Evidence to The United States National Academies, The Organisation for Economic Co-operation and Development, and The Royal Society, Opportunities and Challenges in the Emerging Field of Synthetic Biology' (10 July 2009) Washington DC, 9.

pyrimidines are thymine and cytosine. Two chains of nucleotides are bound together by the interaction of the bases of the individual nucleotides; adenine with thymine and guanine with cytosine. This interaction is called 'base pairing' and results in the two strands of DNA being complementary copies of each other. These ionic bonds between the two strands of DNA form the double helix structure. The order of bases forms the 'code' of the DNA and is called the DNA sequence. The length of the DNA molecule is measured by the number of interacting base pairs; for example, a short length of DNA might be a 100 base pair (bp) sequence. Longer molecules of DNA are described in kilobases (kb).

A gene is a section of DNA encoding instructions for production of a particular protein. Many genes are a few kb in length, but some are megabases long. Enzymes within the cell read the DNA and copy it into ribonucleic acid (RNA), a single-stranded nucleotide chain. Chains of RNA can perform cellular functions in their own right, but they are more commonly transcribed by other cellular enzymes into proteins, themselves long chains of amino acids. Proteins carry out the majority of cellular functions, as enzymes (which catalyse reactions within or outside of the cell), signals, transporters of other molecules, or structural components of the cell.

Since the 1970s scientists have been able to cut and join strands of DNA at chosen points, in order to add or delete genetic material and so change the genes encoded by the DNA. Recombinant DNA technology began to develop soon after the discovery of DNA ligase (the enzyme which joins fragments of DNA) and restriction endonucleases (enzymes which cut strands of DNA into shorter strands).[73] Scientists were then able to take a gene from one organism and insert it into the genome of another. This transfer of DNA between organisms is the basis of classical recombinant DNA technology.

The other technology of the time was sequencing of DNA extracted from the cells of organisms. The first automated DNA sequencing machines were released in the 1980s. Since then, this technology has become faster and cheaper.[74]

Modern DNA technology involves the use of short DNA primers (20–30 bp) to make many copies of much longer lengths of DNA using polymerase chain reaction (PCR) technology. The primers were the first

---

[73] Daniel G Gibson et al, 'Enzymatic Assembly of DNA Molecules up to Several Hundred Kilobases' (2009) **6**(5) *Nature Methods* 343.

[74] See Rob Carlson, *Biology is Technology: The Promise, Peril, and New Business of Engineering Life* (Harvard University Press, 2010), 66–7.

DNA sequences chemically synthesised and represent the beginning of synthetic biology technologies.

Biologists have also developed the ability to use a genetic sequence as a blueprint from which to design specific DNA sequences using computer software and synthesize them using chemical reactions. This means that human-designed DNA sequences, as opposed to those sequences found in nature, can now be built. Sequences of DNA can now be synthesised to order by DNA synthesis companies. It is possible for research teams to outsource the synthesis work, speeding up research projects. Like gene sequencing, gene synthesis capability has also increased rapidly, while the cost has fallen.[75] Now stretches of DNA and even whole genomes can also be designed, using computer software. The improvement of these design tools over time and the falling cost of gene synthesis are opening up new possibilities in synthetic biology.[76] These developments are also aided by our improving understanding of relationships between genetic sequences and the biological functions of the molecules they encode.[77]

A technical challenge has been to produce longer and longer stretches of synthetic DNA.[78] For example, when a long piece of DNA is synthesised, there is a chance of errors being introduced into the sequence.[79] Venter has pointed out that in his team's experiments, one error in one million base pairs can set progress back considerably and stop a synthetic genome from working as expected.[80] Techniques for reliably assembling long strips of DNA, and software to assist in this, have now been developed. Standardised DNA assembly techniques are a key enabling technology in synthetic biology.[81]

---

[75] See ibid.

[76] Michael K Jensen and Jay D Keasling, 'Recent Applications of Synthetic Biology Tools for Yeast Metabolic Engineering' (2015) **15** *FEMS Yeast Research* 1, 6.

[77] Ibid., 1–2.

[78] See, e.g., Daniel G Gibson et al, 'Complete Chemical Synthesis, Assembly and Cloning of a *Mycoplasma genitalium* Genome' (2008) **319** *Science* 1215, 1215.

[79] The types of errors are discussed in ibid., 1218–9.

[80] J Craig Venter, 'First Self-Replicating Synthetic Bacterial Cell' (Speech delivered at J Craig Venter Institute Press Conference 20 May 2010) http://www.jcvi.org/cms/research/projects/first-self-replicating-synthetic-bacterial-cell/video.

[81] Jensen and Keasling (n 76).

(b) Creating a minimal genome

Since the mid-2000s, scientists at the Venter Institute have been trying to identify the genes which are necessary for an organism's very survival, and those which are dispensable. *Mycoplasma genitalium* is a parasitic bacteria and has the shortest genome length (580kb) of any known organism which can be cultured in the laboratory.[82] Using this bacteria, researchers at the Venter Institute have been trying to identify the set of genes without which the organism cannot survive in the lab.[83] The idea is that a bacterium with this minimal genome will be the most efficient vehicle for scientists to manipulate, and will act as a basis to which additional functions such as foreign genes for producing useful proteins can be added.[84]

In 2006, researchers at the Venter Institute published an article setting out the minimum set of genes required for survival of this bacterium in the laboratory and filed a patent application claiming this set of genes.[85]

(c) Building a synthetic genome

The construction of progressively longer molecules of synthetic DNA has been an important technology in the development of synthetic biology.[86] The Venter Institute scientists needed to solve the problems of how to assemble short pieces of DNA into a complete genome, and how to 'get it to work' when put into a cell.[87] In 2003 Venter's team had their first success in synthetic genomics. They were able to construct the genome of a phage (a viral particle with a genome of 5.4 kb) from small lengths of synthesized DNA (oligonucleotides).[88]

In 2008, scientists at the Venter Institute announced that they had been able to chemically synthesize a complete *M. genitalium* genome of 580 kb.[89] This was part of their project to determine which of its genes

---

[82] Gibson et al (n 78).

[83] See ibid.

[84] See Rutz (n 13); Boldt and Muller (n 5) 387.

[85] Glass et al, 'Essential Genes of a Minimal Bacterium' (n 19); John I Glass, et al (2006) 'Minimal Bacterial Genome', United States Patent Application No 11/546364.

[86] See J Craig Venter, 'Evidence to The Presidential Commission for the Study of Bioethical Issues, Synthetic Biology' (8 July 2010) Washington DC.

[87] Venter, 'First Self-Replicating Synthetic Bacterial Cell' (n 80).

[88] Ibid.; Hamilton O Smith et al, 'Generating a Synthetic Genome by Whole Genome Assembly: φX174 Bacteriophage from Synthetic Oligonucleotides' (2003) **100**(26) *Proceedings of the National Academy of Sciences* 15440.

[89] J Craig Venter Institute, 'Venter Institute Scientists Create First Synthetic Bacterial Genome' (press release, 24 January 2008); Smith et al (n 88).

were essential for survival.[90] Producing the synthetic genome presented a technical challenge as it was the largest man-made construct of DNA that had been attempted.[91] The researchers used a new method of assembling pieces of DNA from smaller oligonucleotides.[92] They progressively joined short lengths of DNA to produce longer pieces, which were then joined to each other, eventually producing pieces representing quarters, then halves and finally the whole genome.[93] The resulting complete genome was sequenced to check that the assembled sequence matched the researchers' design, and the results were published in the journal *Science*.[94] Dan Gibson, lead author, said 'this is an exciting advance for our team and the field. However, we continue to work towards the ultimate goal of inserting the synthetic chromosome into a cell and booting it up to create the first synthetic organism'.[95]

The successful assembly of the synthetic genome was important because previous methods involved adding segments of DNA together one by one to produce a genome, which would be much slower than the scheme of assembling small lengths of DNA reported by Gibson and collaborators.[96] They noted that this method was not restricted to DNA that was chemically synthesised, but could also be used to assemble natural DNA fragments or a combination of natural and synthetic fragments.[97] At the time, Venter said the technique was likely to become 'widely used to advance the field of synthetic genomics' and to have applications in biofuels development.[98]

Later in 2008, researchers at the Venter Institute reported successfully using yeast to build a synthetic *M. genitalium* genome from 25 fragments, in one step.[99] This finding provided a much simpler and more

---

[90] Gibson et al, 'Complete Chemical Synthesis, Assembly and Cloning of a *Mycoplasma genitalium* Genome' (n 78) 1215.

[91] Ibid.

[92] Ibid.

[93] Ibid.

[94] Ibid., 1218.

[95] Venter Institute 'Venter Institute Scientists Create First Synthetic Bacterial Genome' (n 89).

[96] Gibson et al, 'Complete Chemical Synthesis, Assembly and Cloning of a *Mycoplasma genitalium* Genome' (n 78) 1219.

[97] Ibid.

[98] Venter Institute 'Venter Institute Scientists Create First Synthetic Bacterial Genome' (n 89).

[99] Daniel G Gibson et al, 'One-Step Assembly in Yeast of 25 Overlapping DNA Fragments to Form a Complete Synthetic Mycoplasma Genitalium Genome' (2008) **105**(51) *PNAS* 20404.

efficient method of genome assembly.[100] The team noted that this could speed up research in synthetic biology.[101]

(d) Genome transplantation

In order to build a synthetic cell, scientists needed to work out how to put a synthetic genome into a cell. Before moving on to inserting a synthetic genome into a cell, an initial step was to develop techniques for transferring the genome from one species of bacterial cell to another.[102]

In 2007, researchers at the Venter Institute successfully transplanted the natural genome of the bacteria *Mycobacterium mycoides* into another mycobacterium species, *Mycoplasma capricolum*, and demonstrated that it functioned in this new environment.[103] This effectively changed one species into the other.[104] The researchers said of their work that 'it may facilitate construction of useful microorganisms with the potential to solve pressing societal problems in energy production, environmental stewardship and medicine'.[105]

In 2009, Venter Institute scientists showed that a synthetic bacterial genome – made from chemically synthesised DNA which was manipulated and assembled in yeast cells – could then be transferred from yeast into a bacterial cell of a different species from the original bacterial species.[106] This provided another step towards the development of a synthetic cell.[107]

(e) Producing the first 'synthetic cell'

On 20 May 2010, scientists at the Venter Institute announced that they had successfully produced the world's first synthetic cell.[108] Following this announcement there was a storm of media activity, with articles

---

100 Ibid.

101 Ibid., 20409.

102 Carol Lartigue et al, 'Genome Transplantation in Bacteria: Changing One Species Into Another' (2007) **317** *Science* 632, 632.

103 Ibid.

104 Ibid.

105 Ibid.

106 Carol Lartigue et al, 'Creating Bacterial Strains from Genomes That Have Been Cloned and Engineered in Yeast' (2009) **325** *Science* 1693.

107 Ibid., 1693.

108 Venter Institute, 'First Self-replicating Synthetic Bacterial Cell' (n 89); Daniel G Gibson et al, 'Creation of a Bacterial Cell Controlled by a Chemically Synthesized Genome' (2010) **329**(5987) *Science* 52.

about the development being published in newspapers worldwide.[109] This achievement essentially represented a combination and extension of previous experiments conducted by the group. The researchers constructed a synthetic genome, modified from the natural version by the inclusion of a 'watermark' in the DNA code to differentiate it from other bacteria.[110] The scientists then inserted this synthetic genome into cells of another bacterial species and observed that it did indeed change the characteristics of the cell.[111] The recipient cell showed the same phenotype (physical characteristics) as those encoded by the inserted synthetic genome.[112]

This achievement was said to provide 'proof of principle for producing cells based upon computer-designed genome sequences'.[113] The researchers noted that in their experiment the synthetic genome used was similar to a natural genome, but that synthesis and transplant of more novel genomes is likely to be possible in future.[114]

(f) 'Design and synthesis of a minimal bacterial genome'

In 2016, the Venter Institute announced a new breakthrough which combined and extended the developments described above. The Institute produced a viable *M. mycoides* cell with a genome that was built in the laboratory *and* contained only the minimum set of genes required for growth in the lab.[115] This represents the newest milestone in research that had begun 20 years earlier, when the genome of *M. genitalium* was first

---

[109] See, e.g, Ewen Callaway and Andy Coghlan, 'The Meaning of Artificial Life' (2010) **206**(2762) *New Scientist* 6; Leigh Dayton, 'Geneticist Creates First Artificial Life', *The Australian* (21 May 2010); Alasdair Palmer, 'Craig Venter's Research is Scary, But Not in the Way You Think', *Telegraph (UK) online* (22 May 2010); Staff Writer, 'And Man Made Life: Artificial Life, the Stuff of Dreams and Nightmares, Has Arrived' (2010) **395**(8638) *The Economist* 11; and Dan Vergano, 'Scientists Create 1st Bacteria Strain from Man-made DNA', *USA Today* (20 May 2010).

[110] J Craig Venter, 'Synthetic Life' (Speech delivered at Ted Med 2009, San Diego, October 2009) http://www.youtube.com/watch?v=dvBV2qnSZwo; G Gibson et al, 'Creation of a Bacterial Cell Controlled by a Chemically Synthesized Genome' (n 108).

[111] Gibson et al, ibid., 52.

[112] Ibid

[113] Ibid., 55.

[114] Ibid.

[115] Clyde A Hutchison et al, 'Design and Synthesis of a Minimal Bacterial Genome' (2016) **351**(6280) *Science* 62531.

sequenced.[116] The new cell, JCVI-syn3.0, is different from the synthetic bacterial cell announced in 2010, JCVI-syn1.0. JCVI-syn1.0 had a genome sequence that was close to the naturally occurring *M. mycoides* sequence.[117] The genome itself was built through chemical synthesis and assembly of DNA cassettes, but the actual DNA sequence was largely unchanged from the natural sequence.[118] By contrast, JCVI-syn3.0 has a genome sequence that has been dramatically altered by the researchers to slim down the set of genes to the minimum requirements.[119]

The development of this cell used a combination of 'whole genome design', chemical synthesis of the genome and transplantation of the genome into cells to see if the genome could 'boot up' and operate to enable the cells to survive.[120] Initially, the researchers used the existing biological literature and knowledge about gene function to develop a genome containing the minimum set of genes they thought would be required. However, they found that cells containing this genome could not survive. The team then developed a system to gradually remove genes and test if the cells could still survive, eventually arriving at the minimal gene set.[121]

The goals of creating JCVI-syn3.0 were both to deepen our understanding of biology, and also to build a chassis that can be used in industrial applications.[122] In creating the cell, the team made some surprising and important discoveries. First, they found that of the 473 genes required for the cell to survive, we only know the function of about two-thirds. The remaining 149 genes have unknown roles. The researchers studied these genes and found that they are present in many other species, but they could not determine their function.[123] The team was surprised by this, with Venter commenting that 'We don't know about a

---

[116] Ewen Callaway, "Minimal" Cell Raises Stakes in Race to Harness Synthetic Life' (2016) **531**(7596) *Nature Reviews Microbiology* 557.

[117] Robert F Service, 'Synthetic Microbe has Fewest genes, but Many Mysteries' (2016) **351**(6280) *Science* 1380, 1380–81.

[118] Ibid.

[119] Ibid.

[120] J Craig Venter Institute, 'First Minimal Synthetic Bacterial Cell Designed and Constructed by Scientists at Venter Institute and Synthetic Genomics, Inc' (Press Release, 24 March 2016).

[121] Service (n 117).

[122] Venter Institute, 'First Minimal Synthetic Bacterial Cell Designed and Constructed by Scientists at Venter Institute and Synthetic Genomics, Inc' (n 120).

[123] Ibid.

third of essential life, and we're trying to sort that out now.'[124] Many of these genes are likely to be involved in similar functions to those carried out by the two-thirds with known function, such as producing proteins to carry out functions in the cell.[125] However, the researchers consider that it is likely that some of the genes with unknown function have roles that have not previously been described, and they now intend to investigate the function of the unfamiliar genes further.

Secondly, the researchers explained that their experiment revealed that some genes are not strictly essential for cells to survive, but they are 'quasi-essential' in that they are required for the cells to grow well in the laboratory.[126] Thirdly, they stated that the experiment showed that some genes with essential functions are present in the genome in 'redundant pairs'.[127] They explained that this means that there may be two identical or similar genes for a particular function. If one is deleted the cell can survive, but it cannot survive without both.

The researchers see this development as a significant milestone in the construction approach: 'This paper signifies a major step toward our ability to design and build synthetic organisms from the bottom up with predictable outcomes. The tools and knowledge gained from this work will be essential to producing next generation production platforms for a wide range of disciplines.'[128]

The researchers see a key outcome of this work as new tools and processes for building synthetic genomes.[129] Some tools, including lab kits and software, are now commercially available. However, some scientists have reflected that whole-genome synthesis, as used in this experiment, is not likely to become a method of choice in synthetic biology. George Church commented that 'genome-editing' techniques, where multiple changes are made to a genetic sequence, will continue to be the 'go-to' method.[130] The key genome editing tool, CRISPR-Cas 9 (CRISPR), was developed by Church and others. CRISPR stands for

---

124 Callaway (n 116).

125 Venter Institute, 'First Minimal Synthetic Bacterial Cell Designed and Constructed by Scientists at Venter Institute and Synthetic Genomics, Inc' (n 120).

126 Hutchison et al (n 115).

127 Ibid., 62534–5.

128 Venter Institute, 'First Minimal Synthetic Bacterial Cell Designed and Constructed by Scientists at Venter Institute and Synthetic Genomics, Inc' (n 120).

129 Ibid.

130 Callaway (n 116) 558.

'clustered, regularly interspaced, short palindromic repeat'. The system allows scientists to make numerous deliberate changes to a DNA sequence, rather than relying on random mutatagenesis as some earlier techniques do. A 'guide' molecule is used to target precise DNA sequences where genes can be added or disrupted.[131] This allows much more efficient genetic research.

Venter disagreed with Church's view that CRISPR would become the dominant technique, stating that genome synthesis tools will quickly improve from here on, and become the preferred tool.[132]

(g) Designer yeast chromosomes

Recently, scientists have been able to synthesise a modified yeast chromosome.[133] This is significant because yeast is a eukaryotic organism, a more complex class of organism than the bacteria described above. Annaluru and colleagues have constructed a synthetic yeast chromosome of about 273 kb.[134] They were able to include hundreds of designed changes from the naturally occurring genetic sequence, without the organism losing significant fitness.[135]

This work is part of a move towards building a completely new synthetic eukaryotic chromosome, rather than making changes to small parts of existing chromosomes.[136] The goal is to eventually build a new eukaryotic genome which could be used to produce useful chemicals.[137] Annaluru and colleagues envisage that it will soon become feasible to construct synthetic designer genomes of more complex eukaryotic organisms such as plants and animals, incorporating specific desirable functions.[138]

An international consortium of scientists is working together to build a synthetic yeast genome comprised of multiple chromosomes.

---

131 Heidi Ledford, 'CRISPR, the Disruptor' (2015) **522** *Nature* 20, 21.

132 Callaway (n 116) 558.

133 Narayana Annaluru et al, 'Total Synthesis of a Functional Designer Eukaryotic Chromosome' (2014) **344**(6179) *Science* 55.

134 Ibid., 56.

135 Ibid., 55–8.

136 Jensen and Keasling (n 76) 6.

137 Jensen and Keasling (n 76).

138 Annaluru et al (n 133).

2.3.3 Development of the Parts Agenda

The parts agenda was conceived of by then-MIT scientists Drew Endy, Randy Rettberg and Tom Knight in the early 2000s.[139] They came from various fields of engineering and wished to apply the principles of engineering to biology.[140] They aimed to use these principles to reduce the unreliability and complexity observed in biological research in order to make 'routine the engineering of synthetic biological systems that behave as expected'.[141] Therefore, 'synthetic biology builds on a small inventory of robust, standardized parts to construct from them new, synthetic systems'.[142]

The parts agenda is led by the BioBricks Foundation, a non-profit organisation founded in 2006 by scientists from MIT, Harvard University and the University of California San Francisco.[143] A 'BioBrick™ standard biological part' is a particular type of biological part, one that 'meets the technical and legal standards set forth by the BioBricks Foundation'.[144] It is a 'nucleic acid encoded-molecular biological function (for example, turn on/off gene expression), along with the associated information defining and describing the part'.[145]

Standard parts are made available to synthetic biologists through the Registry of Standard Biological Parts, and through the BioBricks Foundation. The Registry is like an online catalogue of parts for researchers. In 2016, the Registry contained 20 000 parts. I will call the community of scientists working with these resources the 'standard biological parts community'.

The overarching goal of the BioBricks Foundation and the Registry is to support and facilitate the parts agenda. The BioBricks Foundation's slogan is 'Biotechnology in the public interest'.[146] The goals of the BioBricks Foundation are 'to foster and advance innovation, research,

---

139 Rettberg (n 70).

140 Ibid., 25–6.

141 Endy (n 4) 449–52; see also The Bio Fab Group: David Baker et al, 'Engineering Life: Building a Fab for Biology' (2006) June *Scientific American* 44.

142 Jurgen Pleiss, 'The Promise of Synthetic Biology' (2006) **73** *Applied Microbiological Biotechnology* 735, 736.

143 The BioBricks Foundation, *The BioBricks Foundation* (2013) <http://www.biobricks.org/index.php>.

144 The BioBricks Foundation, *FAQ* <http://www.biobricks.org/FAQ.php>.

145 Registry of Standard Biological Parts, *Help: About the Registry* (n 26); ibid.

146 The BioBricks Foundation, *The BioBricks Foundation* (n 143).

standardization, and education in synthetic biology through the open design, construction, distribution, understanding, and use of BioBrick™ compatible parts, namely standardized genetic materials and associated functional information, in ways that benefit all people and the planet'.[147]

Open-access sharing of parts and materials is a key part of this philosophy: 'The Foundation believes that a free and easy-to-use legal framework for sharing and making use of engineered genetic materials underlies and serves these goals.'[148] The strengths and limitations of this legal mechanism are analysed later in this book.

The parts agenda is also advanced through the BIOFAB, founded in 2009.[149] This is a facility at Stanford University, California. It is a 'biological design-build' facility which will produce standard biological parts and make these freely available for use in synthetic biology research.[150] It aims to 'produce thousands of free standardized DNA parts to shorten the development time and lower the cost of synthetic biology for academic or biotech laboratories'.[151] It is run by Drew Endy and has support from the BioBricks Foundation.

2.3.4 The Emergence of Synthetic Biology as a New Field

Synthetic biology has come to be seen as a new field of biological science. There is a sense among scientists in this field that they belong to a community of researchers. Since 2007 the field has had its own journals, such as Springer's *Systems and Synthetic Biology*. The sense of an emergent community is partly due to the leadership activities of the BioBricks Foundation. It has arranged a series of large international conferences of synthetic biology researchers, titled the Synthetic Biology X.0 Conferences.[152] For example, Synthetic Biology 5.0, at Stanford University in 2011, had more than 700 attendees from all over the world. The BioBricks Foundation has also convened discussion groups on legal and technical standards. It has worked to build a community of synthetic biologists with a shared philosophy and common goals.

---

147 Ibid.

148 The BioBricks Foundation, *The BioBrick™ Public Agreement* (2013) The BioBricks Foundation <http://biobricks.org/bpa/>.

149 BIOFAB, *About the BIOFAB* BIOFAB <http://biofab.org/about>.

150 Ibid.

151 The BIOFAB, *Biofab Launches World's First Biological Design-Build Facility* (2010) BIOFAB <http://biofab.org/datasheet>.

152 See The BioBricks Foundation, *SBx.0 Conference Series* (2013) The BioBricks Foundation <http://biobricks.org/programs/sbx-0-conference-series/>.

The BioBricks Foundation has two current projects in its technical program which are taking this community building forward. First, the Foundation is establishing a global network of BIOFABS by building relationships with other BIOFABS around the world. The goals of this endeavour are to promote development and sharing of parts, contribution of information to the public domain and the adoption of 'shared technical standards and ethical values'.[153] Secondly, the Foundation is working on creating a 'Global Roadmap for Synthetic Biology'. This will be an interdisciplinary effort to evaluate the field's progress thus far, identify goals for the next 15 years, identify resources and consider the incorporation of 'safety, ethical, legal and social impacts' into the synthetic biology research and 'community vision'.[154]

The Registry was formerly located at MIT, where it ran the annual iGEM Competition for undergraduate students. This is a key community building institution in synthetic biology. iGEM has grown from a small competition in 2004 to an international event participated in by over 200 teams from all over the world. In 2012, the iGEM Foundation was formed as an 'independent nonprofit organization', which now runs both the competition and the Registry.[155]

SynBERC undertook work to support the Registry and to improve its computer interface. It has also worked to establish other repositories beyond this, to form a 'Web of Registries'.

The activities of the BioBricks Foundation and the Registry in community building and open-access 'sharing' of synthetic biology research will be discussed further in Chapter 3 on approaches to the regulatory issues raised by synthetic biology, Chapter 8 on patenting and Chapter 9 on the standard biological parts commons.

In addition to this community of scientific researchers, a new community of researchers has emerged to study the ethical, legal and social issues (ELSI) arising from this technology. This is partly due to an early decision in the UK to make inclusion of ELSI researchers a condition of certain funding grants for synthetic biology research.[156] Similarly, a decision was made in the US to require the inclusion of ELSI researchers

---

153 The BioBricks Foundation, *Technical Program* (2016) The BioBricks Foundation <http://biobricks.org/programs/technical-program/>.

154 Ibid.

155 iGEM Foundation, *The iGEM Foundation* iGEM <http://igem.org/About>.

156 Biotechnology and Biosciences Research Council, *Networks in Synthetic Biology* Biotechnology and Biosciences Research Council <http://www.bbsrc.ac.uk/funding/opportunities/2007/synthetic-biology.aspx>.

in the multi-institution SynBERC project.[157] This was an interdisciplinary synthetic biology research consortium which aimed to bring together 'biologists, engineers and human scientists from world-class institutions to produce the tools, techniques, and scientific understanding needed to design and construct a broad range of biological tools for health, energy, environment and, ultimately, human welfare'.[158] SynBERC's vision was to 'develop the foundational understanding and technologies to build biological components and assemble them into integrated systems to accomplish many particular tasks ... to train a new cadre of engineers who will specialize in synthetic biology; and ... engage the public about the benefits and opportunities and challenges of engineering biology of synthetic biology'.[159] SynBERC had four research themes: parts, devices, chassis and 'policy and practices'.[160] The first three themes were made up of biological sciences researchers in areas such as computational biology, biological engineering, pharmaceutical chemistry and genetics. The 'policy and practices' aspect was where the issues beyond the technical science, such as safety and ethics, were considered. The new EBRC will continue to undertake policy and public engagement work.[161]

2.4 APPLICATIONS OF SYNTHETIC BIOLOGY

This section describes some of the key applications being developed in synthetic biology research. It is important to understand the key applications in synthetic biology, and the types of technology that have already been developed, because these give rise to the benefits and risks of synthetic biology research. The controversy over benefits and risks, and how to manage risks while achieving maximum benefit, are key themes in this book.

---

[157] Gaymon Bennet and Paul Rabinow, *Synthetic Biology 2007: From Manifestos to Ramifying Research Programs* Connexions <http://cnx.org/content/m18812/1.2/>, 3.

[158] Synthetic Biology Engineering Research Centre, *Welcome to SynBERC* <http://www.synberc.org/>.

[159] Synthetic Biology Engineering Research Centre, *About SynBERC* <http://www.synberc.org/about>.

[160] Synthetic Biology Engineering Research Centre, *Research* (2013) Synthetic Biology Engineering Research Centre <http://synberc.org/content/research>.

[161] Engineering Biology Research Consortium, *Core Activities* (2017) Engineering Biology Research Consortium <https://www.ebrc.org/core-activities>.

Synthetic biology is a science driven by the goal of developing applications to solve practical problems. Keasling has commented that 'Synthetic biology is the engineering of biology to solve important problems. It is basic research with a focus on application. This emerging field of fundamental science has great potential for developing solutions to large-scale societal challenges.'[162] Keasling has emphasised the potential for well-characterised components to improve the success of biological engineering, noting that the access to such components was one of the key reasons his project to produce an anti-malarial drug precursor could be completed within a relatively short time-frame.[163] He has explained: 'through synthetic biology, we hope to make the engineering of biology more predictable and reliable, thereby reducing the cost to develop medicines and other useful products ranging from chemicals and fuels to consumer and commercial goods'.[164]

The most high-profile of the societal challenges which synthetic biology hopes to address are those in medicine and energy. Many of these applications use the approach of engineering a microorganism such as yeast or the bacterium *E. coli* so that it can be used as a factory to produce a useful substance such as a fuel or a therapeutic agent.[165]

2.4.1 Medicine

Potential medical applications of synthetic biology include synthetic vaccine development, synthetic drug production and use of engineered bacteria or viruses as 'pre-programmed cellular robots'[166] to target diseased cells.

In the most famous medical application of synthetic biology, Keasling and colleagues have used a combination of techniques within synthetic biology including metabolic engineering to produce microorganisms which can generate a precursor to the anti-malarial drug artemesinin. They used the approach of creating engineered microorganisms for use as 'living factories' for the manufacturing of the drug precursor, artemisinic

---

162 Keasling, 'Evidence to United States House of Representatives Committee on Energy and Commerce, Developments in Synthetic Genomics and Implications for Health and Energy' (n 53) 1.

163 Ibid., 2.

164 Ibid., 3.

165 Baldwin et al (n 31) 111–14.

166 Mohan-Ram and Waxman (n 7).

acid.[167] With this method, they used yeast cells to synthetically produce the artemisinic acid.[168] The spin-out company Amyris is now using the same platform technology to develop fuels and other industrial chemicals using engineered microorganisms as factories.[169]

Amyris is also researching the use of engineered yeast cells to produce other compounds for use as pharmaceutical drugs. Its goal is to use these cells to produce compounds that are either brand new or naturally occurring, and which are difficult or impossible to produce through chemical synthesis.

The second major application of synthetic biology in medicine is in the production of synthetic vaccines. This was emphasised when synthetic cells were identified as one of the top ten technologies in 2011 by *Technology Review*.[170] Venter has spoken about the benefits of using synthetic biology to make improved vaccines, stressing that these tools might be helpful in developing vaccines for viruses which mutate quickly, like human immunodeficiency virus (HIV).[171] The use of synthetic vaccines could save a lot of time in responding to an infection, according to Venter.[172] The growing ability to assemble synthetic DNA in an automated fashion aids in this synthetic vaccine development.

In 2010, a new private company called Synthetic Genomics Vaccines Inc. was established to 'develop next generation vaccines' using the genome sequencing and synthetic biology abilities of the Venter Institute and the intellectual property held by Synthetic Genomics Inc. to 'significantly advance and enhance vaccine development'.[173] Synthetic Genomics Vaccines Inc. is aiming to build a bank of synthetic vaccine 'seeds'

---

167 Amyris (n 54); Amyris, *Breakthrough Science* (2013) Amyris <http://198.55.195.109/Innovation/155/BreakthroughScience>.

168 Amyris, *Artemesinin* (n 54); Purnick and Weiss (n 55) 415.

169 Amyris, *Artemesinin* (n 54); Amyris, *Breakthrough Science* (n 167); Amyris, *Fuels* (2013) Amyris <http://198.55.195.109/Products/173/Fuels>.

170 Katherine Bourzac, 'Synthetic Cells' (2011) **114**(3) *Technology Review*, 56.

171 Venter, 'Evidence to The Presidential Commission for the Study of Bioethical Issues, Synthetic Biology' (n 86); see also Bourzac (n 170).

172 Venter, 'Designing and Synthesising Life' (n 20); Venter, 'Evidence to The Presidential Commission for the Study of Bioethical Issues, Synthetic Biology' (n 86).

173 Synthetic Genomics Inc., 'Synthetic Genomics Inc and J Craig Venter Institute Form New Company, Synthetic Genomics Vaccines Inc (SGVI), to Develop Next Generation Vaccines' (Press release, 7 October 2010).

that can be used to produce a vaccine quickly.[174] This strategy could be used for example when a new strain of influenza is identified, or when a pandemic disease threat is identified by the World Health Organization.[175] Synthetic Genomics Inc. has partnered with Novartis to develop synthetic biology-based influenza vaccines. It also collaborates with the US Government's Biomedical Advanced Research and Development Authority.

Scientists at Synthetic Genomics Inc. are also investigating the use of synthetic biology to develop alternative anti-microbial treatments.

Some scientists are trying to harness the natural functions of bacteria, such as sensing certain characteristics of their environment and responding by producing certain chemicals or even invading cells.[176] The goal is to develop engineered bacterial cells which can detect and kill cancer cells or other disease cells.[177] For example, Chris Anderson's lab at the University of California, Berkeley, is working on developing bacterial cells which can detect and destroy tumour cells in the human body.[178] Such techniques are known as 'smart therapeutics'.[179]

In a similar vein, researchers at the Wyss Institute have developed a three-dimensional folded shape from DNA, which can operate as a container for biological fragments with a therapeutic effect.[180] In this case, the DNA container was designed to recognize and open itself in response to disease markers of the surface of cancer cells, and deliver a biological message to the cells to die.[181]

Other synthetic biologists are working on applying the techniques of synthetic biology to tissue engineering. This moves beyond manipulating the behaviour of a single cell to manipulating the way in which cells communicate and interact with each other.[182] The idea is to control formation of patterns of cells.[183] Ron Weiss' research team is one of the

---

174 Synthetic Genomics, *Flu Vaccine* (2016) Synthetic Genomics <https://www.syntheticgenomics.com/flu-vaccine>.

175 Ibid.

176 Baldwin et al (n 31) 115.

177 Ibid.

178 University of California Berkeley, *Jon Christopher Anderson* (2012) Berkely Research <http://vcresearch.berkeley.edu/jon-christopher-anderson>.

179 Baldwin et al (n 31) 115.

180 Wyss Institute for Biologically Inspired Engineering, 'Researchers at Wyss Institute Develop DNA Nanorobot to Trigger Targeted Therapeutic Responses' (Press Release, 17 February 2012).

181 Ibid.

182 Baldwin et al (n 31) 115.

183 Ibid.

groups investigating these ideas. This kind of work could lead to insights that aid in the development of synthetic organs.

In a related area, Synthetic Genomics Inc. is undertaking research on development of 'humanised pig organs' using synthetic biology techniques.[184] In 2014, it entered a multi-year partnership with Lung Biotechnology, a subsidiary of biotech firm United Therapeutics, to conduct this research.

2.4.2 Energy

Several groups are working on using synthetic biology to produce biofuels, and significant private investment is occurring in this arena. Research groups are using engineered microorganisms with synthetic metabolic pathways, to make compounds that are not found in nature, or are modified, and have desirable characteristics as fuels.[185]

Synthetic Genomics Inc. is working on a range of biofuels projects and has stated that it is 'developing breakthrough genomic-driven solutions for major global issues. We're focusing first on energy and chemical solutions because we believe there is a great need for cleaner, greener sources for these products'.[186] It is aiming to develop biofuels which can be produced from non-food feedstocks and have superior properties to traditional biofuels.[187] At the US House of Representatives Committee on Energy and Commerce (the House Committee) hearing on synthetic biology, Venter noted that his research team is also trying to use microorganisms to capture carbon dioxide and convert it into fuel that could replace oil.[188]

Synthetic Genomics Inc. is researching the engineering of photosynthetic algae to produce fuel from carbon dioxide and sunlight.[189] It has a partnership with ExxonMobil Research and Engineering Company to

---

184 Synthetic Genomics Inc., *Novel Applications* (2015) Synthetic Genomics <http://www.syntheticgenomics.com/novelapplications.html>.

185 Baldwin et al (n 31) 112–3.

186 Synthetic Genomics Inc., *What We Do* (2012) Synthetic Genomics Inc. <http://www.syntheticgenomics.com/what/>.

187 Synthetic Genomics Inc., *Next Generation Fuels and Chemicals* (2012) Synthetic Genomics <http://www.syntheticgenomics.com/what/renewablefuels.html>.

188 Venter, 'Evidence to United States House of Representatives Committee on Energy and Commerce, Effects of Developments in Synthetic Genomics and Implications for Energy and Health' (n 8).

189 Synthetic Genomics Inc., *Next Generation Fuels and Chemicals* (n 187).

develop synthetic photosynthetic algae which produce biofuels.[190] Venter has explained:

> We believe that the ability to write the genetic code, as we describe synthetic genomics, will enable a better understanding of the fundamentals of living cells. It will also enable us to direct cells and organisms to perform jobs, such as producing clean water or new biofuels that natural species cannot currently do to the needed scale and efficiencies.[191]

Synthetic Genomics Inc. is also trying to use the techniques of synthetic biology to engineer microorganisms that are able to efficiently convert plant biomass into biofuels.[192]

Keasling's company Amyris is also working on producing renewable energy as an alternative to fossil fuels. Amyris has stated that its 'renewable fuels are designed to be optimal transportation fuels ... Cost competitive replacements for petroleum-derived fuels, compatible with existing engines yet with superior performance'.[193] Amyris is using control of metabolic pathways in microorganisms to turn them into 'living factories' to convert plant sugars into useful chemicals.[194] These plant feedstocks, such as Brazilian sugarcane, are described as renewable.[195]

Amyris has a key molecule, 'biofene' (the brand name for the chemical farnesene) which is produced using Amyris' microbial platform technology.[196] Biofene is described as a 'building block' which has a wide range of uses in producing chemicals for use not only as fuels but also in cosmetics, fragrances, industrial lubricants and other industrial chemicals.[197] Amyris is already commercially producing renewable diesel and jet fuels. It has a commercial production facility in Brazil and has partnered with the energy company Total to produce renewable diesel and jet fuels with improved performance.

---

190 ExxonMobil, *Algae Biofuels* <http://www.exxonmobil.com/Corporate/energy_climate_con_vehicle_algae.aspx>.

191 Venter, Prepared Statement 'Evidence to United States House of Representatives Committee on Energy and Commerce, Effects of Developments in Synthetic Genomics and Implications for Energy and Health' (n 8).

192 Synthetic Genomics Inc., *Next Generation Fuels and Chemicals* (n 189).

193 Amyris, *Fuels* (n 169).

194 Amyris, *Breakthrough Science* (n 167).

195 Ibid.; Amyris, *Business Strategy* Amyris <http://www.amyris.com/Company/151/BusinessStrategy>.

196 Amyris, *Breakthrough Science* (n 167).

197 Ibid.

LS9 was a San Francisco-based synthetic biology company developing 'high-performance chemicals and advanced biofuels'.[198] It had a grand vision: to use the life sciences to facilitate the transition to renewable energy.[199] LS9 was originally co-founded by Church, Keasling and others, and has now been sold to another biofuels company.

2.4.3 Environmental Remediation

Synthetic biology has potential applications in environmental monitoring and remediation. For example, bacteria could be engineered to behave as biosensors which detect pollutants in the environment and degrade these.[200] A 2012 investigation of environmental and industrial applications of synthetic biology found that in environmental remediation, the synthetic biology applications with greatest potential impact are biosensors, 'synthetic biology-based carbon capture' and solid waste treatment.[201] There are also possible uses of synthetic biology in water desalination and soil remediation.[202]

In an example of biosensing, synthetic biology may be used to turn plants into very sensitive sensors for detection of pollutants or biohazards.[203] Researchers have developed engineered plants with leaves that change from green to white when very small concentrations of certain chemicals are detected by their roots in the soil.[204] They are looking into developing these into 'a simple and inexpensive means to monitor human surroundings for substances such as pollutants, explosives, or chemical agents'.[205]

2.4.4 Agricultural Applications

Synthetic biology also has agricultural applications in the engineering of crops for improved performance and yield in areas such as drought and

---

198 LS9 (n 49).

199 LS9, *Vision* (2011) LS9 <http://ls9.com/about-us/vision>.

200 Baldwin et al (n 31) 116.

201 Markus Schmidt, 'Executive Summary' in Markus Schmidt (ed.), *Synthetic Biology: Industrial and Environmental Applications* (Wiley-Blackwell, 2012) xxi, xxii–xxiv.

202 Ibid., xxii–xxiv.

203 Mauricio S Antunes et al, 'Programmable Ligand Detection System in Plants through a Synthetic Signal Transduction Pathway' (2011) **6**(1) *PLoS ONE* e16292, e16292.

204 Ibid.

205 Ibid.

salinity tolerance.[206] Synthetic Genomics Inc. has a research program in food and nutritional products. It established a company called Agradis to develop more environmentally friendly, sustainable food crops as well as plants which can be used as feedstocks for biofuel production.[207] Agradis has since been acquired by Monsanto.

Synthetic biology techniques are also being used in research into crop protection chemicals. For example, Dow Agrosciences has a collaboration with biotechnology company Radiant Genomics to develop new chemicals for this purpose.

It has also been proposed that synthetic biology techniques can be used to improve the protein content of crops used as animal feed. As meat consumption is increasing, there is a need for larger amounts of high-quality, protein-rich animal feed.[208] Currently the protein content of animal feed is largely derived from soy beans, but alternative protein-rich animal feed crops will also be needed as soy beans can only be grown in certain regions.[209] Possibilities include developing animal feed crops which can be grown in cooler areas and engineering other crops to give a protein yield similar to soy beans.[210]

2.4.5 Chemical Production

As well as applications in producing fuels, synthetic biology can also be used to produce 'commodity chemicals'.[211] These are basic chemicals such as ethylene and propylene which are the foundation for many different types of chemical production.[212] For example, DuPont has produced a commodity chemical using engineered *E. coli*, using less energy than the conventional process for producing that chemical.[213]

---

206 Baldwin et al (n 31) 116.

207 Synthetic Genomics Inc., 'Synthetic Genomics Inc and Plenus, SA de CV Form New Sustainable Agriculture Company, Agradis, Inc to Develop and Commercialize Products Using Genomic Technologies' (Press release, 24 October 2011).

208 Eliot Herman, 'The Challenge to Meet Global Need for Protein Sources for Animal Production' (Paper presented at the 21st Century Borders/Synthetic Biology: Focus on Responsibility and Governance, 2013) <http://www.scienceforglobalpolicy.org/LinkClick.aspx?fileticket=VqAc-vBzvMk%3d&tabid=133>, 45–9.

209 Ibid.

210 Ibid.

211 Baldwin et al (n 31) 113.

212 Ibid.

213 Ibid.

Many other chemicals can also be produced with synthetic biology. For example, Amyris is developing products for use in cosmetics, industrial lubricants, fragrances and other areas. Amyris states that the industrial oils and lubricants produced from biofene have better performance in terms of biodegradability, toxicity and renewability than traditional products.[214] Amyris has established a joint venture with Cosan, a Brazilian company, to produce renewable industrial lubricants from biofene. The American Refining Group has invested in this venture.

Genomatica is a biotechnology firm which is using microorganisms with modified metabolic pathways to produce butanediol from plant feedstocks. This can be used to produce a variety of industrial chemicals. Genomatica has many patents, with more applications pending.

2.4.6 Military/Security Applications

Synthetic biology has a range of potential defensive military and security applications.[215] For example, the UK Government has offered funding for research into:

> innovative proof-of-concept and/or preliminary speculative scientific investigations that seek to explore the potential applicability of synthetic biology to meeting the UK's current and future national security and defence needs. This may include … research that can underpin novel sensors, detectors and diagnostic approaches, including in field devices; development of new materials and methods for decontamination and trauma care … smart materials and coatings … and improvements to bioprocess or metabolic engineering for the production of biologics or small molecules to be used in such technologies.[216]

The US Department of Defense has been changing its traditional focus to encompass fields such as biology.[217] The Department of Defense's interests include developing methods of detecting explosives or pollutants based on a better understanding of how organisms detect the presence of

---

214 Amyris, *Industrial* (2016) Amyris <https://amyris.com/product-category/industrial/>.

215 See Mukunda et al (n 44).

216 Biotechnology and Biological Sciences Research Council, *Joint Synthetic Biology Initiative: Call for Proposals* (2011) <http://www.bbsrc.ac.uk/web/FILES/Guidelines/rc-dstl-synbio-call.pdf>.

217 Sharon Weinberger, 'Pentagon Turns to "Softer" Sciences' (2010) **464** *Nature* 970, 970.

different substances in their environment, and 'tools that will allow us to detect adversarial uses of synthetic biology'.[218]

Along with defensive applications, synthetic biology also has the potential to be used in weapons development. The regulatory issues raised by this will be discussed in Chapter 7.

2.5 MAPPING RESEARCH IN SYNTHETIC BIOLOGY

This section considers where research in synthetic biology is occurring and how it is funded. The bulk of synthetic biology research is occurring in the US. Research is also occurring in the EU and the UK, as well as some in Asia.

Large investments are being made in both publicly funded and commercially driven research in synthetic biology. The nature of funding models for synthetic biology research impacts on various aspects of its social and scientific context. For example, there is a tension between proactive patenting of synthetic biology research and open sharing of research. There are diverging cultures, interests and expectations regarding intellectual property between for-profit and academic researchers in synthetic biology.[219] There is also related concern about how to obtain societal benefit from publicly funded research, and how to promote responsibility of scientists to those who provide the funding for their research.[220] These are important themes throughout this book.

2.5.1 United States

(a) Government-funded research

A number of US Government agencies are funding research in synthetic biology. In the House Committee hearing, Keasling discussed the important role federal funding has played and should continue to play in this field, telling the Committee that: 'Your actions in [and] the support of

---

218 Ibid.

219 Sibylle Gaisser and Thomas Reiss, 'Shaping the Science-Industry-Policy Interface In Synthetic Biology' (2009) **3**(1–4) *Systems and Synthetic Biology* 109, 110, 112.

220 On these issues see The Presidential Commission for the Study of Bioethical Issues, 'New Directions: The Ethics of Synthetic Biology and Emerging Technologies' (n 3) 6–7, 113–22, 141.

Congress will determine whether the efforts described today are ultimately successful.'[221]

The major providers of public funding for synthetic biology research in the US include DARPA, the Department of Defence, the NIH, the National Science Foundation and the National Aeronautics and Space Administration (NASA).[222] Funding is also provided by the DOE and the Department of Agriculture.

Synthetic biology was identified by the Obama administration as a research area which can contribute to the development of a 'bioeconomy', 'a large and rapidly growing segment of the world economy that provides substantial public benefit'.[223] The *National Bioeconomy Blueprint* set out the US Government's strategy for realising the potential of the bioeconomy.[224] This included supporting research and development and promoting the transfer of inventions from the laboratory to the market.[225] One of the strategic objectives was 'unlocking the potential of synthetic biology'.[226] Under this objective, the DOE provided $30 million for synthetic biology research to 'identify biological design principles that will provide understandings of plant and microbial systems to enable synthetic redesign', with the goal of aiding development of clean energy.[227]

The development of advanced biofuels was a major priority for the Obama administration, which aimed to reduce the dependence of the US on oil.[228] Significant public funding has been devoted to this area. For example, in 2012, Obama announced that $30 million would be provided to match private investments in biofuels and $32 million would be

---

221 Keasling, 'Evidence to United States House of Representatives Committee on Energy and Commerce, Developments in Synthetic Genomics and Implications for Health and Energy' (n 53) 37.

222 See Woodrow Wilson Center, 'US Trends in Synthetic Biology Research Funding' (September 2015) <http://www.synbioproject.org/site/assets/files/1386/final_web_print_sept2015.pdf?>, 10.

223 'National Bioeconomy Blueprint' (The White House, 2012) <http://www.whitehouse.gov/sites/default/files/microsites/ostp/national_bioeconomy_blueprint_april_2012.pdf>, 1, 15.

224 Ibid., 2.

225 Ibid., 2–3.

226 Ibid., 22.

227 Ibid.

228 Department of Energy, 'Obama Administration Announces New Investments to Advance Biofuels Industry and Enhance America's Energy Security' (Press release, 2 July 2012).

provided for earlier stage biofuels innovation, including $12 million for synthetic biology.[229]

The DOE also funds the Joint Bioenergy Institute (JBEI), a Bioenergy Research Centre. Its mission is to use biotechnology to convert non-food biomass into renewable biofuels for transport. JBEI's research includes using synthetic biology for fuel synthesis in yeast and other microbes.

The NSF has also funded the $28 million interdisciplinary multi-institution research centre SynBERC, among other projects.[230] SynBERC's NSF funding ended in 2016. The new EBRC was established in order to fill the resulting gap and provide a central organisation to bring together synthetic biologists and allow them to work through research challenges and set a strategic plan for the advancement of the field in the US.[231] It is a non-profit organisation and will be supported with a combination of government funding (including NSF funding) and membership fees.

The NIH provides funding for a range of synthetic biology projects and funds research into ethical issues associated with emerging technologies such as synthetic biology. However, according to the Woodrow Wilson Center, which has been studying government funding of synthetic biology, the overall federal funding for research into risk and ELSI matters raised by synthetic biology is relatively low.[232] The centre's findings indicate that funding for ELSI is about 1 per cent of total federal funding for synthetic biology research, significantly lower than for other major emerging technology initiatives such as the Human Genome Project.[233]

---

229 Ibid.

230 Synthetic Biology Engineering Research Centre, *About SynBERC* (n 159); National Science Foundation, *University of California Berkeley* (2013) Gen-3 Engineering Research Centers (ERC): (ERC) Active Awards <http://www.nsf.gov/awards/award_visualization.jsp?org=NSF&pims_id=5502&ProgEleCode=1480&RestrictActive=on&BooleanElement=true&BooleanRef=true&from=fund#region=US-CA&instId=0013128000>.

231 Engineering Biology Research Consortium, *Frequently Asked Questions* (2017) Engineering Biology Research Consortium <https://www.ebrc.org/faq>.

232 Woodrow Wilson Center (n 222) 10.

233 Woodrow Wilson Center, 'US Trends in Synthetic Biology Research Funding' (n 222).

In relation to research funded by defence agencies, DARPA is taking a particularly keen interest in pushing synthetic biology research forward.[234] It is now the largest government funder of synthetic biology in the US. According to the Woodrow Wilson Center, DARPA's funding for synthetic biology research has increased from just under $20 million in 2011 to over $100 million in 2015.[235] Since it became interested in synthetic biology in 2011, DARPA has been funding a program called 'Living Foundries'. This program aims to:

> apply an engineering framework to biology to harness its use as a technology and drive its advance as a manufacturing platform ... Living Foundries aims to enable on-demand production of new and high-value materials, devices and capabilities for the Department of Defense and establish a new manufacturing capability for the United States.[236]

Tens of millions of dollars have now been awarded under this program. In 2015, DARPA awarded $32 million to MIT and the Broad Institute of MIT and Harvard to support a 'foundry' for the assembly of genetic constructs. DARPA had initially provided seed funding for the establishment of this facility. The development of the synthetic minimal bacterial cell, announced by the Venter Institute in 2016, was also partly funded under this program.

(b) Private sector research

It is important to note that in the US synthetic biology is not only a research endeavour of universities and government. There is significant private investment in synthetic biology companies, such as Synthetic Genomics Inc. and Amyris, which have a range of commercial partnerships.

The Venter Institute is a dominant player in synthetic biology research. It is non-profit and works with a mixture of government and private funding. The Venter Institute is investigating applications of synthetic biology in the areas of energy and medicine. In addition to scientific projects, the Venter Institute also conducts research into the ethical and social issues raised by synthetic biology, viewing itself as a driver of

---

[234] See Alicia Jackson, 'What Do We Need From Synthetic Biology?' (Paper presented at the Fifth International Conference on Synthetic Biology, Stanford University, 2011).

[235] Woodrow Wilson Center (n 222) 4.

[236] Defense Advanced Research Projects Agency, *Living Foundries Industry Day* (3 June 2011) FedBizOpps.gov <https://www.fbo.gov/index?s=opportunity&mode=form&id=77b10102cba5c98c6be6b5c3d7a11387&tab=core&_cview=0>.

discussions in these areas.[237] Synthetic Genomics Inc., the company that commercialises the Venter Institute's research, is a key player in the private sector synthetic biology environment. There are various other private biofuels companies working in synthetic biology as noted above, including Joule and the Renewable Energy Group.

In addition to synthetic biology research companies, there are also companies which provide services for this research, such as gene synthesis. Companies such as DNA2.0, Blue Heron Bio and Gen 9 and SG-DNA, a subsidiary of Synthetic Genomics, provide DNA synthesis in the US. This industry is also expanding internationally.

(c) Philanthropic funding

Philanthropic funding is also playing a role in the synthetic biology research environment in the US. For example, the Wyss Institute at Harvard University is a biological research centre established with a $125 million gift to Harvard in 2009.[238] This gift was doubled to $250 million in 2013. Synthetic biology is a key aspect of the Wyss Institute's research, which aims to 'discover the engineering principles that Nature uses to build living things, and harness these insights to create biologically inspired materials and devices that will revolutionize healthcare and create a more sustainable world'.[239]

The Bill and Melinda Gates Foundation (the Gates Foundation) also has a presence in the synthetic biology research environment. It has provided funding for synthetic biology research and its representatives have spoken at the large Synthetic Biology X.0 conferences. The Gates Foundation collaborated with Amyris on the artemisinin project and provided funding.[240] Amyris and the Gates Foundation continue to work together in a public-private partnership to provide more affordable malaria treatment.

---

237 Venter, Prepared Statement 'Evidence to United States House of Representatives Committee on Energy and Commerce, Effects of Developments in Synthetic Genomics and Implications for Energy and Health' (n 8).

238 Wyss Institute for Biologically Inspired Engineering, *About Us* (2013) Wyss Institute for Biologically Inspired Engineering <http://wyss.harvard.edu/viewpage/about-us/about-us>.

239 Ibid.

240 Amyris, *Artemesinin* (n 54).

Since 2011, the Gates Foundation is funding use of synthetic biology to address global health problems through its Grand Challenges program.[241] This is available to research institutions all over the world. Initial grants are $100 000, with the possibility of a further $1 million for successful projects.

2.5.2 European Union

Synthetic biology has formed the focus of a great deal of reporting, scoping and interdisciplinary research in Europe since 2005. In the EU research is funded through the Framework Programmes for Research and Technological Development. Under the Sixth Framework an initiative called New and Emerging Science and Technology (NEST) was funded. A range of synthetic biology projects were funded including projects on specific applications such as antibody production, creating synthetic membranes, and engineering photosynthetic bacteria for energy production.[242] Other projects have included ethical and social aspects of synthetic biology. For example, SYNBIOSAFE was a project to examine the ethics and safety of synthetic biology research and 'stimulate a debate on these issues'.[243] The European Group on Ethics in Science and New Technologies to the European Commission has also looked into synthetic biology. In 2009, the Group produced a report on ethical, safety, security and regulatory issues relating to synthetic biology.[244]

Community building has also been a focus, for example the SYNBIOCOMM project aimed to promote establishment of a synthetic biology community by hosting a European conference on synthetic

---

[241] Bill and Melinda Gates Foundation, *Apply Synthetic Biology to Global Health Challenges* (2011) Grand Challenges in Global Health <http://www.grandchallenges.org/Explorations/Topics/Pages/SyntheticBiologyRound7.aspx>.

[242] European Commission, 'Synthetic Biology' (European Commission Directorate General for Research, 2007) <http://ec.europa.eu/research/fp6/nest/pdf/nest_synth_bio.pdf>, 6, 18, 20.

[243] SYNBIOSAFE, *Safety and Ethical Aspects of Synthetic Biology* SYNBIOSAFE <http://www.synbiosafe.eu/>.

[244] European Group on Ethics in Science and New Technologies to the European Commission, 'Ethical Aspects of Synthetic Biology: Proceedings of the Roundtable Debate' (European Commission, 19 May 2009) <http://ec.europa.eu/bepa/european-group-ethics/docs/publications/round_table_ethical_aspects_of_synthetic_biology.pdf>.

biology and funding European universities to participate in the iGEM Competition.[245]

More recently, the program ERASynBio was established under the Seventh Framework to fund synthetic biology research.[246] The goals are to strengthen the research community and 'provide an evolving line of European transnational projects and facilitate the identification of barriers in Synthetic Biology and needs of the stakeholders'.[247] Research under this program must incorporate ethical, social and regulatory issues.[248] From 2015, ERASynBio is self-sustaining.

The EU Commission has funded a new initiative titled MGR-Grammar as part of its Horizon 2020 program. This project will use synthetic biology and gene synthesis technologies to study how the expression of genes is regulated. This is expected to lead to improved genetic devices, and eventual applications in fuels, medicine and other areas.[249]

2.5.3 United Kingdom

Synthetic biology is one of the strategic research priorities of the UK's Biotechnology and Biosciences Research Council (BBSRC) and the Royal Society.[250] Areas relating to synthetic biology, such as bioenergy, are also research priorities.[251] The UK Government has identified the life sciences, including synthetic biology, as one of the 'eight great technologies of the future' in which Britain can be a leader.[252]

---

[245] European Commission, 'Synthetic Biology' (n 242); SynBioComm, *SynBioComm: Towards a European Synthetic Biology Community* SYNBIOSAFE <http://www.synbiosafe.eu/index.php?page=synbiocomm>.

[246] ERASynBio, *ERASynBio* ERASynBio <http://www.erasynbio.eu/project>.

[247] ERASynBio, 'ERASynBio 1st Joint Call for Transnational Research Projects: Building Synthetic Biology Capacity through Innovative Transnational Projects' (ERASynBio, 2013) <http://www.erasynbio.eu/lw_resource/datapool/_items/item_40/erasynbio1_call_text_130422_final.pdf>, 3.

[248] Ibid., 5.

[249] European Commission, *What are the Rules of Gene Regulation?* (2016) Horizon 2020 <https://ec.europa.eu/programmes/horizon2020/en/news/what-are-rules-gene-regulation>.

[250] Biotechnology and Biosciences Research Council, *Strategic Priorities* (2013) Biotechnology and Biosciences Research Council <http://www.bbsrc.ac.uk/funding/priorities/priorities-index.aspx>.

[251] Ibid.

[252] Biotechnology and Biological Sciences Research Council, *Business Secretary Announces £40M for UK Synthetic Biology* (2015) Biotechnology and Biological Sciences Research Council <http://www.bbsrc.ac.uk/news/policy/2015/150129-pr-business-secretary-40m-for-synbio/>.

There has also been substantial research into the ethical and social issues associated with synthetic biology. For example, the Royal Academy of Engineering undertook an enquiry into synthetic biology and produced two reports in 2009: 'Synthetic Biology: Scope, Applications and Implications' and 'Synthetic Biology: Public Dialogue on Synthetic Biology'.[253]

The BBSRC and the Engineering and Physical Sciences Research Council (EPSRC) have sought proposals for interdisciplinary networks for synthetic biology research since 2008. Initially, seven networks of synthetic biology researchers were funded for three years. These networks incorporated researchers from various institutions. They went beyond purely scientific research. Scientists were required to incorporate consideration of associated ethical, legal and social issues by including humanities researchers in the research teams.[254] Some of the networks have continued on with further funding.

The UK Research Councils also fund the £102 million Synthetic Biology for Growth Programme.[255] This programme funds six interdisciplinary centres for synthetic biology research, to the tune of £70 million over five years. This programme has also invested £18 million for capacity building in DNA synthesis over five years. The programme provides funding for doctoral training in synthetic biology, promoting the continued growth of the field in the UK.

A new Centre for Synthetic Biology and Innovation was established in 2009, when the sociology and anthropology researchers within the BIOS Centre for the Study of Bioscience, Biomedicine, Biotechnology and Society at the London School of Economics moved to Imperial College London to work in partnership with life scientists there. The centre is funded with an EPSRC Science and Innovation Award 'that aims to build

---

[253] The Royal Academy of Engineering, 'Synthetic Biology: Public Dialogue on Synthetic Biology' (The Royal Academy of Engineering, 2009) <http://www.raeng.org.uk/news/publications/list/reports/syn_bio_dialogue_report.pdf>; The Royal Academy of Engineering, 'Synthetic Biology: Scope, Applications and Implications' (The Royal Academy of Engineering, 2009) <http://www.raeng.org.uk/news/publications/list/reports/Synthetic_biology.pdf>.

[254] Biotechnology and Biosciences Research Council, *Networks in Synthetic Biology* Biotechnology and Biosciences Research Council <http://www.bbsrc.ac.uk/funding/opportunities/2007/synthetic-biology.aspx>.

[255] Biotechnology and Biological Sciences Research Council, *Synthetic Biology for Growth Programme* (2016) Biotechnology and Biological Sciences Research Council <http://www.bbsrc.ac.uk/research/programmes-networks/synthetic-biology-growth-programme/>.

new activity in areas of national strategic importance'.[256] The social science researchers aim 'to influence the trajectory of this emerging field of science and technology and contribute to the development of an appropriate governance regime'.[257]

The UK has a multidisciplinary programme of research into national security and defence applications of synthetic biology. A joint synthetic biology initiative was established in 2011 as a partnership between the BBSRC, the Defence Science and Technology Laboratory, the Engineering and Physical Sciences Research Council and the Medical Research Council.[258] This initiative provided £2.4 million to fund research.

2.5.4 Do-it-yourself Synthetic Biology Research

Synthetic biology research is also occurring outside traditional institutional and company settings. A 'do-it-yourself' (DIY) biology movement has emerged, mainly in the US but also in other jurisdictions. These are biologists who wish to experiment on their own projects. They are also known as 'bio punks' and 'biohackers'. DIYBio is an international group 'dedicated to making biology an accessible pursuit for citizen scientists, amateur biologists and biological engineers who value openness and safety'.[259] It has local groups in North America, Europe and Asia. DIYBio is engaged in considering issues of safety, security and intellectual property. For example, its website states that the vision of making biology accessible requires 'mechanisms for amateurs to increase their knowledge and skills, access to a community of experts, the development of a code of ethics, responsible oversight, and leadership on issues that are unique to doing biology outside of traditional professional settings'.[260] The existence of this 'do-it-yourself' biology community raises particular regulatory issues which will be examined in later chapters.

---

[256] Centre for Synthetic Biology and Innovation, *About CSynBI* (2016) Kings College London <http://www.kcl.ac.uk/sspp/departments/sshm/research/CSynBI/Index.aspx>.

[257] Ibid.

[258] Biotechnology and Biological Sciences Research Council, *Joint Synthetic Biology Initiative* (2011) <http://www.bbsrc.ac.uk/jointsyntheticbiology>.

[259] DIYbio, *An Institution for the Do-It-Yourself Biologist* DIYbio <http://diybio.org/>.

[260] Ibid.

2.6 CONCLUSION

This chapter has analysed the scientific context of the emerging field of synthetic biology. The key points are that synthetic biology is a term which describes several approaches to biological engineering: the construction approach, the parts agenda and metabolic engineering. These approaches are underpinned by common enabling technologies including DNA synthesis and assembly, software for assisting in those processes, and genome editing. The enabling technologies of synthetic biology and the various scientific approaches within the field are key scientific aspects which impact upon the types of risks posed and the regulatory challenges these create.

Secondly, the key scientific figures in synthetic biology research are J Craig Venter, Drew Endy, George Church and Jay Keasling. These are important scientists and thinkers working in the US, which dominates the research scene. These scientists have been influential in creating the scientific context of synthetic biology and have facilitated its emergence as a new field. They have also been influential in debates about the regulatory challenges posed by synthetic biology.

Thirdly, synthetic biology is a science driven by the need to develop solutions to real-world challenges. It has a wide range of significant applications in biofuels, medicine, agriculture, environmental remediation, defence and security. Research in these areas will potentially yield transformative applications. Some beneficial outcomes of synthetic biology research can already be seen in medicine and energy. The nature of these applications is also relevant to the nature of the risks posed by synthetic biology.

Large amounts of public funding are being devoted to these applications, particularly those in therapeutics and biofuels development. This underscores the need to promote societally beneficial technology development in synthetic biology. This is one of the key regulatory roles identified in the Introduction. In addition, significant private investment is occurring in key companies such as Synthetic Genomics Inc. and Amyris.

There are also individuals interested in pursuing synthetic biology outside universities and companies. This do-it-yourself biology movement presents unique regulatory challenges.

For all this, the field of synthetic biology is still young, having only been around for ten–15 years. If synthetic biology's development as a field was compared to the timeline for development of microchips for computers, we would now be at about 1960.[261] Challenges that lie ahead for the field include moving towards more real world applications, moving into multicellular rather than single-cell systems, constructing more complex genetic circuits, and working out how to use synthetic genetic circuits safely in mammals.[262] While our understanding of the relationship between genetic sequences and biological functions is improving and the costs of gene synthesis are falling, our understanding of biology has been said to be well behind what would be needed to create a completely synthetic organism.[263]

---

261 Jeffrey C Way et al, 'Integrating Biological Redesign: Where Synthetic Biology Came From and Where It Needs to Go' (2014) **157**(March 27) *Cell* 151.

262 George M Church et al, 'Realizing the Potential of Synthetic Biology' (2014) **15**(4) *Nature Reviews Molecular Cell Biology* 289.

263 Cameron et al (n 47) 389.

3. Science meets politics, activists and governance: evaluating key positions in synthetic biology debates

3.1 INTRODUCTION

The Venter Institute's bold announcement of the creation of a synthetic cell in 2010 provided a major catalyst for debate about the potential benefits and risks of synthetic biology and how it should be regulated. The United States' Presidential Commission for the Study of Bioethical Issues was charged to investigate synthetic biology in 2010. Its consultations brought together some of the key arguments and actors in this debate. Debates have also occurred in academic journals, newspapers and through various hearings and conferences. The debate has been highly contested, and there is a lack of consensus on issues of risk, benefit and regulation.

This chapter considers how key actors in this debate have approached the regulatory issues raised by synthetic biology and evaluates these approaches. The major scientific actors include the charismatic, dynamic individuals, J Craig Venter and Drew Endy. While both are strong believers in the powerful potential of synthetic biology, they have differing philosophies on issues of intellectual property, safety, security and regulation. The Canadian ETC Group – a pro-environmental, anti-technology group – is the most visible, vocal actor in opposition to synthetic biology. These actors have helped to bring the topic of synthetic biology and the issues it raises for regulation onto the public stage. The various issues raised by synthetic biology have also been analysed from a sociological viewpoint, as explored in this chapter.

3.2. J CRAIG VENTER'S NON-INTERVENTIONIST VISION OF REGULATION

Venter is a scientist and entrepreneur. He is a controversial, polarising figure in the biology world, at times cast as biology's villain or a maverick scientist.[1] He tends to cause issues to explode on to the agenda by making bold claims which are widely publicised. Venter calls for a non-interventionist approach to the regulation of synthetic biology involving self-regulation and light-handed external regulation.

3.2.1 Boosterism

Venter's involvement in the announcement of the development of a synthetic cell demonstrates his boosterism in relation to scientific developments. Venter's Institute announced the creation of a man-made bacterial cell by publishing an article in *Science* entitled, 'Creation of a Bacterial Cell Controlled by a Chemically Synthesized Genome.'[2] This was presented as a ground-breaking achievement, the 'first synthetic species'.[3]

Some outlandish comparisons and interpretations appeared during the ensuing coverage. In *The Economist*, references to Frankenstein and Jurassic Park appeared, along with biblical references.[4] In an article entitled, 'And man made life,' it was announced that 'creating life is no longer the prerogative of the Gods'.[5] Venter was described as having 'mastered' life and being able to control it.[6]

This reaction was unwarranted. In the original paper, the scientists concede that although they have referred to the resulting bacteria as a 'synthetic cell', only the DNA was synthetic; the cell cytoplasm and other parts of the cell were in fact not synthetic. This would not necessarily be clear to a layperson reading the original press release. Indeed, the terms of the announcement were seen by some scientists as misleading. For example, biologist Bonnie Bassler told the Commission: 'The authors did not create … The title does not represent the scientific

---

[1] See, e.g., Jocelyn Kaiser, 'Attempt to Patent Artificial Organism Draws a Protest' (2007) **316**(15 June 2007) *Science* 1557.

[2] Daniel G Gibson et al, 'Creation of a Bacterial Cell Controlled by a Chemically Synthesized Genome' (2010) **329**(5987) *Science* 52.

[3] J Craig Venter, 'Evidence to United States House of Representatives Committee on Energy and Commerce, Effects of Developments in Synthetic Genomics and Implications for Energy and Health' (27 May 2010) Washington DC, 28.

[4] Staff Writer, 'Genesis Redux' (2010) **395**(8683) *The Economist* 81; Staff Writer, 'And Man Made Life: Artificial Life, the Stuff of Dreams and Nightmares, Has Arrived' (2010) **395**(8638) *The Economist* 11.

[5] Staff Writer (n 4).

[6] Ibid.

findings in the paper. The title has unnecessarily alarmed people'.[7] Bassler also raised concern that the announcement could distract attention from the actual science of synthetic biology and the real issues.

A week after the original announcement, a more circumspect description of the achievement appeared in a *New Scientist* editorial. Venter and co-authors described it as a 'small, but highly significant step in synthetic genomics'.[8] They wrote: 'we did not create life from scratch: we transformed existing life into new life … the result is not an artificial life form; it is a living, self-replicating cell that most microbiologists would find hard to distinguish from its progenitor cell, unless they sequenced its DNA'.

While the announcement was described in the media as 'momentous',[9] it was in fact an incremental step forward from the previous research.[10] Venter's team was already able to produce, in the laboratory, a man-made bacterial genome.[11] The advance in this experiment was that the team had been able to cause a cell to take up this genome, and carry out the genetic instructions it contained.[12] The team was also already able to cause a cell to absorb a novel genome comprised of natural DNA.[13] Here the difference was that the DNA making up the genome introduced into the cell had been made by humans in the laboratory, rather than occurring in nature.

Venter also gave evidence to the House Committee. He explained that:

> What it is not, it is not life from scratch. We used a living cell and converted it into a new cell, based on this synthetic genome. But it is the first cell to

---

7 Bonnie Bassler, 'Evidence to The Presidential Commission for the Study of Bioethical Issues, Synthetic Biology' (8 July 2010) Washington DC. A similar view was expressed by professor of biomedical engineering Jim Collins: Mark Bedau et al, 'Life After the Synthetic Cell' (2010) **465**(7297) *Nature* 422, 424.

8 J Craig Venter et al, 'Synthetic Genomics: Where Next?' (2010) **2762** *New Scientist* 3.

9 Staff Writer (n 4).

10 Bassler, 'Evidence to The Presidential Commission for the Study of Bioethical Issues, Synthetic Biology' (n 7).

11 See Daniel G Gibson et al, 'Complete Chemical Synthesis, Assembly and Cloning of a *Mycoplasma genitalium* Genome' (2008) **319** *Science* 1215; Daniel G Gibson et al, 'One Step Assembly in Yeast of 25 Overlapping DNA Fragments to Form a Complete Synthetic Mycoplasma genitalium Genome' (2008) **105**(51) *PNAS* 20404.

12 See Gibson et al, 'Creation of a Bacterial Cell Controlled by a Chemically Synthesized Genome' (n 2).

13 Carol Lartigue et al, 'Genome Transplantation in Bacteria: Changing One Species Into Another' (2007) **317** *Science* 632.

> have its parent being a computer … we can now start in the digital computer and go out and write new software of life.[14]

Venter has at times minimised the actual technical scientific change or advance that was achieved with the creation of the synthetic cell, while at other times promoting the potential beneficial impact of this development and synthetic biology generally. First, Venter has argued that synthetic biology is important in terms of the progress of science and knowledge. For example, he has characterised the synthetic cell as 'a new understanding of life', and the tools of synthetic biology as 'powerful tools that give us a new way to look at the world'.[15] He has also called it 'a giant philosophical change in how we view life'.[16] Meanwhile he has sometimes downplayed the actual technical advance, describing it as a 'baby step'.[17] The hope underpinning such statements may be that they will help to calm fears and reduce the concern that followed the announcement of the synthetic cell. Venter has stated that 'the thing that can kill this [field of research] is the idea that this is so scary that we shouldn't do this'.[18]

Second, Venter has maintained that synthetic biology can address grand challenges such as the energy crisis and climate change. He sees a critical role for synthetic biology in reducing our dependence on fossil fuels.[19]

Third, Venter has contended that synthetic biology can play a role in improving healthcare, such as improving the development of vaccines.

---

14 Venter, 'Evidence to United States House of Representatives Committee on Energy and Commerce' (n 3) 28–9.

15 Ibid.

16 J Craig Venter, 'First Self-Replicating Synthetic Bacterial Cell' (Speech delivered at J Craig Venter Institute Press Conference 20 May 2010) <http://www.jcvi.org/cms/research/projects/first-self-replicating-synthetic-bacterial-cell/video>.

17 Prepared Statement of J Craig Venter, 'Evidence to United States House of Representatives Committee on Energy and Commerce, Effects of Developments in Synthetic Genomics and Implications for Energy and Health' (27 May 2010) Washington DC, 78.

18 J Craig Venter, 'Craig Venter is on the Verge of Creating Synthetic Life' (Speech delivered at Ted Talks, February 2008) <http://www.ted.com/talks/craig_venter_is_on_the_verge_of_creating_synthetic_life.html>.

19 J Craig Venter, 'Evidence to The Presidential Commission for the Study of Bioethical Issues, Synthetic Biology' (8 July 2010) Washington DC; Venter, 'Evidence to United States House of Representatives Committee on Energy and Commerce, Effects of Developments in Synthetic Genomics and Implications for Energy and Health' (n 17).

He predicted that synthetic vaccines will be developed much faster than is possible with conventional techniques, improving our ability to respond to emerging health threats.[20] Synthetic vaccine technology would help to manage new influenza viruses and to produce vaccines for rapidly mutating viruses such as HIV.[21]

Finally, Venter has argued that synthetic biology will have benefits to the economy, and indeed may help to establish a new industry. He is not concerned about intellectual property being used to limit access to knowledge and goods in synthetic biology; he is more worried about lack of commercial exploitation of the opportunities associated with synthetic biology.[22] Venter is in favour of using the intellectual property system to propel advances in synthetic biology and promote funding of this research by the private sector.[23] His company is thus pursuing several families of patent applications in synthetic biology, as discussed in later chapters. He regards the availability of federal funding as critical for the advancement of synthetic biology, suggesting that a failure to provide funding would be 'tragic' and noting that there is an opportunity for federal funding to be involved in research at the forefront of the field.[24]

3.2.2 Risks of Synthetic Biology

Venter has been concerned to show that he is aware of the concerns that synthetic biology raises in terms of biosecurity and environmental risk. He thinks 'having an intelligent ethical legal framework for this new science to emerge in is absolutely critical'.[25] He has pointed out that the Venter Institute has, from an early stage, been 'trying to push the envelope on the policy side'.[26] He has commented that at the Venter Institute these issues are considered to be 'as important as the scientific research'.[27] This is a change from Venter's earlier approach to the issues raised by his research: 'So, I think we set a precedent in this field,

20 Ibid.

21 Ibid.

22 Venter, 'Evidence to The Presidential Commission for the Study of Bioethical Issues, Synthetic Biology' (n 19).

23 Ibid.

24 Venter, 'Evidence to United States House of Representatives Committee on Energy and Commerce' (n 3) 95.

25 Venter, 'Evidence to The Presidential Commission for the Study of Bioethical Issues, Synthetic Biology' (n 19).

26 Venter, 'First Self-Replicating Synthetic Bacterial Cell' (n 16).

27 Venter, 'Evidence to United States House of Representatives Committee on Energy and Commerce' (n 3) 11.

actually going against my usual trend by asking for permission first, rather than asking forgiveness afterwards, for moving things ahead.'[28] He told the House Committee that his group had 'asked for ethical review of this research before we did the first experiments. This was back in 1997 ... We have had ongoing discussion, trying to drive the discussion ... So the scientists, I think, not only are being responsible, we are asking the questions before anybody else has'.[29]

The Venter Institute has been involved in projects regarding the ethical issues surrounding synthetic biology, including projects with the Hastings Centre and the Woodrow Wilson Centre. The Venter Institute clearly regards public engagement and outreach as important. They conduct lectures, congressional education and other media activities.

In his efforts to demonstrate his Institute's role in driving discussion of ethical and governance issues, Venter has frequently referred to two studies. These are an early study of ethical issues commissioned by him in the 1990s which resulted in the publication of an article by Stanford bioethicist Mildred Cho and collaborators, 'Ethical Considerations in Synthesizing a Minimal Genome' in 1999,[30] and a 2007 report prepared by a group of experts, including staff of the Venter Institute, 'Synthetic Genomics: Options for Governance'.[31]

The article by Cho and collaborators of 1999 pointed out that there was, at that early stage in the development of minimal genome technologies, 'an opportunity for proactive identification and debate of the associated ethical issues'.[32] Overall the paper was positive towards the continued development of minimal genomes. It did not make specific recommendations, but identified the key issues and raised questions that

---

[28] Venter, 'Evidence to The Presidential Commission for the Study of Bioethical Issues, Synthetic Biology' (n 19).

[29] Venter, 'Evidence to United States House of Representatives Committee on Energy and Commerce' (n 3).

[30] Mildred K Cho et al, 'Ethical Considerations in Synthesizing a Minimal Genome' (1999) **286**(5447) *Science* 2087. Among the authors is The Ethics of Genomics Group, comprised of Robert Aronowitz, Charles Bosk, A L Caplan, M K Cho, Mary Lynn Dell, M Susan Lindee, D Magnus, D McGee, Glenn McGee, Jon F Merz, Michael Orsi, Claire Pouncey, Pamela Sankar, Gerald I Wolpe, and Paul R Wolpe.

[31] M S Garfinkel et al, 'Synthetic Genomics: Options for Governance' (The J Craig Venter Institute, Massachusetts Institute of Technology, Center for Strategic and International Studies, October 2007) <http://www.jcvi.org/cms/fileadmin/site/research/projects/synthetic-genomics-report/synthetic-genomics-report.pdf>.

[32] Cho et al (n 30) 2087.

were perceived as important for further consideration. The questions included: 'How can the technology be used for the benefit of all, and what can be done in law and social policy to ensure that outcome?'[33]

Cho and collaborators identified three major groups of regulatory issues: intellectual property, biosecurity and biosafety. They identified the issue of potential impact of intellectual property and commercialisation in synthetic biology on 'the ability of both industry and academia to continue developing the technology for the public good'.[34]

The authors also identified the possibility for the techniques of gene sequencing, DNA synthesis and study of the genes associated with diseases to be used for developing biological weapons.[35] They stopped short of recommending that the research not continue or proposing specific limitations, instead suggesting that national and international oversight should be considered.

The 2007 report 'Synthetic Genomics: Options for Governance' was produced by two policy researchers at Venter's Institute, Michele Garfinkel and Robert M Friedman, in collaboration with Drew Endy, then at MIT and Gerald L Epstein at the Centre for Strategic and International Studies. Venter was one of the group of experts which contributed to the study. The approach taken in this study was to identify policy solutions which would '*both* minimize the risks from nefarious uses *and* minimize the impediments to beneficial uses of the technology'.[36] The report set out a range of options for addressing the risks relating to bioterrorism, safety of laboratory workers and the public, and harm to the environment. Options were assessed for their effectiveness in reducing risks relating to bioterrorism, biosafety and environmental harm, as well as the 'cost and burden' imposed on research.[37]

3.2.3 Managing the Risks

Venter's position suggests that the risks posed by synthetic biology are manageable. He has downplayed the security risks of synthetic biology at

---

33 Ibid., 2090.

34 Ibid., 2089.

35 Ibid.

36 Garfinkel et al (n 31) 17 (emphasis in original).

37 M S Garfinkel et al, 'Synthetic Genomics: Options for Governance' (The J Craig Venter Institute, Massachusetts Institute of Technology, Center for Strategic and International Studies, October 2007) <http://www.jcvi.org/cms/fileadmin/site/research/projects/synthetic-genomics-report/synthetic-genomics-report.pdf> iii.

times, reassuring the Presidential Commission that 'some of that can be readily prevented by some pretty straightforward regulations'.[38] Nevertheless Venter does perceive a need for screening of orders for synthetic DNA and checking the identity of those ordering:

> I think those [risks] can be pretty much eliminated by requiring companies to screen against A-list agents [a prescribed list of dangerous micro-organisms such as Ebola] and requiring bona fide institutions to be doing this work versus being done in somebody's garage … we think there's some guidelines that are needed.[39]

Venter appears to support voluntary regulations for synthetic biology. He approves of the voluntary guidelines for screening of gene synthesis that had been proposed by the US Department of Health and Human Services.[40] These guide DNA synthesis companies on how to monitor orders for synthetic DNA and check them for genes from dangerous organisms, in order to manage biosecurity risks of DNA synthesis activities.[41] Venter called them 'a great step in the direction to prohibit these frivolous uses'.[42] It is not clear whether he would support these as enforceable legal requirements, or voluntary requirements only. Venter has emphasised the track record of molecular biologists in developing and complying with appropriate guidelines.[43]

Overall, it is fair to characterise his position as a preference for light-handed regulation. Venter has suggested that the science of synthetic biology itself will provide answers to some of the risks. For example, he has stressed that technology can also limit the environmental risks associated with synthetic biology, advocating technical scientific measures such as engineering 'suicide genes' into organisms so that they are unable to survive in the environment.[44] This could involve, for example,

---

[38] Venter, 'Evidence to The Presidential Commission for the Study of Bioethical Issues, Synthetic Biology' (n 19).

[39] Ibid.

[40] Venter, 'Evidence to United States House of Representatives Committee on Energy and Commerce' (n 3) 78.

[41] See the guidelines which have now been introduced: *Screening Framework Guidance for Providers of Synthetic Double Stranded DNA*, Federal Register 75(197) 62820 (13 October 2010).

[42] Venter, 'Evidence to United States House of Representatives Committee on Energy and Commerce' (n 3) 78.

[43] Venter, 'Evidence to The Presidential Commission for the Study of Bioethical Issues, Synthetic Biology' (n 19).

[44] Venter, 'Evidence to United States House of Representatives Committee on Energy and Commerce' (n 3) 84.

engineering microbes to be dependent on an artificially supplied nutrient, so that they would die off if they were released into nature.[45] The message implicit in these comments is that the risks can be managed.

More recently, a somewhat different approach to environmental risk was taken in the Venter Institute report entitled 'Synthetic Biology and the US Regulatory System: Challenges and Options'.[46] This 2014 report focused on the safety and environmental risks of organisms produced with synthetic biology techniques. The Institute saw this research as important because some synthetic biology products are intended for use in the environment, such as algae that produce biofuels.[47] The report stated that 'while such products do not necessarily pose more or different environmental *risks* than the organisms engineered by other means, the novelty of the traits made possible by the technology may create more *uncertainty* about the organism's environmental impact'.[48] This seems a significant concession, given some of Venter's earlier comments regarding risk. The report further noted that the range of engineered microbes that can be produced will become more diverse and the organisms will become more complex, as our understanding of gene circuits improves.[49] These factors will make risk assessment more difficult.[50]

In 2015, the Venter Institute published a further report entitled 'DNA Synthesis and Biosecurity: Lessons Learned and Options for the Future'.[51] This report investigates the compliance with the voluntary government guidelines for the gene synthesis industry on checking synthetic DNA orders for similarity with pathogenic sequences in order to mitigate biosecurity risks. The report points out some changes in the nature of the industry that are likely to make these guidelines less effective in future. It canvasses a range of options, with an emphasis on voluntary measures.

---

45 See Gerd HG Moe-Behrens et al, 'Preparing Synthetic Biology for the World' (25 January 2013) **4** *Fronters in Microbiology* 5 <http://www.ncbi.nlm.nih.gov/pmc/articles/PMC3554958/pdf/fmicb-04-00005.pdf>.

46 Sarah R Carter et al, 'Synthetic Biology and the US Biotechnology Regulatory System: Challenges and Options' (J Craig Venter Institute, 2014) <http://www.jcvi.org/cms/fileadmin/site/research/projects/synthetic-biology-and-the-us-regulatory-system/full-report.pdf>.

47 Ibid., 6.

48 Ibid., 23.

49 Ibid., 5, 9.

50 Ibid., 5.

51 Sarah R Carter and Robert M Friedman, 'DNA Synthesis and Biosecurity: Lessons Learned and Options for the Future' (J Craig Venter Institute, 2015).

3.2.4 Governance

Venter's statements at times suggest that it is not necessary for government or other law reform bodies such as the Commission to intervene in a heavy-handed way in managing the issues the scientific research raises. This can also be seen in his comments about the good track record of biologists in complying with guidelines, and the lack of accidents and problems occurring as a result of this.[52]

Venter has noted there has been review at the highest level of government, emphasising that he briefed Congress and the White House before the synthetic cell announcement.[53] He pointed out that work on the synthetic cell was reviewed by the various government agencies and officials had the chance to consider the development prior to publication.[54] Venter has also pointed out that he has briefed the Department of Energy, Congress and the White House on his previous work in synthetic biology.[55] These comments may suggest that Venter hopes to demonstrate that the scientists are already being heavily reviewed, and that a new regulatory approach is not required.

Venter is understandably keen that regulation should not impede the ability of scientists to push synthetic biology forward. He has identified the opportunity to avoid repeating the experience in embryonic stem cell research, where an overly restrictive federal policy caused embryonic stem cell research in the US to lag behind advances in other countries.[56] Showing this preference for minimal external regulation, Venter supported the recommendations of the Presidential Commission, stating that they are 'wise, warranted and *restrained* which will help to ensure that this young field of research will flourish in a positive manner'.[57]

However, a more moderate approach is taken in the Venter Institute's 2014 report on options for regulation of synthetic biology products. There seems to be an implicit assumption that there are circumstances in

---

52 See Venter, 'Evidence to The Presidential Commission for the Study of Bioethical Issues, Synthetic Biology' (n 19).

53 Venter, 'Evidence to United States House of Representatives Committee on Energy and Commerce' (n 3) 31.

54 Ibid., 27.

55 Ibid., 31.

56 Venter, 'Evidence to The Presidential Commission for the Study of Bioethical Issues, Synthetic Biology' (n 19).

57 J Craig Venter, 'Statement from J Craig Venter, PhD, on the Presidential Commission for the Study of Bioethical Issues Report, New Directions: The Ethics of Synthetic Biology and Emerging Technologies' (Released Statement, 16 December 2010) (emphasis added).

which a degree of external oversight is desirable to manage risks. The report investigated whether the regulatory authorities within the US Coordinated Framework for the Regulation of Biotechnology (Coordinated Framework) have sufficient authority to regulate the end-products of synthetic biology research. The investigators searched for new regulatory problems created by synthetic biology, noting that the existing regulatory system was developed to manage organisms engineered with earlier genetic engineering techniques.[58]

The 2014 report identified two key regulatory challenges. First, it noted that plants engineered using synthetic biology are likely to be outside the regulatory authority of the US Department of Agriculture's Animal and Plant Health Inspection Service (APHIS). This is because earlier genetic engineering techniques used 'plant pests' to introduce new genes into plants, and APHIS' regulatory authority depends on the use of such plant pests.[59] By contrast, newer techniques do not require the use of pests.[60] The report canvassed options to address this gap, ranging from self-regulation and industry standards to changing the scope of regulatory authority of government agencies. The pros and cons of each option were set out, without making recommendations for particular options to be adopted. The report did advocate to some extent for voluntary processes to regulate plant products produced with synthetic biology, but also pointed out the downsides of such an approach. It noted that APHIS has played a useful role in limiting the environmental risks posed by engineered plants, and that if plants produced with newer techniques were not subject to such oversight this would be a significant change in regulatory approach.[61]

The second regulatory challenge noted in the report is the likely increase in workload of the US Environmental Protection Agency (EPA). As discussed in Chapter 5, the EPA regulates engineered microbes. The Venter Institute investigators considered that new synthetic biology techniques are likely to increase the number of applications to the EPA for experimental release and raised concern about adequate funding and expertise as well as potential delays in the approvals process. They considered options such as increasing funding for the EPA and changing its regulatory authority to place more onus on product developers to report on environmental and safety risk.

---

58 Carter et al (n 46) 4.
59 Ibid., 24–5.
60 Ibid., 25–6.
61 Ibid., 25–7.

The 2015 Venter Institute report on DNA synthesis and biosecurity concluded that the partnership between industry and government to implement voluntary gene synthesis screening procedures has had reasonable success.[62] The industry is said to be 'highly motivated' to comply with the voluntary procedures provided by the Department of Health and Human Services, due to the need to maintain good public opinion and avoid potential legal liability.[63] However, the heavy administrative burden of compliance is emphasised and enforceable government oversight is not contemplated.[64] The report does note an option to strengthen the partnership arrangement by imposing a contractual requirement for researchers using federal funding to only use those DNA synthesis providers who comply with the voluntary procedures.[65]

3.2.5 Critique of Venter's Approach

As discussed, Venter has referred to two early studies of ethical and regulatory issues to demonstrate his Institute's role in driving discussion. The early study by Cho and colleagues identified issues but did not make recommendations. The 2007 report 'Synthetic Genomics: Options for Governance' also did not provide overall conclusions as to the best package of policy interventions, but discussed the effectiveness of each option. A key technical conclusion from the research commissioned for the study was that at that time, constructing a virus in a laboratory was more difficult than obtaining a natural virus, but that this situation could be reversed within ten years (by 2017).[66] This conclusion is updated in the 2015 Venter Institute report, to state that construction of viral genomes is now much easier, though construction of bacterial genomes remains highly challenging.[67]

The 2007 report emphasised by Venter has limitations. For instance, while the report discusses 'requiring' gene synthesis companies to screen orders for synthetic DNA, implement customer verification and store data about these orders, it does not consider the question of whether this should be a binding, legal requirement or a softer tool such as voluntary guidelines or a code of practice.

---

62 Carter and Friedman (n 51) 8.
63 Ibid., 8.
64 Ibid., e.g., 11–13.
65 Ibid., 17.
66 Garfinkel et al (n 31) 16.
67 Carter and Friedman (n 51) 5.

What can be drawn from the report in terms of a regulatory approach is a conclusion about the technical feasibility of using synthetic biology techniques to develop biological weapons, comparison of different options in terms of policy risk reduction and cost burden, and an emphasis on the need for regulations to keep pace with the science. This report's acknowledgement that programs of regulation developed today will need to be reconsidered in several years is, however, useful. An adaptive approach is mentioned as a consideration, but is not specifically recommended.

While it was useful to engage experts to identify key ethical issues early, and to explore oversight options for biosecurity, producing two reports over the decade before the announcement of the synthetic cell in 2010 does not seem to amount to driving the ethical debate, as Venter's comments claim. The 2014 and 2015 Venter Institute studies also do not make specific recommendations for action, but the inclusive and consultative approach is constructive. These studies show more of a trend towards the Institute taking a leadership role because they involve broad consultation with the relevant stakeholders in government, industry and academia and identification of existing and likely future gaps in regulation.

Venter is involved in large-scale privately funded research in synthetic biology (as well as his publicly funded work) and hence he advocates for use of the intellectual property system to drive synthetic biology research forward. While Venter may stress the goals of his Institute in pursuing synthetic biology research for the benefit of society, there are reasons to scrutinise the way in which intellectual property is used in synthetic biology, and the extent to which public benefit will be achieved. Venter's approach to intellectual property has been controversial since the early 1990s, in relation to technology such as expressed sequence tags and human genome sequence data.[68] Venter's approach to ownership of and access to the human genome data was seen as damaging to the integrity

---

68 Though I acknowledge that Venter has stated that he did not support patents on expressed sequence tags or commercial monopoly over all of the human genome data: J Craig Venter, *A Life Decoded, My Genome: My Life* (Viking, 2007), 130–33, 135–7, 150, 158. He has stated that he wished the human genome data to be publicly available, and patents to be limited to those genes which clearly had therapeutic or diagnostic value: ibid., 180, 260. However, in relation to the controversy about Venter's approach to intellectual property see, e.g., Robert Cook-Deegan, *The Gene Wars: Science, Politics and the Human Genome* (Norton, 1994); Bruce Alberts and Aaron Klug, 'The Human Genome Itself Must be Freely Available to all Humankind' (2000) **404**(6776) *Nature* 325; Declan Butler, 'US/UK Statement on Genome Data Prompts Debate on "Free Access"' (2000) **404**(6776) *Nature* 324; Matthew Rimmer, 'The

and progress of scientific research.[69] There was also concern that there was inadequate attention to questions of ethics and access in relation to Venter's Sorcerer II expedition, in which Venter took water samples from various oceans around the world and used shotgun sequencing to sequence the genomes of microbes contained in these waters, looking for industrially useful genes.[70] While Venter stated that he would put the data into the public domain, concerns were raised in various countries that proper access and benefit sharing arrangements had not been negotiated with local governments, and that in some cases sufficient local laws had not been developed to manage the issues.[71] Venter was seen by some to be dismissive of objections to his approach and questions raised about access arrangements.[72] These examples suggest that there is reason to consider the engagement with ethical questions and access concerns relating to synthetic biology research.

The studies of ethical and regulatory issues by Venter and his Institute are not disinterested; they are prepared in part by people who have a strong interest in synthetic biology proceeding with minimal oversight from government. For example, Venter was one of the core group of researchers who worked on the 2007 biosecurity project, along with 17 others. Venter appears to prefer that scientists set the agenda and resolve issues with technical measures. There may be an underlying technocratic approach to policy making, involving deference to scientific experts.[73] From this perspective, failure to make the 'right' regulatory decision can be seen as being due to technical incompetence, or lack of consultation with scientific experts.[74] There is an ongoing debate about the utility of technocratic processes and the contrasting democratic approach which emphasises public participation. For example, the various errors 'ordinary people' (including legislators) tend to make in their thinking about risk have led environmental and administrative law

---

Sorcerer II Expedition: Intellectual Property and Biodiscovery' (2009) **6** *Macquarie Journal of Intellectual Property* 147.

69 See, e.g., Editorial, 'Slippery slopes?' (2001) **19**(1) *Nature Biotechnology* 1; Alberts and Klug (n 68).

70 See Rimmer (n 68).

71 See ibid., 150–51, 161–71.

72 Ibid., 171, 184.

73 See discussion of the technocratic approach in Sheila Jasanoff, *The Fifth Branch: Science Advisors as Policy Makers* (Harvard University Press, 1990), vii.

74 Ibid., 15–16.

scholar Cass Sunstein to argue for a 'large' role for experts.[75] However, there is literature in the sociology of scientific knowledge and science and technology studies observing the imperfect objectivity of scientific knowledge, which is not created in a vacuum free of political factors, values and interests.[76] Scientific knowledge is not immune to these influences when given as advice to government. Indeed, 'science policy cannot be trusted to scientists' because some scientists may, at times be more interested in the pursuit of scientific knowledge than other concerns such as public safety; the goals of science and policy are not the same.[77] Scientific expertise should certainly be sought and taken into account in developing regulatory interventions for emerging technologies. The experience of scientists in interacting with the legal system and its real-life impact on their work are critical considerations. But governments cannot outsource science policy to scientists. To do so would be to abdicate their public responsibilities, and would result in regulatory interventions lacking in legitimacy.

As discussed, Venter's comments have at times both minimised the actual technical scientific change or advance that was achieved with the creation of the synthetic cell and emphasised the potential beneficial impact of this development and synthetic biology generally. Indeed, it is in Venter's interests to suggest that when the risks of synthetic biology are being discussed, synthetic biology is a continuation of existing biological research. That is: 'For regulatory and intellectual property purposes it is in the interest of practitioners to argue that synthetic biology is no more than a cumulative advance on what is already being done. Consequently, only existing rules and doctrines need apply.'[78]

---

[75] See, e.g., Cass R Sunstein, *Risk and Reason: Safety, Law and the Environment* (Cambridge University Press, 2002), 113; Cass R Sunstein, *Laws of Fear: Beyond the Precautionary Principle* (Cambridge University Press, 2005), 126.

[76] See, e.g., Bruce Bimber and David H Guston, 'Politics by the Same Means: Government and Science in the United States' in Sheila Jasanoff et al (eds), *Handbook of Science and Technology Studies* (SAGE Publications Inc, Revised ed, 1995) 554. See also David S Caudill, 'Synthetic Science: A Response to Rabinow' (2009) **21**(3) *Law and Literature* 431, 441, where Caudill points out that synthetic biologists at SynBERC are 'inevitably, engaged in a social, political, institutional and ethical enterprise'.

[77] Richard A Posner, *Catastrophe: Risk and Response* (Oxford University Press Inc, 2004), 98–9.

[78] Graham Dutfield, 'Patent Law, Biotechnologies and Language in Subject Matter Expansionism' in Matthew Rimmer and Alison McLennan (eds), *Intellectual Property and Emerging Technologies: The New Biology* (Edward Elgar,

Venter formed a dim view of government agencies and bureaucracy during his career as an NIH researcher, where his early efforts to begin sequencing the human genome were thwarted.[79] He also encountered opposition from 'senior government officials' when trying to publish the sequence of the smallpox genome, which his research team sequenced in the early 90s.[80] Venter thought the smallpox genome sequence should be published despite his view that it would be possible to 'resurrect' the virus from this sequence information within several years.[81] It may be that progress and dissemination of scientific information are primary considerations, and that the security implications and the role of government in oversight may be secondary at times.

Some scientists take the view 'that freedom of scientific research should enjoy absolute priority over every other social value'.[82] However, such an approach would lack procedural legitimacy. Public participation is a key element of procedural legitimacy and it is important that perspectives beyond those of scientists are considered.[83]

While the goal of scientific progress is appropriate in a scientific leader, Venter's approach has limitations because it can tend to reduce debate about the risks of synthetic biology research and how it might be regulated.[84] A more constructive approach would take into account the role of government in decision-making about synthetic biology and the objections and concerns held by other players, including the public. This

---

2012) 117, 119. See also Joy Y Zhang et al, 'The Transnational Governance of Synthetic Biology: Scientific Uncertainty, Cross-borderness and the "Art" of Governance' (BIOS Working Paper No: 4, BIOS (Centre for the Study of Bioscience, Biomedicine, Biotechnology and Society), May 2011) <http://www2.lse.ac.uk/BIOS/publications/Working%20Papers/WP4TransnationalGovernance.pdf>, 7.

[79] Venter, *A Life Decoded, My Genome: My Life* (n 68) 89–153.

[80] Ibid., 173–4.

[81] Ibid.

[82] See Posner (n 77) 99.

[83] As explained in the Introduction, see Roger Brownsword and Han Somsen, 'Law, Innovation and Technology: Before We Fast Forward – A Forum for Debate' (2009) **1** *Law, Innovation and Technology* 1, 18–19; Roger Brownsword and Morag Goodwin, *Law and the Technologies of the Twenty-First Century: Text and Materials* (Cambridge University Press, 2012), 48.

[84] In a similar vein, social science researchers Joy Zhang and collaborators have identified a tendency for a 'closing down' of discussion about synthetic biology. Discussion tends to revolve around questions about specific risks, and is 'closed down' by saying that all risks are covered by existing regulations, which means that indeterminate, unknown aspects are not analysed, nor are the social changes that occur with scientific development: Zhang et al (n 78) 16–17.

would engender a regulatory approach with greater legitimacy. There are moves towards this type of approach in the Venter Institute's 2014 and 2015 reports, which involved broad consultation with policy makers, industry, the DIY community and others.

3.3 DREW ENDY'S REGULATION USING COMMUNAL NORMS

Drew Endy is a leading scientist in synthetic biology working in the US. Endy's approach is characterised by a preference for a 'do it together'[85] approach to both the scientific research itself and discussions about the issues it raises. This means he has encouraged community building in the field. In terms of intellectual property, it means he advocates sharing of data and tools rather than use of intellectual property rights. In the context of safety and security, it means he pro-actively raises and discusses these questions, and proposes regulatory approaches which involve scientists and law enforcement agencies in working together.

In 2005, Endy published his seminal article *Foundations for Engineering Biology*, which set out his vision for the development of the new field. He emphasised the potentially game-changing nature of applying engineering principles to biology through the parts approach and set out what he saw at that time as the challenges that needed to be overcome in order for the emerging field to be successful.[86] His other leadership activities have included co-founding key institutions in synthetic biology and organising the first major conference in the Synthetic Biology X.0 series held in 2004.

Endy has had a high level of involvement in discussions of the ethical and regulatory issues around synthetic biology. Endy has driven the development of models for the management of intellectual property and the regulation of biosecurity risks associated with synthetic biology. As early as 2003, when the field of synthetic biology was in its earliest stages, Endy was raising questions about the 'biological risk landscape'

---

[85] 'Do it together' is a phrase Drew Endy used in his evidence to the Presidential Commission in preference to the 'Do It Yourself' label describing some individuals working in synthetic biology outside institutional structures. See Drew Endy, 'Evidence to The Presidential Commission for the Study of Bioethical Issues, Synthetic Biology' (8 July 2010) Washington DC.

[86] Drew Endy, 'Foundations for Engineering Biology' (2005) **438**(7067) *Nature* 449.

and investigating strategies for response to these risks.[87] In 2003, he prepared some exploratory material regarding the risks of synthetic biology as part of a study commissioned by DARPA.[88] In 2007 he joined with leading synthetic biologist George Church and scientists at various gene synthesis companies and officers of the Federal Bureau of Investigation to publish an article in 'Nature Biotechnology' proposing an oversight framework for addressing security issues relating to DNA synthesis.[89] Endy was also involved (with Venter and others) in the study which led to the 2007 report, 'Synthetic Genomics: Options for Governance', discussed above. He has given evidence to various government bodies, including the Presidential Commission's 2010 inquiry into synthetic biology.

Endy has also been involved in the broader issues around synthetic biology in the US through the BioBricks Foundation, which has held workshops and convened ongoing discussions on the development of legal standards for sharing of BioBricks research. The BioBricks Foundation has also included discussion of ethical, legal and social issues in its iGEM program. Endy was also involved with the now-closed interdisciplinary project SynBERC. He directed the 'Practices' thrust of research which investigated safety and security issues as well as 'ownership, sharing and innovation' issues.[90] The activities of this research team also included promoting 'dialog between regulators, policy makers, and researchers that leads to shared understanding of needs and requirements regarding the engineering of microbes for release beyond controlled laboratory environments'.[91] A new Engineering Biology Research Consortium (EBRC) has now grown from the former SynBERC, and will continue policy and public engagement activities.

Along with Venter, Endy has pointed out the beneficial applications of synthetic biology. However, Endy's focus has been more on explaining the science and emphasising the nature of the change in biological engineering that is occurring. He has suggested that very significant

---

87 Drew Endy, 'Synthetic Biology' (2003) <http://dspace.mit.edu/bitstream/handle/1721.1/38455/SyntheticBiologyStudy.pdf>.

88 Drew Endy, 'Strategy for Biological Risk and Security' (2003) <http://dspace.mit.edu/bitstream/handle/1721.1/30595/BioRisk.v2.pdf?sequence=1>; Endy, 'Synthetic Biology' (n 87).

89 Hans Bügl et al, 'DNA Synthesis and Biological Security' (2007) **25**(6) *Nature Biotechnology* 627.

90 Synthetic Biology Engineering Research Centre, *Practices* <http://synberc.org/practices>.

91 Ibid.

technical advances have occurred and continue to occur, and that their potential for moving science forward should not be underestimated. He told the Presidential Commission:

> I believe that the capacity to synthesize and construct DNA is … currently the number one and most important technology of the 21st Century … I'd agree technically, scientifically, that synthetic biology represents an outgrowth of genetic engineering. But I would … urge you to recognize that changes in process, pace, and scope can lead to very significant transitions … Please don't underestimate the consequences of changes in process, even if the technology is [on] an incremental development path.[92]

Endy suggested that there were new issues for the Commission to consider, raised by synthetic biology: 'I think the change of capacity here is very important and means that the committee is unlikely, in my opinion, to find sufficient all of the existing and past bioethics work in the subject. I think there is new stuff happening that warrants additional attention.'[93]

3.3.1 Norms of the Standard Biological Parts Community

I will describe the BioBricks community's regulation-related norms as norms of 'sharing', 'positive contribution' and 'engagement'.

(a) Sharing

The norm of 'sharing' refers to the importance of making scientific knowledge 'freely available for ethical, open innovation'.[94] A program of the BioBricks Foundation aims to 'improve the process of scientific discovery by lowering the technical barriers to sharing and dissemination of knowledge in biological research, building a community of researchers that values, practices, and innovates the open sharing of information'.[95]

Endy is a longstanding supporter of openness in biology. He set out the problems associated with intellectual property and the need for sharing mechanisms in his 2005 article.[96] He has even been credited with

---

92 Endy, 'Evidence to The Presidential Commission for the Study of Bioethical Issues, Synthetic Biology' (n 85).

93 Ibid.

94 The BioBricks Foundation, *About* (2012) <http://biobricks.org/about-foundation/>.

95 The BioBricks Foundation, *Education Program* (2012) <http://biobricks.org/programs/education-program/>.

96 Endy, 'Foundations for Engineering Biology' (n 86).

creating the term 'open source biology'.[97] Endy's view is that the patent system creates difficulties for synthetic biology due to the financial investment required to obtain patent protection.[98] The patent system does not match 'the scale and pace of work'.[99] Further, Endy sees negotiating intellectual property access as time consuming and expensive. He has stated that the ability to develop an improved ownership framework will impact on the development and acceptance of synthetic biology.[100]

Endy assisted in drafting the BioBricks Public Agreement, which created a formal legal mechanism for sharing of BioBricks biological parts via contract. Endy's leadership through the BioBricks Foundation has been a large part of the development of a community of synthetic biologists and a 'commons' for the sharing of synthetic biology research and information.[101]

(b) Positive contribution

The norm of positive contribution refers to a general desire of the synthetic biology community to use synthetic biology to develop beneficial applications to solve various problems in the real world. Endy has emphasised the importance of keeping synthetic biology 'overwhelmingly constructive' since as early as 2005.[102] The BioBricks Foundation styles itself as a 'public benefit organisation', founded by 'scientists and engineers who recognised that synthetic biology had the potential to produce big impacts on people and the planet and who wanted to ensure

---

[97] Nathanael Johnson, 'Steal This Genome!' (March 30, 2005) *East Bay Express* <http://www.eastbayexpress.com/ebx/steal-this-genome/Content?oid=1077148>.

[98] Endy, 'Evidence to The Presidential Commission for the Study of Bioethical Issues, Synthetic Biology' (n 85).

[99] Ibid.

[100] Prepared Statement of Drew Endy, 'Evidence to United States House of Representatives Committee on Energy and Commerce, Effects of Developments in Synthetic Genomics and Implications for Energy and Health' (27 May 2010) Washington DC, 4.

[101] This commons is examined in Chapter 9. The existence of a 'commons' in synthetic biology has been pointed out by James Boyle in Boyle, *The Public Domain: Enclosing the Commons of the Mind* (Yale University Press, 2008) 176–8; Sapna Kumar and Arti Rai, 'Synthetic Biology: The Intellectual Property Puzzle' (2007) **85** *Texas Law Review* 1745, 1763–7; Arti Rai and James Boyle, 'Synthetic Biology: Caught Between Property Rights, the Public Domain and the Commons' (2007) **5**(3) *PLoS Biology* 0389, 0391–2.

[102] Endy, 'Foundations for Engineering Biology' (n 86) 453.

that this emerging field would serve the public interest'.[103] It envisions synthetic biology being a 'force for good' in fields such as medicine, agriculture and environmental remediation.[104] This rhetoric of making a positive contribution through the applications of synthetic biology has been prominent in discussions at the Synthetic Biology X.0 conferences.

The BioBricks Foundation's emphasis on promoting biosafety and biosecurity in synthetic biology research is also part of this desire to make a positive contribution.[105] In addition to the intellectual property conditions, the BioBrick Public Agreement contains terms requiring that the parties 'comply with laws and regulations applicable to the Materials, including laws and regulations governing export control and safety'.[106] The User Agreement within the BioBrick Public Agreement further stipulates that the 'User will refrain from using the Materials in connection with any intentionally harmful, negligent, or unsafe uses'.[107] Partly because of the inclusion of such terms in the BioBrick Public Agreement, that agreement has been seen as an initial 'constitution' for the field of synthetic biology.[108]

The notions of complying with applicable laws and using materials only for safe, legal and beneficial uses are thus important principles which guide the research within the standard biological parts community.

(c) Engagement

This norm refers to the high level of awareness of ethical and regulatory issues within the standard biological parts community, and willingness to engage in discussions about these. This can be seen in the activities of Endy himself as well as the synthetic biology community generally. The importance of this norm is clear from Endy's participation in the policy discussions as noted above. This norm is also found in statements of the BioBricks Foundation. For example, the Foundation's website states 'We

---

103 The BioBricks Foundation, *About* (n 94).

104 Ibid.

105 Ibid.

106 The BioBricks Foundation, *The BioBrick™ Public Agreement* (2013) The BioBricks Foundation <http://biobricks.org/bpa/>, sample Contributor Agreement, § 7.

107 Ibid., §§ 4 and 5.

108 Andrew W Torrance, 'Synthesizing Law for Synthetic Biology' (2010) **11**(2) *Minnesota Journal of Law, Science and Technology* 629, 663.

believe biosecurity, biosafety, bioethics, environmental health, and sustainability must be integrated with scientific research and applied technology'.[109]

3.3.2 Endy's Views on Safety and Security Regulation

Like Venter, Endy has pointed to the success of biosafety regulation based on recombinant DNA technology, in particular the 'past practical successes' of the Recombinant DNA Advisory Committee and The Office of Biotechnology Assessment at the NIH.[110] He has emphasised that biosafety matters have been taken on board in iGEM, where teams have to investigate biosafety issues associated with their project.[111] This is seen as a way of 'practically promulgating best practices as they exist around the topic of biosafety'.[112] Endy participated in production of the report *Synthetic Genomics: Options for Governance*, along with Venter.

In 2007, Endy and others set out a tiered model for biosecurity regulation relating to DNA synthesis. The authors highlighted the potential for DNA synthesis technology to be 'purposefully misapplied' and took the view that oversight at that time was not sufficient.[113] They thought that the regulation designed for recombinant DNA technology had worked well partly because the research only occurred in settings where there were biosafety frameworks. Endy told the Commission that he sees the major regulatory challenge as how to regulate those working on synthetic biology *outside* such institutions.[114]

3.3.3 Critique of Endy's Regulation Through Communal Norms

Endy's approach to the regulatory issues raised by synthetic biology has been to take leadership in discussions within the synthetic biology community about these issues and participate in discussions outside this community.

In terms of intellectual property, the BioBrick Public Agreement model advocated by Endy will help to ensure open access to the basic building

---

109 The BioBricks Foundation, *About* (n 94).

110 Endy, 'Evidence to The Presidential Commission for the Study of Bioethical Issues, Synthetic Biology' (n 85).

111 Ibid.

112 Ibid.

113 Bügl et al (n 89) 627.

114 Endy, 'Evidence to The Presidential Commission for the Study of Bioethical Issues, Synthetic Biology' (n 85).

blocks of synthetic biology. The major regulatory limitation of the BioBrick Public Agreement model is that the requirements cannot be enforced by the BioBricks Foundation. It has no authority to have any involvement in either the enforcement of intellectual property terms of the BioBrick Public Agreement or the regulatory and safety term.

Thus while both the awareness of the synthetic biologists of the broader issues and the desire to encourage compliance with laws and promote responsibility are admirable, the regulatory term cannot have practical effect, because it cannot be enforced. This is different from the role of a biobank, which does have some regulatory functions. A biobank can also have an important role monitoring ethical and other regulatory requirements.[115] For example, the Steering Committee of the UK Stem Cell Bank oversees research on stem cell lines that have already been established, including those which have not been deposited in the bank, and ensures that research is done ethically and transparently.[116] Under its Code of Practice, the Steering Committee ensures that all ethical requirements such as donor consent have been met before stem cells are deposited in the bank.[117] Further, researchers wishing to access stem cells have to apply to the Steering Committee, providing details of the value of their research and any peer review. Access is given on the condition that use of the cells must comply with all applicable laws, and approval is only granted for the particular research specified. In a similar way, the UK Biobank ensures that all consents and approvals are in place before deposit and access to material.[118] It retains ownership of the material and the 'right to take legal action against any unauthorized use or abuse of the database or samples'.[119]

Thus the self-regulation through communal norms espoused by Endy has limitations in that it cannot be enforced. While this approach is not sufficient in itself as a regulatory strategy for synthetic biology, it has a

---

[115] See discussion of biobank roles in Dianne Nicol and Richard Gold, 'Standards for Biobank Access and Intellectual Property' in Matthew Rimmer and Alison McLennan (eds), *Intellectual Property and Emerging Technologies: The New Biology* (Edward Elgar, 2012).

[116] United Kingdom Stem Cell Bank, 'Code of Practice for the Use of Human Stem Cell Lines' (UK Stem Cell Bank, 2010) <http://www.nibsc.org/asset.ashx?assetid=36019c52-bba2-497b-903f-df33e5c39dbc>, 22, 30–32.

[117] Ibid., 23–5.

[118] UK Biobank, 'UK Biobank Ethics and Governance Framework' (UK Biobank, 2007) <http://www.ukbiobank.ac.uk/wp-content/uploads/2011/05/EGF20082.pdf?phpMyAdmin=trmKQlYdjjnQIgJ%2CfAzikMhEnx6> 4–14.

[119] Ibid., 12.

number of positive elements. Endy's engagement with the issues surrounding synthetic biology is more comprehensive than Venter's. Endy has posed questions regarding safety, security, intellectual property and ethics. He has mentioned that synthetic biology brings together different scientific disciplines, and also creates an opportunity for these scientists to come together with ethicists and others to discuss how to make the best decisions about synthetic biology.[120]

It is positive for the development of a regulatory regime for synthetic biology that the standard biological parts community is so highly aware of ethical and regulatory issues, and enthusiastically engages in discussions with government regarding these. This openness to discussion should assist to develop a regulatory system that maximises regulatory effectiveness and connectedness.

Effective regulation is regulation which is 'fit for purpose' and is 'making a difference'.[121] Endy's approach – particularly the norm of engagement within the general synthetic biology community that he has fostered – is likely to assist policy makers to access the best quality, most relevant scientific information possible. This should enhance the understanding of the issues and risks by regulators, contributing to their ability to develop effective regulatory mechanisms to respond to synthetic biology. Of course, the scientific perspective should not be the only perspective considered.[122] But scientific information should be taken into account as an important aspect of regulatory policy making.[123] Having access to high-quality scientific information will also assist in promoting regulatory connection. The information will enable regulatory instruments to be modified over time in order to remain 'connected' to the stage of development of synthetic biology.[124]

---

[120] Drew Endy, 'Evidence to The United States National Academies, The Organisation for Economic Co-operation and Development, and The Royal Society, Opportunities and Challenges in the Emerging Field of Synthetic Biology' (9 July 2009) Washington DC, 96; Endy, 'Synthetic Biology' (n 87).

[121] See Brownsword and Goodwin (n 83) 277; Brownsword and Somsen (n 83) 11.

[122] Brownsword and Somsen (n 83) 18–19.

[123] D J Galligan, 'Citizen's Rights and Participation in the Regulation of Biotechnology' in Francesco Francioni (ed.), *Biotechnologies and International Human Rights* (Hart Publishing Limited, 2007) 342, 343–4.

[124] For further explanation of connection see Brownsword and Goodwin (n 83) 372.

This willingness to engage in discussion of security regulation is tempered by concern about the impact of overregulation or security fears on the future progress and funding of synthetic biology research. For example, Endy and his collaborators argue that a 'governance framework that stymies the open commercial development of synthesis technology will retard research and make the challenge of responsibly developing the technology more difficult'.[125] However, his approach encompasses a role for both scientists and regulators, and emphasises the need for public discussion.[126]

Another strength of Endy's approach is that it has been unusually forward-thinking. In 2003, he and colleagues asked 'what steps can be taken now, at the beginning of the field, to minimize the number of individuals who could or would act to cause harm by a future biological technology?'[127] That early study concluded that 'any meaningful strategy for minimizing future biological risk requires that the development of technologies for engineering biology proceed alongside the development of non-technical approaches to risk management'.[128] Subsequently, the article, 'DNA Synthesis and Biological Security', by Endy and others, identified pressure points on the regulatory system and stated that 'steps should be taken now' to address these.[129] This forward-thinking approach is also likely to promote regulatory connection, as potential future regulatory problems are considered in advance and seen as important.

Further, there was acknowledgement of the role of law enforcement authorities in overseeing DNA synthesis, and the suggestion that obligations regarding responsible behaviour of those using DNA synthesis technology should initially be voluntary and later *mandatory*.[130] This goes further than Venter's non-interventionist approach.

3.4 SOCIOLOGICAL APPROACHES

The problem of regulating synthetic biology has also been considered through a sociological lens. In 2011, a working paper on the 'art of governance' of synthetic biology was published by sociologists Zhang,

---

[125] Bügl et al (n 89) 628.

[126] See Endy, 'Evidence to The Presidential Commission for the Study of Bioethical Issues, Synthetic Biology' (n 85).

[127] Endy, 'Synthetic Biology' (n 87) 16.

[128] Ibid., 3.

[129] Bügl et al (n 89) 638.

[130] Ibid., 627–29.

Marris and Rose of BIOS, the former Centre for the Study of Bioscience, Biomedicine, Biotechnology and Society at the London School of Economics.[131]

The BIOS paper 'does not seek to identify common principles or conventions which should form the basis of synthetic biology regulations'.[132] It proposes that there is a need for an 'art of governance': 'an approach which employs proactive, open-ended regulatory styles able to work with uncertainty and change, to make links across borders, and to adapt to evolving relations among changing stakeholders, including researchers, research funders, industry, and multiple publics'.[133]

The paper emphasised four key concepts: the importance of moving beyond a focus on reducing or eliminating uncertainty; 'external accountability'; the cross-border nature of synthetic biology; and the 'fragmentation of social authorities'.[134] External accountability refers to the need for accountability across the various disciplines which make up synthetic biology, and outside science to 'all those who may be affected'.[135] Cross-borderness refers to both the way in which synthetic biology research crosses traditional disciplinary boundaries, as well as its international nature. Fragmentation of social authorities refers to the fact that no regulatory authority or social actor has capacity to oversee all aspects of synthetic biology.

A strength of the paper is that it acknowledges the pervasiveness of uncertainty in relation to synthetic biology's impacts. It proposes reflexiveness in legislation, and that synthetic biology regulation should not be a 'rigid regulatory regime' but a 'flexible and evolving "art of governance"'.[136] This approach has echoes of planned adaptation, which is considered in Chapter 5.

The main limitation of the BIOS paper is that it does not propose a specific approach to the regulatory challenges raised by synthetic biology; this is not its intention.[137] Instead, the authors aimed to understand the 'technical and social context which allows the paradoxical views [that synthetic biology is revolutionary but that no new regulatory mechanisms

---

[131] This research group is now part of the interdisciplinary Centre for Synthetic Biology and Innovation at Imperial College London.

[132] Zhang et al (n 78) 21.

[133] Ibid., 4.

[134] Ibid.

[135] Ibid.

[136] Ibid., 3.

[137] Ibid., 21.

are needed] to co-exist'.[138] Thus the regulatory problem is approached from a sociological viewpoint. I draw on aspects of this 'art of governance' approach from time to time in subsequent chapters. In particular, the concept of 'external accountability' is helpful in explaining why self-regulatory approaches in synthetic biology lack legitimacy.

3.5 ETC GROUP'S CALL FOR A MORATORIUM

The ETC Group is a Canada-based civil society organisation which campaigns about the development and impact of new technologies. It is a small outfit with offices in several countries. It works with other civil society organisations, particularly in developing countries. The ETC Group originally began its work focusing on agricultural biodiversity, food security and the impact of new technologies on poor people.[139] The group later expanded its research to consider 'social and environmental concerns related to biotechnology and nanotechnology'.[140] It claims a 'unique ability to transform research into public policy and institutional change', such as success in using techniques such as objecting to patents and mobilising public opinion to alter the biotechnology patent landscape.[141]

'ETC' stands for erosion, technology and concentration. 'Erosion' is used to refer to the erosion of living species and the natural environment, as well as erosion of knowledge and rights. 'Concentration' refers to the 're-organization of economic power into the hands of high-tech global oligopolies'.[142] Intellectual property monopolies are seen as a tool in this concentration of power.

The ETC Group is concerned with the impact of new technologies on poorer countries, and believes that when major new technologies are introduced into a society which is not 'just', this will 'exacerbate the gap between the rich and the poor'. It promotes transparency in the development of new technologies, and equitable sharing of benefits. It has drawn attention to the lack of consultation with people in the global South who will be affected by these new technologies.

---

138 Ibid., 7.

139 ETC Group, *ETC Group: A Brief History* ETC Group <http://www.etcgroup.org/en/about/History_of_etcgroup_page>.

140 Ibid.

141 Ibid.

142 ETC Group, *The Issues* ETC Group <http://www.etcgroup.org/en/issues>.

3.5.1 ETC Group's Criticism of Synthetic Biology and its Regulation

The ETC Group has been vocal in its objections to synthetic biology since 2006, releasing a number of press releases and reports and giving evidence to the Presidential Commission. The group also joined with other civil society organisations to object to the Venter Institute and Sloan Foundation's report, 'Synthetic Genomics: Options for Governance' in 2007 and the Commission's recommendations in 2010.[143]

The ETC Group uses press releases and reports to put forward its position in the debate. It uses attention-grabbing language to draw attention to the risks of synthetic biology. It has come up with several colourful labels to describe particular concerns, such as 'pandora's bug' to describe the potential for a synthetic biology to unleash danger, and 'microbesoft' to describe concern about intellectual property monopolies in this field.[144]

ETC Group's position is largely related to concerns about the new 'bioeconomy', a term used to refer to the economy it sees developing in which 'living plant material [biomass] rather than fossil plant material is the key feedstock of production'.[145] The group has argued that the need for biomass to drive the bioeconomy will change the way in which land is currently used in biomass-rich areas, with adverse impacts for local people and ecosystems, particularly in the global South.[146] The group has pointed with alarm to the current interest of large companies in the land and resources of biomass-rich countries. It has emphasised that synthetic biology companies such as Synthetic Genomics Inc. have partnerships

---

[143] ETC Group, 'Syns of Omission: Civil Society Organizations Respond to Report on Synthetic Biology Governance from the J Craig Venter Institute and Alfred P Sloan Foundation' (News Release, 17 October 2007) http://www.etcgroup.org/en/node/654; A group of Non-government Organisations, *Letter to Dr Amy Gutmann, Chair, Presidential Commission of the Study of Bioethical Issues*, (16 December 2010) <http://www.etcgroup.org/upload/publication/pdf_file/Civil%20Society%20Letter%20to%20Presidents%20Commission%20on%20Synthetic%20Biology_0.pdf>.

[144] ETC Group, *Patenting Pandora's Bug: Goodbye, Dolly…Hello, Synthia! J. Craig Venter Institute Seeks Monopoly Patents on the World's First-Ever Human-Made Life Form* (7 June 2007) <http://www.etcgroup.org/en/node/631>.

[145] Jim Thomas, 'Evidence to The Presidential Commission for the Study of Bioethical Issues, Synthetic Biology' (8 July 2010) Washington DC.

[146] ETC Group, 'The New Biomassters: Synthetic Biology and the Next Assault on Biodiversity and Livelihoods' (ETC Group, 2010) <http://www.etcgroup.org/upload/publication/pdf_file/biomassters_27feb2011.pdf>.

with large multinational corporations such as British Petroleum. It has called the attempts of large companies to buy land for growing biomass a 'land grab'.[147] The ETC Group has argued that there is not necessarily enough biomass to power a bioeconomy, and that biomass is not really a renewable resource.[148] Further, the group has been sceptical about the real benefits of products such as algal biofuels. Overall, the group has referred to the emergence of the new bioeconomy as an 'earth grab'.[149]

These objections apply to all industrial biotechnologies that could enable the development of this bioeconomy. Synthetic biology is seen as a key enabling technology. ETC Group staff member Jim Thomas submitted to the Commission that 'If you don't look at the broader economic shifts that are at play, you're going to miss the real socio-economic impact of synthetic biology. The bioeconomy will reshape the world and impact rights and potentially fuel inequalities'.[150]

The ETC Group is concerned that there has not been enough debate and investigation regarding synthetic biology, and is alarmed at the speed with which it is progressing. It has argued that synthetic biology products are being deployed in advance of proper discussion and investigation of their efficacy and safety, and without oversight.[151] Thomas told the Commission that 'I hope this really opens up a debate that will allow for the proper regulations and oversight of this technology'.[152] He urged the Commission to seize the opportunity 'to learn from history, rather than to be doomed to repeat it'. These are concerns about the legitimacy of regulatory interventions which apply to synthetic biology and their effectiveness in terms of adequate risk management.

In relation to risks raised by synthetic biology, the ETC Group advocates a precautionary approach.[153] It is concerned about the risks associated with accidental or deliberate release of organisms engineered with synthetic biology into the environment, pointing out that the impact of such release is unknown. It sees current risk assessment methods as inappropriate for synthetic biology and is not convinced by scientists' 'technofixes' for these risks, such as inclusion of suicide genes.

---

[147] Ibid., 15.

[148] Ibid., 25–6, 31.

[149] Ibid., 55.

[150] Thomas (n 145).

[151] ETC Group (n 146) 37.

[152] Thomas (n 145).

[153] ETC Group et al, 'The Principles for the Oversight of Synthetic Biology' (ETC Group, 13 March 2012) <http://libcloud.s3.amazonaws.com/93/ae/9/2287/1/Principles_for_the_oversight_of_synthetic_biology.pdf>, 3.

The ETC Group also opposes synthetic biology research on the grounds that it could be used to create biological weapons.[154] The group has demanded 'strong legal measures and enforcement' in relation to these risks, rejecting proposals for voluntary guidelines.[155] This is a concern that self-regulatory or voluntary measures will lack regulatory effectiveness.

The ETC Group also has concerns about the impact of intellectual property rights in synthetic biology, including overly broad patents. The Venter Institute's patent applications through Synthetic Genomics Inc. have been strongly opposed by the Group, with warnings that Venter is trying to patent 'life itself' and that the patents could lead to a problematic monopoly.[156] This is also a concern about regulatory effectiveness, in that the ETC Group is concerned that the patent system may be used in a way that is contrary to public interests.

The ETC Group has advocated both a total global halt to synthetic biology research, and a moratorium on environmental release and commercial use of organisms produced using it.[157] In 2012, along with other civil society organisations, it demanded a moratorium on commercial use and environmental release until certain oversight principles are implemented.[158] These included employing the precautionary principle, introducing 'mandatory synthetic biology-specific regulations', protecting the environment and guaranteeing 'the right-to-know and democratic participation'.[159]

The ETC Group responded to Venter's synthetic cell announcement by warning that 'there still remains no proper national or international oversight of new high-risk technologies that carry vast implications for

---

154 ETC Group, 'Extreme Genetic Engineering: An Introduction to Synthetic Biology' (ETC Group, January 2007) <http://www.etcgroup.org/upload/publication/602/01/synbioreportweb.pdf>, 23–6.

155 Ibid., 48–50.

156 ETC Group (n 144); ETC Group, *Extreme Monopoly: Venter's Team Makes Vast Patent Grab on Synthetic Genomes* (8 December 2007) <http://www.etcgroup.org/en/node/663>.

157 ETC Group, 'Synthia is Alive … and Breeding: Panacea or Pandora's Box?' (News Release, 20 May 2010); Thomas (n 145); NGO, *Letter to Dr Amy Gutmann* (n 143); ETC Group et al, 'The Principles for the Oversight of Synthetic Biology' (n 153).

158 ETC Group et al (n 153).

159 ETC Group et al, 'The Principles for the Oversight of Synthetic Biology' (n 153) 1.

humanity and the natural world'.[160] The group of civil society organisations responded negatively to the Presidential Commission's recommendations and demanded a 'moratorium on the release and commercial use of synthetic organisms until a thorough study of all the environmental and socio-economic impacts of this emerging technology has taken place'.[161]

In 2012 the ETC Group and other civil society organisations released a report demanding that the precautionary principle be employed and regulations specific to synthetic biology be introduced.[162] The group was very critical of self-regulation by scientists and argued that this amounts to *no* regulation at all. It argued that self-regulation lacks transparency and penalties for breaches.

The ETC Group also strongly objected to efforts of the synthetic biology community to develop a guiding declaration in 2006, seeing these as an attempt to prevent public debate and pre-empt government regulation.[163] These objections effectively shut down consideration of a code of conduct or ethics within the framework of the Synthetic Biology X.0 conferences.[164] The Group alleges that the recommendations of the 1975 Asilomar Conference on Recombinant DNA Molecules was similarly intentioned, drawing parallels between the two initiatives.

The 2010 Presidential Commission report failed to ameliorate the ETC Group's concerns, with the group and partners responding that: 'These recommendations give industry a free pass, while failing to ensure that the environment and public health are protected. We need stronger, more transparent regulation for synthetic biology, not less.'[165]

The ETC Group also criticised the Commission for 'ignoring' the precautionary principle, arguing, along with other NGOs, that 'business as usual has won out over precaution'.[166]

---

[160] ETC Group, 'Synthia is Alive … and Breeding: Panacea or Pandora's Box?' (n 157).

[161] A group of Non-government Organisations (n 143).

[162] ETC Group et al 'The Principles for the Oversight of Synthetic Biology' (n 153) 2.

[163] ETC Group, 'Global Coalition Sounds the Alarm on Synthetic Biology, Demands Oversight and Societal Debate' (News Release, 19 May 2006); ETC Group, 'Extreme Genetic Engineering: An Introduction to Synthetic Biology' (n 154) 46–8; ETC Group, 'The G(e)nomes of Zurich: Civil Society Calls for Urgent Controls on Synthetic Life' (News Release, 25 June 2007).

[164] See Zhang et al (n 78) 19.

[165] ETC Group, *NGOs Blast US Presidential Commission on Bioethics* (16 December 2010) <http://www.etcgroup.org/en/node/5244>.

[166] Ibid.

3.5.2 Critique of the ETC Group's Moratorium Approach

The ETC Group espouses a 'democratic' approach to regulation of synthetic biology, in contrast to technocratic perspectives. In a democratic model, public participation is seen as important in regulatory decision-making relating to science and technology, and poor decision-making is seen to occur when regulatory agencies do not 'incorporate a full enough range of values into their decision-making'.[167] The ETC Group's concerns relate to both procedural legitimacy and legitimacy of regulatory purpose.

The major limitation of ETC Group's approach is that it comprises a wholesale rejection of the technology of synthetic biology – there is no room for compromise in the narrative of the evils of the technology espoused by the group. It does not want this research to go ahead. A sweeping moratorium might allow some of the risks of synthetic biology to be managed. However, it would reduce the ability of synthetic biology research to generate beneficial applications. This would thwart the facilitative role of emerging technology regulation identified in the Introduction.

The ETC Group's approach is also not adequately informed by science and there is no engagement with other perspectives such as those of ethicists, sociologists and lawyers.

The ETC Group's criticism that the Commission ignored the precautionary principle is largely valid. I explore this critique further in Chapter 5. Similarly, the group's point that there is insufficient oversight of synthetic biology is valid. However, while the group demands 'proper regulation and oversight' of synthetic biology,[168] it has not suggested what kind of regulation is appropriate or how it should be developed. For example, there would be difficulty in defining what organisms would be covered by the demanded moratorium on environmental release and commercial use of organisms produced using synthetic biology. This is because there is no agreed definition of synthetic biology, and the field actually comprises a broad range of approaches to biological engineering.

The ETC Group is right to argue that self-regulation by scientists is insufficient to manage the regulatory issues and various risks raised by synthetic biology. This is because scientists are not always completely objective, as discussed above, and may not have the requisite regulatory or policy expertise. Appraisal from outside the synthetic biology research

---

167 Jasanoff (n 73) vii, 16.

168 Thomas (n 145).

field and 'external accountability' are important.[169] These are needed for engendering public trust in regulation efforts. Self-regulation can create problems of procedural legitimacy as well as regulatory effectiveness.[170] These problems arise because where self-regulation is relied upon, regulatees are able to develop self-interested regulation.[171] For example, companies may develop a code of practice which allows them to publicly declare their responsible attitude, but which is in fact cheap to implement and not sufficiently effective.

However, given the strong engagement of leaders such as Endy, it is unlikely the self-regulation efforts were designed to supplant external legal regulation. For that reason, the shutting down of these discussions by the ETC Group was not constructive.

It is valid for the ETC Group to raise questions concerning the potentially damaging impact of intellectual property monopolies on synthetic biology research, given the issues that have arisen due to patenting in other areas such as gene technology and embryonic stem cell research.[172] However, as I demonstrate in Chapter 8, the patent system in the US is, in the sample studied, working well to limit the scope of some of the overly broad patent applications which have been filed for synthetic biology inventions. As I show in Chapter 9, the BioBricks Foundation is working to keep basic building blocks in synthetic biology research openly available. Thus regulatory ineffectiveness in relation to the intellectual property mechanisms applicable to synthetic biology is not clearly demonstrated.

3.6 CONCLUSION

This chapter has evaluated the strengths and weaknesses of the approaches of Venter, Endy, sociologists Zhang, Marris and Rose and the ETC Group to the regulatory issues raised by synthetic biology.

---

169 See discussion of external accountability in Zhang et al (n 78) 4, 17–18.

170 Roger Brownsword, *Rights, Regulation and the Technological Revolution* (Oxford University Press, 2008), 153–5.

171 Ibid.

172 In relation to patent controversies in other areas of biotechnology see, e.g., Matthew Rimmer, *Intellectual Property and Biotechnology: Biological Inventions* (Edward Elgar Publishing, 2008); Matthew Rimmer and Alison McLennan (eds), *Intellectual Property and Emerging Technologies: The New Biology* (Edward Elgar Publishing, 2012).

I have proposed that Venter's approach leaves room for a stronger acknowledgement of the role for government and other players, including the public. There are merits to Endy's approach based on communal norms which represents a constructive, cooperative engagement with the issues. This should assist in developing effective, connected regulation. The high level of openness to discussion should also help to promote regulatory legitimacy.

The concepts proposed by Zhang, Marris and Rose are valuable in mapping the regulatory challenges posed by synthetic biology, though they were not directed at proposing regulatory solutions. I will draw on these concepts in subsequent chapters.

Lastly, I have agreed with the ETC Group that self-regulation by scientists is insufficient to address the regulatory challenges of synthetic biology, but that its demand for a moratorium is too extreme.

4. New directions? The United States Presidential Commission's investigation of synthetic biology's risks, benefits and oversight

4.1 INTRODUCTION

On 20 May 2010, the same day on which the successful creation of a synthetic bacterial cell was announced by the J Craig Venter Institute, President Obama asked his Presidential Commission for the Study of Bioethical Issues to consider the potential benefits for medicine, the environment, security and other areas, as well as considering the possible health, security and other risks.[1] The Commission was directed to develop recommendations for 'actions the Federal Government should take to ensure that America reaps the benefit of this developing field of science while identifying appropriate ethical boundaries and minimizing identified risks'.[2] The Commission's Report, 'New Directions: The Ethics of Synthetic Biology and Emerging Technologies' was released late in 2010.

This chapter examines and critiques the Commission's approach and recommendations in the context of the focus of this book on the regulatory challenges of synthetic biology.

1 The Presidential Commission for the Study of Bioethical Issues, 'New Directions: The Ethics of Synthetic Biology and Emerging Technologies' (The Presidential Commission for the Study of Bioethical Issues, December 2010) <http://bioethics.gov/sites/default/files/PCSBI-Synthetic-Biology-Report-12.16.10_0.pdf>, vi.

2 Ibid.

4.2 BACKGROUND TO THE PRESIDENTIAL COMMISSION AND ITS REVIEW OF SYNTHETIC BIOLOGY

4.2.1 The Presidential Commission

The Commission was an expert advisory body established by Obama by Executive Order in November 2009.[3] In establishing this Commission, Obama continued a tradition of bioethics advisory councils stretching back to the 1970s.

The Commission was an interdisciplinary panel of 13 experts appointed by Obama. The members were drawn heavily from academia, particularly medical experts and those with expertise in philosophy and bioethics. They did not receive remuneration. The Commission was chaired by Amy Gutmann, President of the University of Pennsylvania and a political scientist and philosopher.

The Commission's work was framed in broad and permissive terms under the Executive Order, by reference to non-exhaustive lists of issues and considerations. The Commission was authorised to undertake a broad range of activities such as conducting original research (including empirical research), conducting hearings and producing reports or other materials. There were no limits on what it could recommend, and it could propose 'any legal, regulatory or policy actions it deemed appropriate'.[4] The Commission was required to determine its priorities with regard to the 'connection of the issues to the goal of Federal advancement of science and technology'.[5] This was the only requirement to consider the agenda of the government of the day.

4.2.2 The Presidential Commission's Review of Synthetic Biology

Obama directed the Commission to report on synthetic biology within six months. In gathering information for its investigation into synthetic biology, the Commission consulted widely and received verbal evidence from 37 experts including scientists, ethicists and other humanities

---

[3] *Establishing the Presidential Commission for the Study of Bioethical Issues*, Executive Order 13521 of 24 November 2009, Federal Register 74(228) 30 November 2009.

[4] Ibid., s. 2 (a) (ii).

[5] Ibid., s. 2 (e).

experts, legal experts, industry representatives, government officials and civil society groups. It also sought written submissions.

The Commission held a series of three public meetings in the second half of 2010. They included presentations and question sessions with invited experts, as well as time for the public to ask questions or make comments. The Commission received evidence from scientists working in synthetic biology and related fields, including Venter, Endy and the ETC Group.

4.3 PRINCIPLES PROPOSED BY THE PRESIDENTIAL COMMISSION FOR ASSESSING EMERGING TECHNOLOGIES

The approach of the Presidential Commission to the regulatory challenges raised by synthetic biology is analysed here, drawing on transcripts of Commission consultations, its Report and publications by Chair Amy Gutmann.

The Commission found that existing analysis of the issues surrounding synthetic biology did not provide 'a broad-based ethical framework' to underpin approaches to addressing the issues.[6] Therefore it developed five ethical principles: public beneficence; responsible stewardship; intellectual freedom and responsibility; democratic deliberation; and justice and fairness. It intended these 'principles to illuminate and guide public policy choices to ensure that new technologies, including synthetic biology, can be developed in an ethically responsible manner'.[7]

The 'New Directions' report focused on risk, with an emphasis on risk identification and management as well as oversight of research in order to minimise risk. The Commission wanted to ensure the government was 'attentive to the risks related to synthetic biology, including risks that may emerge as the field matures'.[8] In response to the question of 'whether this field brings unique concerns that are so novel or serious that special restrictions are warranted at this time', the Presidential Commission decided that special restrictions were not required, but that prudent vigilance should be implemented.[9]

---

[6] Amy Gutmann, 'The Ethics of Synthetic Biology: Guiding Principles for Emerging Technologies' (2011) **41**(4) *Hastings Center Report* 17, 18.

[7] The Presidential Commission for the Study of Bioethical Issues (n 1) 4.

[8] Gutmann (n 6) 21–2.

[9] Ibid., 21.

It concluded that: 'A moratorium at this time on synthetic biology research would inappropriately limit intellectual freedom.'[10] Instead of a moratorium or new regulations, the Presidential Commission recommended 'ongoing government monitoring and dialogue between the private and public sectors'.[11]

The Commission did not see a need for a new agency to regulate synthetic biology at the time of its Report.[12]

4.3.1 Public Beneficence

The Commission stated that the goal of public beneficence is 'to act to maximize public benefits and minimize public harm. This principle encompasses the duty of a society and its government to promote individual activities and institutional practices, including scientific and biomedical research, that have great potential to improve the public's well-being'.[13]

The key elements of this principle for developing regulation are the goals of minimising public harm while maximising public benefit, and the requirement that policies be reviewed. The Commission stated that 'public beneficence requires that when seeking the benefits of synthetic biology, the public and its representatives be vigilant about risks and harms, standing ready to revise policies that pursue potential benefits with insufficient caution'.[14] Under this principle, the Commission recommended a review of public funding for synthetic biology, public investment in synthetic biology research that is likely to be publicly beneficial, and an investigation into licensing and sharing practices in synthetic biology research.[15]

4.3.2 Responsible Stewardship

The Presidential Commission explained that the principle of responsible stewardship 'reflects a shared obligation among members of the domestic and global communities to act in ways that demonstrate concern for those who are not in a position to represent themselves (e.g., children and future generations) and for the environment in which future generations

---

10 The Presidential Commission for the Study of Bioethical Issues (n 1) 12.
11 Gutmann (n 6) 19.
12 The Presidential Commission for the Study of Bioethical Issues (n 1) 8.
13 Ibid., 4.
14 Ibid.
15 Ibid., 6–7.

will flourish or suffer'.[16] Gutmann stated that this principle is applied through 'open decision-making processes informed by the best available science'.[17] Responsible stewardship encompasses the concepts of 'prudent vigilance' and 'coordinated regulation'. Prudent vigilance refers to the establishment of processes for assessing risks and benefits before scientific projects are undertaken as well as after.[18] It also encompasses the need to 'continue to assess safety and security as technologies develop' and to limit their use as necessary.[19] This includes evaluating safety and security as technologies 'diffuse into public and private sectors'.[20]

Under the principle of responsible stewardship, the Presidential Commission advocated a consistent and coordinated approach to regulation of synthetic biology. It recommended that: 'A mechanism or body should be identified to: (1) leverage existing resources by providing ongoing and coordinated review of developments in synthetic biology, (2) ensure that regulatory requirements are consistent and non-contradictory, and (3) periodically and on a timely basis inform the public of its findings.'[21] The concepts of responsible stewardship and prudent vigilance also led the Presidential Commission to recommend initial review of risk assessment policies.[22] It proposed an interagency discussion of risk assessment activities, including identifying strategies for harmonising these and identifying any gaps in current processes for assessing the risks associated with release of synthetic organisms.[23]

The Commission also proposed 'ongoing review of the ability of synthetic organisms to multiply in their natural environment' and identification 'as needed, [of] reliable containment and control mechanisms'.[24]

4.3.3 Intellectual Freedom and Responsibility

The principle of intellectual freedom and responsibility 'asserts that restrictions on research, whether by self-regulation by scientists or by

---

16 Ibid., 25.
17 Gutmann (n 6) 19.
18 The Presidential Commission for the Study of Bioethical Issues (n 1) 4.
19 Ibid.
20 Gutmann (n 6) 19.
21 The Presidential Commission for the Study of Bioethical Issues (n 1) 8.
22 Ibid., 9.
23 Ibid.
24 Ibid.

government intervention, should limit the free pursuit of knowledge only when the perceived risk is too great to proceed without limit'.[25]

The Presidential Commission recognised the importance of scientific creativity and experimentation to progress.[26] It argued that 'a robust public policy regarding the responsible conduct of science must promote the creative spirit of scientists and unambiguously protect their intellectual freedom'.[27]

The responsibility element reflects the recognition that scientists have obligations beyond merely furthering creative scientific enquiry, because:

> Society as a whole has a stake in what scientists and engineers do. Scientists and engineers should recognize the potential impact of their research on those who will experience both its benefits and burdens and their responsibility to those who provide the means, directly or indirectly, for their research.[28]

The Commission saw a role for self-regulation by the synthetic biology research community, and considered supporting this to be part of fostering responsibility.[29] It went on to state: 'Creating a culture of responsibility in the synthetic biology community could do more to promote responsible stewardship in synthetic biology than any other single strategy.'[30] The emphasis on creating a culture of responsibility and on supporting self-regulation may be recognition of the fact that there is some synthetic biology research being done outside traditional institutional settings, such as in the do-it-yourself biology community. This reality is noted repeatedly throughout the Report.

The concept of 'enhanced watchfulness' which forms part of the recommendation 'requires the scientific community to recognize the varied risks associated with synthetic biology and develop internal processes to identify and respond to potential threats rapidly and effectively'.[31] Thus self-regulation and communal norms were emphasised by the Presidential Commission in preference to government regulation.

The principle of intellectual freedom and responsibility also encompasses the ideal of 'regulatory parsimony'. Regulatory parsimony requires 'only as much oversight as is truly necessary to ensure justice,

---

25 Ibid., 12.
26 Ibid., 27.
27 Ibid.
28 Ibid., 141.
29 Ibid., 145.
30 Ibid., 11.
31 Ibid., 145.

fairness, security, and safety while pursuing the public good'.[32] The Commission elaborated:

> Regulatory parsimony is particularly important in emerging technologies, which by their very definition are still in formation and are *not always well-suited for sharply specified limitations*. The blunt instruments of statutory and regulatory restraint may not only inhibit the distribution of new benefits, but they can be counterproductive to security and safety by preventing researchers from developing effective safeguards.[33]

Under the principle of 'promoting intellectual freedom and responsibility', the Commission recommended a continued 'periodic assessment of security and safety risks' and that if the initial review revealed 'significant unmanaged security or safety concerns', additional controls should be considered.[34]

4.3.4 Democratic Deliberation

The fourth principle, democratic deliberation, relates to open and respectful debate and inclusive processes for public discussion.[35] Amy Gutmann and political scientist Dennis Thompson explain that:

> At the core of deliberative democracy is the idea that citizens and officials must justify any demands for collective action by giving reasons that can be accepted by those who are bound by the action. When citizens morally disagree about public policy, they therefore should deliberate with one another, seeking moral agreement when they can, and maintaining mutual respect when they cannot.[36]

The Commission stated that 'through democratic deliberation, questions about synthetic biology can be explored and evaluated on an ongoing basis in a manner that welcomes the respectful exchange of opposing views'.[37] Therefore, it 'sought to provide an inclusive forum for discussion' through its inquiry into synthetic biology.[38]

---

32 Ibid., 5.
33 Ibid., 28 (emphasis added).
34 Ibid., 13–14.
35 Ibid., 5.
36 Amy Gutmann and Dennis Thompson, 'Deliberating about Bioethics' (1997) **27**(3) *The Hastings Center Report* 38, 38.
37 The Presidential Commission for the Study of Bioethical Issues (n 1) 14.
38 Gutmann (n 6) 21.

Under this principle, the Commission encouraged an ongoing exchange about synthetic biology between various groups.[39] It also encouraged the use of 'clear and accurate language' about synthetic biology, and recommended establishment of a 'publicly accessible fact check mechanism' to enable impartial verification of scientific claims about synthetic biology.[40] In addition, it recommended expansion of education activities and identification of strategies to promote scientific and ethical literacy.[41]

4.3.5 Justice and Fairness

The fifth principle, justice and fairness, relates to promoting availability of information and fair distribution of the risks and benefits of research in synthetic biology.[42] While justice and fairness are relevant considerations regardless of the funding source for research, the Commission pointed out that 'fair distribution of the benefits of synthetic biology is an especially important consideration for government funded research'.[43] The Commission noted the potential for synthetic biology to cause environmental harm or damage to human health.[44] In some cases 'special consideration and safety reviews may be needed'.[45] Where there is potential for 'increased risk of harmful environmental exposures in specific locations', justice and fairness requires 'vigilant attention to these environmental risks'.[46]

To promote justice and fairness, the Commission recommended an evaluation of mechanisms that ensure the risks of synthetic biology research are 'not unfairly or unnecessarily distributed'.[47] It proposed that commercial manufacturers should 'ensure that risks and potential benefits to communities and the environment are assessed and managed so that the most serious risks … are not unfairly or unnecessarily borne by

---

39 The Presidential Commission for the Study of Bioethical Issues (n 1) 15.
40 Ibid., 15.
41 Ibid., 16, 157.
42 Ibid., 5.
43 Ibid., 162.
44 Ibid., 163.
45 Ibid.
46 Ibid., 164.
47 Ibid.

certain individuals, subgroups or populations'.[48] It further recommended evaluation of regulatory requirements regarding risk and benefit distribution.[49]

4.4 CRITIQUE OF THE COMMISSION'S APPROACH

Among the five ethical principles, the key principles for responding to the regulatory challenges raised by synthetic biology are public beneficence, responsible stewardship, intellectual freedom and responsibility, and democratic deliberation. This section provides a critique focusing on those principles.

4.4.1 Lost Opportunity to Make Recommendations for Policy Action

Presidential Commissions can perform a number of functions in a democratic society. These functions have become a field of study in themselves. Political scientist Amy Zegart points out that scholarship shows that Presidential Commissions can increase public support for a presidentially supported initiative, demonstrate high-level concern about an issue or a particular crisis, gather facts and information about an issue, investigate policy alternatives and promote consensus, among other functions.[50] Specifically, bioethics commissions have been demonstrated to stimulate debate, provide fora for the public debate of bioethical issues, gather and analyse information, disseminate information and create 'an instructive picture of the scientific, moral, legal and political landscape'.[51]

The Presidential Commission's inquiry into synthetic biology in 2010 had several functions. First, it had a 'symbolic reassurance' function in that it demonstrated a response to the 'crisis' of the Venter Institute's synthetic cell announcement. In a case of symbolic reassurance, a commission's establishment shows presidential awareness of an issue and

48 Ibid., 164–5.

49 Ibid., 165.

50 Amy B Zegart, 'Blue Ribbons, Black Boxes: Toward a Better Understanding of Presidential Commissions' (2004) **34**(2) *Presidential Studies Quarterly* 366, 372.

51 Margaret Foster Riley and Richard A Merrill, 'Regulating Reproductive Genetics: A Review of American Bioethics Commissions and Comparison to the British Human Fertilisation and Embryology Authority' (2005) **6** *Columbia Science & Techology Law Review* 1, 94–98.

public concern, and can also be used to 'inform the public and its leaders and to reassure the disaffected that they are being "taken seriously"'.[52] It responds to a sense that the president 'must *do something*'.[53] The impetus created by the synthetic cell announcement is shown in the fact that the reference was issued to the Commission on the same day as the Venter Institute's press release was issued, 20 May 2010. The Executive Office of the President is likely to have had advance notice of the impending announcement of the synthetic cell, in keeping with Venter's practice of briefing the White House before publishing major developments. The office would have recognised that the announcement could generate significant concern about the risks of synthetic biology, as indeed it did in the following days and weeks. It was important for the government to demonstrate that the matter was being considered.

The Commission's second function was to gather and digest information and present policy advice, one of the key functions identified by scholars of Presidential Commissions. For example, Zegart's typology of commissions includes both proactive and reactive 'information commissions', which aim to influence government officials in order to improve policy.[54] Their proactive function is to present facts and analysis as well as to identify policy challenges and options.[55] These commissions 'identify new policy problems looming on the horizon'.[56] They are designed to 'get in front of a policy problem', investigate possible future developments and propose how the government might respond.[57]

Gutmann regarded the Presidential Commission's work as proactive and noted that its process differed from previous bioethics advisory bodies, which took several years to study and debate issues before making recommendations.[58] She observed:

> The Commission had a unique opportunity to contribute proactively to a field of scientific enquiry that is relatively young … Its most promising potential benefits and most worrisome risks are not yet upon us, allowing time for

---

[52] See David Flitner Jr, *The Politics of Presidential Commissions* (Transnational Pub, 1986), 17–19.

[53] Ibid., 19.

[54] Zegart (n 50) 375.

[55] Ibid., 376.

[56] Ibid., 375.

[57] Ibid., 376.

[58] Gutmann (n 6) 20.

efforts to publicly consider and recommend safe development of the field for the good of all.[59]

Gutmann saw the reference from Obama as 'an exceptional opportunity to look forward instead of to merely react and to lead a proactive review of this emerging field'.[60]

However, this opportunity was not altogether seized. The Commission analysed the verbal evidence, submissions and some of the relevant literature in order to perform the functions of synthesising and digesting this information, presenting facts and identifying policy issues. However, the Commission was not able to go beyond this to provide regulatory policy *options* for assessment by the government. Instead, the Commission was limited to clarifying the nature of scientific developments in synthetic biology, identifying key risks and potentially beneficial outcomes and pinpointing the issues requiring further investigation. Many of its recommendations identify areas where more information is required.

The Commission did identify regulatory issues relatively early in the development of the technology of synthetic biology. However, this is only the beginning of a policy discussion. The way in which the Commission's principled approach would impact upon the development of regulatory strategies is not clear. There is a need to go beyond its survey of issues to move towards developing regulatory mechanisms.

The Presidential Commission was instructed to report within six months of the reference being issued.[61] This appears to have created limitations on what could be achieved. Presidential Commissions *can* provide advice as to specific policy action and legislative proposals. For example, President George W Bush's President's Council on Bioethics made specific recommendations concerning the use and creation of human embryos in research (though these were not implemented through legislative change).[62] Bush's Council considered the issues raised by embryonic stem cell research for a much longer time than the Presidential Commission studied synthetic biology: two years as opposed to six months. Also, the issue of embryo research can be seen as a narrower

---

59 Ibid., 17.

60 Ibid., 18.

61 Barack Obama, 'Letter from President Barack Obama to Dr Amy Gutmann' (20 May 2010); Barack Obama, 'Memorandum for Dr Amy Gutmann' (24 November 2010).

62 See recommendations in The President's Council on Bioethics, 'Reproduction and Responsibility: The Regulation of New Biotechnologies' (March 2004) <http://bioethics.georgetown.edu/pcbc/reports/reproductionandresponsibility/index.html>, 205–24.

scope of inquiry than all the ethical, regulatory and social issues raised by synthetic biology and its diverse applications. Thus the Bush Council was able to make more concrete policy proposals.

Perhaps the Commission shied away from making more specific recommendations because in addition to lacking time, it did not have sufficient information upon which to base these, or any specific regulatory proposals to assess. For some of the more complex issues, no positive models were put to the Commission. For example, no specific model for addressing biosecurity risk was put to the Commission in its consultations. Further, the Commission was not able to conduct in-depth empirical research into the issues raised by synthetic biology. Its review of the relevant literature was limited. It did not go in depth into the impact of synthetic biology on existing regulatory structures and the regulatory challenges which are raised. Although bioethics commissions often synthesise information, sometimes they have conducted or commissioned empirical or in-depth research.[63] For its research into human subject protection, the same Presidential Commission assembled an International Research Panel of experts which provided it with advice.[64] It also asked 18 government agencies which conduct research on human subjects for information, and constructed a database representing the research landscape.[65] An Empirical Advisory Group of experts assisted the Commission in this empirical work.

Thus while the Presidential Commission identified the key regulatory issues, it was not able to propose how to address these. While the principles espoused by the Commission have logical appeal, they are vague in the sense that they do not lead the reader to any guidance about how to make decisions, regulations and policy. More concrete regulatory recommendations could have been made, particularly in the areas of

---

[63] Alexander Morgan Capron, 'Governmental Bioethics Commissions: The Nature of the Beast' (2002) **9**(3) *Medical Ethics* 4.

[64] See The Presidential Commission for the Study of Bioethical Issues, 'Moral Science: Protecting Participants in Human Subjects Research' (The Presidential Commission for the Study of Bioethical Issues, December 2011) <http://bioethics.gov/cms/sites/default/files/Moral%20Science.pdf>, 4; International Research Panel of the Presidential Commission for the Study of Bioethical Issues, 'Research Across Borders: International Research Panel of the Presidential Commission for the Study of Bioethical Issues' (September 2011) <http://bioethics.gov/sites/default/files/IRP-%20Research%20Across%20Borders.pdf>.

[65] See The Presidential Commission for the Study of Bioethical Issues, 'Moral Science: Protecting Participants in Human Subjects Research' (n 64) 4–5.

intellectual property and biosecurity. While the report focuses on environmental risk, much of the evidence raised in consultations actually concerned these other two issues.

The Commission had an opportunity to push discussion forward, to move beyond raising the issues and canvassing the different points of view, and suggesting that someone should keep an eye on things. This analysis suggests that the short time frame and the complexity of the issues have limited the ability of the Commission to present policy proposals.[66]

4.4.2 Similar Concerns?

In its overall conclusion, the Presidential Commission stated that synthetic biology raises largely similar concerns to previous technologies, and that there are many oversight bodies and mechanisms that are well situated to respond to these. However, it acknowledged that some risks warrant special attention, particularly due to the uncertainty regarding the biological properties of synthetic organisms, and the do-it-yourself biology movement.[67]

The Commission concluded that the risks appeared to be 'appropriately managed', but recommended that the Executive Office of the President should 'evaluate and re-evaluate periodically the effectiveness of current research oversight mechanisms' and 'the government should continue to assess specific security and safety risks'.[68] More specific recommendations could have been made about managing these risks, as is argued in subsequent chapters.

Social science researcher Zhang and collaborators have identified a tendency for a 'closing down' of discussion about synthetic biology.[69]

---

66 Of course there are also more cynical explanations, such as that the success of such commissions is often measured by implementation of their recommendations, and therefore they are likely to recommend only policy actions seen as attractive to the President or already being taken: see W David Snowball, 'Rhetorical Constraints on Government Commissions' in Christine M Miller and Bruce C McKinney (eds), *Government Commission Communication* (Praeger, 1993) 31, 44–7.

67 The Presidential Commission for the Study of Bioethical Issues (n 1) 125.

68 Ibid., 13.

69 Joy Y Zhang et al, 'The Transnational Governance of Synthetic Biology: Scientific Uncertainty, Cross-borderness and the "Art" of Governance' (BIOS Working Paper No: 4, BIOS (Centre for the Study of Bioscience, Biomedicine, Biotechnology and Society), May 2011) <http://www2.lse.ac.uk/BIOS/publications/Working%20Papers/WP4TransnationalGovernance.pdf>, 16.

Discussion tends to revolve around questions about specific risks, and is 'closed down' by saying that all risks are covered by existing regulations.[70] This means that indeterminate, unknown aspects are not analysed, nor are the social changes that occur with scientific development.[71] Simply stating that synthetic biology's risks are covered by regulation for genetically modified organisms ignores the problems, controversies and inadequacies of that regulatory regime.[72] This closing down is also reflected in the Commission's report. For example, it concluded that in the areas of agriculture, food and environment, the environmental risks of synthetic biology are similar to genetic modification technologies.[73] This discussion did not consider the *differences* between the social and scientific context and regulatory challenges of genetic modification technologies and those of synthetic biology. As I establish in Chapter 5, there are important differences between these, which impact on the need for new regulatory strategies.

The Presidential Commission developed principles for the assessment of emerging technologies generally. There is certainly a role for looking at similarities between synthetic biology and earlier technologies. Indeed, law traditionally works by drawing analogies between new cases or present circumstances, and those to which the law has previously applied.[74] However, if this analysis closes off discussion of the distinctiveness of the newly emerging technology, this is problematic. While acknowledging the importance of developing general ideas and principles to apply to the development of regulation of emerging technologies where possible, it is *also* critical to look at what is distinctive about a particular technology. It is important to take into account the specific features of the technology.[75] This analysis should consider distinctiveness both in terms of the science itself, its social and scientific context and its regulatory challenges. This would promote the goals of regulatory

---

[70] Ibid.

[71] Ibid.

[72] Ibid., 17.

[73] The Presidential Commission for the Study of Bioethical Issues, 'New Directions: The Ethics of Synthetic Biology and Emerging Technologies' (n 1) 70.

[74] Roger Brownsword, 'So What Does the World Need Now? Reflections on Regulating Technologies' in Roger Brownsword and Karen Yeung (eds), *Regulating Technologies: Legal Futures, Regulatory Frames, and Technological Fixes* (2008) 23, 30–32.

[75] Ibid.

legitimacy, effectiveness and connection, and the dual regulatory roles of maximising benefit and minimising harm.

4.4.3 Regulatory Parsimony

As discussed, the concept of regulatory parsimony requires that there should only be 'as much oversight as is truly necessary to ensure justice, fairness, security, and safety while pursuing the public good'.[76] But how is the question of how much regulation is truly necessary to be answered? The principles and discussion set out in the Presidential Commission's Report do not aid in answering this question and developing a regulatory response. More guidance in relation to a regulatory strategy is required, for example in the area of biosecurity, which the Commission recognises is 'one of the most widely voiced risks' of synthetic biology research.[77]

Gutmann argues that this principle of regulatory parsimony is particularly important when evaluating emerging technologies 'where the temptation to stifle innovation on the basis of uncertainty and fear of the unknown is particularly great'.[78] She describes statutory and regulatory measures as 'blunt instruments', suggesting that they could limit the distribution of beneficial outcomes of the technology, while being 'counter-productive to security and safety by preventing researchers from developing effective safeguards'.[79]

While Gutmann's concerns are valid observations about regulating emerging technologies, synthetic biology requires differentiated consideration because of the novelty of the risks, the concomitant lack of current information to assess and manage these, and the severity of harm that could be caused if risks are realised. The technical feasibility of constructing a novel pathogen using synthetic biology techniques may be science fiction for now, but it could occur in the near future. There needs to be enforceable, accountable control of access to sequences of synthetic DNA from dangerous pathogens. This could be introduced now, before problems are observed under the approach of the Presidential Commission.

Gutmann has also argued that the Commission's recommendations bring to the fore the role of those outside government in addressing

---

76 Gutmann (n 6) 19.

77 The Presidential Commission for the Study of Bioethical Issues, 'New Directions: The Ethics of Synthetic Biology and Emerging Technologies' (n 1) 72.

78 Gutmann (n 6) 19.

79 Ibid.

emerging technologies, particularly citizens, experts and the scientific community.[80] The preference for less regulation seems to have led the Commission to prefer *self*-regulation in some areas of risk management, comparing the field of synthetic biology to self-oversight and caution in the recombinant DNA field. It stated that: 'While self-governance is not a sufficient means to mitigate all risks, it is likely an effective way to control many of the risks associated with emerging technologies, including synthetic biology, particularly at this early stage.'[81]

However, there are significant problems associated with self-regulation, as discussed in the previous chapter. Self-regulation can produce self-interested regulation.[82] There are issues of legitimacy of the regulation,[83] lack of public trust and the prospect of overly techno-optimistic regulation. These matters are particularly problematic in the synthetic biology context due to the high level of uncertainty regarding the risks of environmental harm, biosafety problems and biosecurity risk. This is particularly the case where scientists are outside the usual venues for scientific research, as is the case in the emerging field of do-it-yourself biology.

4.4.4 Approach to Risk

The Presidential Commission chose to develop the new concepts of responsible stewardship and prudent vigilance for the assessment and management of risks posed by synthetic biology. In general, the Commission's approach to risks raised by synthetic biology, based on prudent vigilance and intellectual freedom and responsibility, is not sufficiently cautious.[84]

In subsequent chapters I argue that the Presidential Commission should have recommended greater regulatory action to address environmental, biosafety and biosecurity risks.

---

[80] Ibid., 20.

[81] The Presidential Commission for the Study of Bioethical Issues, 'New Directions: The Ethics of Synthetic Biology and Emerging Technologies' (n 1) 129.

[82] Roger Brownsword, *Rights, Regulation and the Technological Revolution* (Oxford University Press, 2008), 153–5.

[83] Ibid.

[84] Per Sandin, 'A New Virtue Based Understanding of the Precautionary Principle' in Mark A Bedau and Emily C Parke (eds), *The Ethics of Protocells: Moral and Social Implications of Creating Life in the Laboratory* (MIT Press, 2009) 89, 98.

4.4.5 Strengths of the Presidential Commission's Approach

There are a number of useful aspects of the approach taken in the Commission's Report. These include the recommendation to engage in a broad gap-finding exercise in the regulatory 'patchwork', and the recognition that circumstances will change over time which may require regulatory measures to be changed.

(a) Adaptation

A theme running through the recommendations was the importance of reviewing various matters on an ongoing basis and updating regulatory processes where necessary. This includes review of scientific information, risks and risk assessment processes. In relation to promoting responsible stewardship, the Commission recommended continuing review and updating of regulatory processes, 'to ensure that regulators have adequate information'.[85] However, a specific process for adaptation was not suggested.

The principle of public beneficence espoused by the Presidential Commission also encompasses 'being ready to revise policies that pursue public benefit with insufficient caution'.[86] The Commission noted that the existing regulatory environment for synthetic biology in the US comprises a 'patchwork quilt of measures [which] is built on long-standing practices that have adapted to new technologies over time'.[87] These measures include regulation by many government agencies at the level of the lab, field, environment and marketplace, and cover specific areas of risks and specific applications of synthetic biology.[88] The Presidential Commission observed the need for adapting regulation in this field.[89]

The theme of ongoing review continued in the Commission's application of the principle of intellectual freedom and responsibility. It recommended that there should be evaluation and periodic re-evaluation of the 'effectiveness of current research oversight mechanisms and determine what, if any, additional steps should be taken to foster

---

[85] The Presidential Commission for the Study of Bioethical Issues, 'New Directions: The Ethics of Synthetic Biology and Emerging Technologies' (n 1) 9.
[86] Ibid., 4.
[87] Ibid., 101.
[88] Ibid., 79–111.
[89] Ibid., 102.

accountability at the institutional level without unduly limiting intellectual freedom'.[90]

While not expressly stated, it seems that the concept of prudent vigilance was preferred to other approaches because it was seen as allowing adaptable regulation. For example, Commission Member John Arras explained that the concept 'includes ongoing assessments of risk as the technology develops … Because of the difficulty of risk analysis in the face of uncertainty, particularly for low probability/high impact events, ongoing assessments will be needed as the field progresses. Regulatory systems should be updated as necessary'.[91] Thus ongoing evaluation of risks is a requirement of prudent vigilance.[92] This is the most useful aspect of this principle. If policies are adaptable, synthetic biology regulation has a better chance of remaining 'connected' to the state of the science, an important characteristic of regimes for regulating emerging technologies.[93] Further, policies that are not working well can be identified and adapted. This allows regulators to identify and address the problem of regulatory 'disconnection' and to find ways to 'reconnect' the regulatory regime to the state of development of the technology.[94] For example, in relation to intellectual property, if 'agency learning'[95] can occur through review of the patent environment and patent granting practices in synthetic biology, potential unintended adverse impacts of the patent system may be mitigated or avoided. These mechanisms will also assist in promoting the goal of regulatory effectiveness.

(b) Coordination

The Presidential Commission's recommendations regarding coordination among federal agencies are also helpful. Maximising coordination should

---

[90] Ibid., 13.

[91] John Arras 'Comments in Commission Deliberations, The Presidential Commission for the Study of Bioethical Issues, Synthetic Biology' 16 November 2010.

[92] The Presidential Commission for the Study of Bioethical Issues, 'New Directions: The Ethics of Synthetic Biology and Emerging Technologies' (n 1) 27.

[93] As explained in the Introduction, see Brownsword, 'So What Does the World Need Now?' (n 74).

[94] See Roger Brownsword and Han Somsen, 'Law, Innovation and Technology: Before We Fast Forward – A Forum for Debate' (2009) **1** *Law, Innovation and Technology* 1, 26–8.

[95] See Lawrence C McCray et al, 'Planned Adaptation in Risk Regulation: An Initial Survey of US Environmental, Health, and Safety Regulation' (2010) **77** *Technological Forecasting & Social Change* 951, 952.

help to reduce the burden on scientists and costs of implementing regulatory measures. The issue of poor regulatory coordination has been noted in discussion regarding synthetic biology regulation.[96] Similarly, the Commission heeded the evidence it repeatedly received that synthetic biology research is a truly international field. The recommended goals of promoting consistency of regulation across jurisdictions and participating in engagement and discussion at the international level are therefore practical and important.

(c) Democratic deliberation and legitimacy

A key benefit of the Presidential Commission's inquiry into synthetic biology was that of fostering debate and providing an inclusive forum for this to occur. The creation of a forum for discussion was part of the Commission's goal of facilitating of democratic deliberation. Facilitating deliberative democracy can be seen as a function of law reform bodies generally.[97] This includes bioethics commissions, as a particular species of law reform body.

The Commission furthered the goals of democratic deliberation by adopting a process that encouraged wide public participation and was transparent. For example, in conducting its consultations, the Presidential Commission sought written comment from *any* interested party. It conducted its inquiry in a very publicly accessible way, holding consultations in public, allowing public questions to the invited experts and Commission Members and posting transcripts of these sessions on its website.

The Commission provided a venue for discussion by experts in the field and also members of the public, allowing contribution to the debates from a broader range of actors than had previously occurred.

The Commission tried to be a broker in the sense that democratic deliberation requires looking for common ground between perspectives as much as possible.[98] Gutmann stated that, 'When common ground was

---

[96] See summary of discussion at synthetic biology conference in '21st Century Borders/Synthetic Biology: Focus on Responsibility and Governance' (Institute on Science for Global Policy, 2013) <http://www.scienceforglobalpolicy.org/LinkClick.aspx?fileticket=VqAc-vBzvMk%3d&tabid=133> 72, 74–5, 95.

[97] See e.g., discussion of law reform commissions in jurisdictions using the Westminster system of government in Marcia Neave, 'Institutional Law Reform In Australia: The Past and the Future' (2005) **23** *Windsor Year Book of Access to Justice* 343, 362.

[98] See Gutmann (n 6) 19.

impossible to find, the Commission cultivated mutual respect through active engagement with differing views.'[99] Thus the Commission sought to bring together the various stakeholders involved, find a middle ground between techno-optimism and techno-phobia, present an unbiased account of the issues and offer reasonable recommendations.

This wide consultation process should help to promote regulatory legitimacy, identified as one of the desirable criteria for strategies for regulating emerging technologies.[100] As the Commission pointed out, 'Public deliberation is particularly valuable while the field is still young.'[101] Through this consultative approach, the Commission also showed awareness of the importance of public trust in both the scientific community and the regulatory systems that are applied to synthetic biology.[102]

The Commission's focus on democratic deliberation and public benefit will also assist in furthering a principle identified in sociological review of synthetic biology regulation, the principle of external accountability.[103] This refers to the need for accountability of the researchers to those outside science, to all who may be affected by the research.[104]

4.5 CONCLUSION

This chapter has considered the role and the recommendations of the Commission in light of the criteria of regulatory legitimacy, effectiveness and connection introduced in Chapter 1. The facilitation of democratic deliberation by the Commission should assist in addressing the plurality of views about the risks and benefits of synthetic biology.[105] The process

---

99 The Presidential Commission for the Study of Bioethical Issues, 'New Directions: The Ethics of Synthetic Biology and Emerging Technologies' (n 1) 21.

100 See Brownsword, 'So What Does the World Need Now' (n 74) 31–2; see also Cass R Sunstein, *Risk and Reason: Safety, Law and the Environment* (Cambridge University Press, 2002), 294.

101 The Presidential Commission for the Study of Bioethical Issues (n 1) 152.

102 For discussion of the importance of public trust in relation to emerging technologies and their regulation generally see Gregory N Mandel, 'Regulating Emerging Technologies' (2009) **1** *Law, Innovation & Technology* 75, 90.

103 Zhang et al (n 69) 4.

104 Ibid.

105 See Roger Brownsword and Morag Goodwin, *Law and the Technologies of the Twenty-First Century: Text and Materials* (Cambridge University Press, 2012), 59.

adopted by the Commission has improved procedural legitimacy of regulatory mechanisms which are implemented based on its recommendations.[106] In its acknowledgement of the need for adaptive regulation, it has promoted connection between regulatory instruments and the stage of development of synthetic biology.

Overall, the Commission's Report had a helpful emphasis on becoming informed, rather than panicking about synthetic biology. It was realistic about the nature of what was achieved by the Venter group. Nonetheless, the Commission could have done more to seize its opportunity to make recommendations early in the development of an emerging technology. It did not put forward regulatory proposals for assessment by the government. Its recommendations do not form the basis of regulation of synthetic biology that could fulfil the dual regulatory roles of managing risks and facilitating desirable innovation.[107] Its role was limited to 'symbolic reassurance', creating an inclusive forum for public debate and synthesising information.

The substantive recommendations are problematic in relation to regulatory effectiveness. Further, the Commission emphasised what was *similar* about the risk profile of synthetic biology and recombinant DNA technology, but the *differences* in the regulatory challenges posed were not explored. The Presidential Commission's focus on similarities has limited its ability to develop effective regulatory proposals for synthetic biology, both in relation to the intellectual property issues posed by synthetic biology and the matters of environmental, biosafety and security risks. Subsequent chapters examine the distinctiveness of synthetic biology in these areas in greater depth.

---

[106] The Commission itself pointed out that 'public participation frequently fosters the perceived political legitimacy of the policies and practices that are ultimately chosen': The Presidential Commission for the Study of Bioethical Issues (n 1) 152.

[107] These regulatory roles are explained in Chapter 1 and are used as a starting point for the analysis in this book, drawing on Brownsword, *Rights, Regulation and the Technological Revolution* (n 82); Brownsword and Somsen (n 94); Brownsword and Goodwin (n 105).

PART II

Regulating for the risks

5. Environmental risk: uncertainty, precaution, prudent vigilance and adaptation

5.1 INTRODUCTION

Risk and uncertainty give rise to some of synthetic biology's major regulatory challenges. Two key areas of risk generated by synthetic biology research are intentional bioterrorist attack on the one hand and accidental damage to human health or to the environment on the other. Uncertainty regarding these risks is pervasive in debates about synthetic biology. This chapter explores the issues posed by potential damage to the environment and biodiversity.

Synthetic biology's applications in relation to environmental remediation, energy and agriculture pose significant environmental safety risk, as they involve living cells as opposed to non-living compositions of biological parts, and could be released into the open environment.[1] Harm to the environment could arise in several ways. Synthetic organisms, or highly modified organisms, could be released into the environment either intentionally, such as in a commercial field trial, or accidentally. They will be different from naturally occurring organisms in more complex and substantial ways than previously developed recombinant DNA organisms, which may create greater uncertainty about their environmental impacts and make regulation more difficult.[2] Synthetic or highly engineered organisms could survive and persist in the natural environment rather than dying off.[3] They might reproduce and spread, impacting upon

---

[1] Jennifer Kuzma and Todd Tanji, 'Unpackaging Synthetic Biology: Identification of Oversight Policy Problems and Options' (2010) **4**(1) *Regulation & Governance* 92, 103–6.

[2] Genya V Dana et al, 'Synthetic Biology: Four Steps to Avoid a Synthetic-biology Disaster' (2012) **483**(7387) *Nature* 29; Kuzma and Tanji, ibid., 108.

[3] Dana et al, ibid.

natural ecosystems.[4] They may be able to adapt to the natural environment and change their characteristics over time, possibly out-competing natural organisms and damaging biodiversity.[5] Damage to existing ecosystems could also occur if these organisms transfer their DNA to natural organisms.[6] Genes can 'flow' between microorganisms like bacteria in several ways, such as through viruses, and it is difficult to predict how this will impact upon natural bacteria.[7] The release of synthetic organisms could also produce novel pests or invasive species.[8] There are many areas where more information about the behaviour of synthetic or highly modified organisms in the environment is needed.[9] There is also uncertainty about how effective proposed technical methods of enhancing the safety of these organisms will be.[10]

Environmental remediation and renewable technologies are areas in which synthetic biology promises some of the most transformative benefits.[11] Synthetic biology companies are investing in research using

---

4 Ibid.

5 Ibid.

6 Denise Caruso, 'Synthetic Biology: An Overview and Recommendations for Anticipating and Addressing Emerging Risks' (Science Progress, 2008) <http://scienceprogress.org/2008/11/synthetic-biology/>, 2–3; The Woodrow Wilson International Center for Scholars: Science Technology & Innovation Program, *Comprehensive Environmental Assessment and Synthetic Biology Applications Workshop: Conference Notes Version 1.0* (2011) <http://www.synbioproject.org/process/assets/files/6609/_draft/final_cea_and_syn_bio_workshop_notes_wilson_center.pdf>, 4-6.

7 The Woodrow Wilson International Center for Scholars: Science Technology & Innovation Program, *Comprehensive Environmental Assessment and Synthetic Biology Applications Workshop: Background Materials* (2011) <http://www.synbioproject.org/process/assets/files/6609/_draft/background_packet_for_cea_and_syn_bio_workshop_jul_28_2011.pdf>, 15; Kent H Redford et al, 'Synthetic Biology and Conservation of Nature: Wicked Problems and Wicked Solutions' (2013) **11**(4) *PLoS Biol* e1001530.

8 The Presidential Commission for the Study of Bioethical Issues, 'New Directions: The Ethics of Synthetic Biology and Emerging Technologies' (The Presidential Commission for the Study of Bioethical Issues, December 2010) <http://bioethics.gov/sites/default/files/PCSBI-Synthetic-Biology-Report-12.16.10_0.pdf>, 70.

9 Redford et al (n 7) 3; Markus Schmidt et al, 'A Priority Paper for the Societal and Ethical Aspects of Synthetic Biology' (2009) **3**(1) *Systems and Synthetic Biology* 3, 4.

10 Caruso (n 6) 3.

11 Markus Schmidt, 'Executive Summary' in Markus Schmidt (ed.), *Synthetic Biology: Industrial and Environmental Applications* (Wiley-Blackwell, 2012) xxi, xxi.

synthetic biology to produce cleaner technologies, such as biofuels, as discussed in Chapter 2. Synthetic biology may also have applications in pollutant detection and environmental remediation. The question is how to draw the balance between minimising the risk of damage to the environment and health from synthetic biology products, while also capturing the potential benefits of such products. These are difficult issues to reconcile, but doing so is important as it is the role of regulation to both promote the beneficial innovation and minimise the potential harms of synthetic biology.[12]

This chapter considers legal approaches to managing uncertainty, before looking at how these issues are playing out in the synthetic biology context. It critiques the Presidential Commission's focus on the concept of prudent vigilance. This chapter considers the existing regulation of environmental risk arising from synthetic biology products. It then explores an approach to managing uncertainty drawn from political science and environmental regulation, 'planned adaptation', and evaluates its usefulness in the synthetic biology context. The final part of this chapter argues that the Commission should have adopted a precautionary approach and taken less comfort from the possibility of technical measures to prevent environmental harm.

5.2 RISK, UNCERTAINTY AND PRECAUTION

5.2.1 Risk and Uncertainty

Environmental degradation and public health are areas which have been discussed in terms of risk, uncertainty and precaution for several decades. More recently, since changes to the political and social environment following the attacks of September 2001, terrorism has also come within this discourse.[13]

---

[12] As discussed in Chapter 1, addressing issues of risk is one of the roles of regulation of emerging technologies set out in Roger Brownsword and Han Somsen, 'Law, Innovation and Technology: Before We Fast Forward – A Forum for Debate' (2009) **1** *Law, Innovation and Technology* 1, 2–3; see also Roger Brownsword and Morag Goodwin, *Law and the Technologies of the Twenty-First Century: Text and Materials* (Cambridge University Press, 2012), 23, 46.

[13] See, e.g., the discussion of bioterrorism in Michael McFadden et al, 'Uncertainty, Terrorism and Law Enforcement' in Gabriele Bammer and Michael Smithson (eds), *Uncertainty and Risk: Multidisciplinary Perspectives* (Earthscan, 2008) 261.

There is a distinction between risk and uncertainty. Risk refers to an event, the probability of which can be estimated using either theory, experience or both, whereas uncertainty arises when the probability of a risk eventuating cannot be estimated by either of these methods.[14] Scientific uncertainty involves 'controversies over the possibility of adverse effects to the environment or human health, their scope or their degree of seriousness'.[15] There is intractable scientific and other uncertainty regarding the biosafety and bioterrorism risks associated with synthetic biology. A bioterrorist attack is an event for which the probability cannot be estimated as it is not possible to obtain all the necessary intelligence information.[16] The likelihood of a bioterrorist attack involving synthetic biology techniques is uncertain because we do not know when it will become feasible for scientists to use synthetic biology to produce a dangerous organism in a laboratory, or at what point this would become easier than procuring naturally occurring dangerous organisms. There are also uncertainties about precisely what resources in terms of skills, equipment and financial resources would be needed. Risks from accidentally generated biosafety threats are also uncertain, because it is not possible to know what precisely do-it-yourself biologists are working on, how well contained these activities are and what the impact on human health would be if there was accidental generation and release of a dangerous organism.

Scientists cannot predict how biodiversity would be affected by organisms produced with synthetic biology, due to their novelty. We do not have experience of similar organisms which we can draw upon to make predictions about the environmental impact of synthetic organisms, or genetically engineered algae and bacteria, the main vehicles for current synthetic biology research.[17] This means that it is not possible to predict the nature of all potential environmental harm. We are also unable to estimate the probability or severity of any particular kind of environmental harm which could be caused by synthetic biology. Regulators are

---

[14] Richard A Posner, *Catastrophe: Risk and Response* (Oxford University Press Inc, 2004), 171.

[15] Rene von Schomberg, 'The Precautionary Principle and its Normative Challenges' in Elizabeth Fisher, Judith Jones and Rene von Schomberg (eds), *Implementing the Precautionary Principle: Perspectives and Prospects* (Edward Elgar Publishing, 2006) 19, 23.

[16] Posner (n 14) 174.

[17] Allison Snow, 'Evidence to The Presidential Commission for the Study of Bioethical Issues, Synthetic Biology' (8 July 2010) Washington DC.

in a situation of 'ignorance', that is they are 'unable to specify either the probability of bad outcomes or their nature'.[18]

The Commission acknowledged this problem of uncertainty in its deliberations on synthetic biology. For example, Commission member John Arras commented:

> synthetic biology is a new and very energetic field. We're witnessing very rapid development. It's very difficult to predict ahead of time exactly what sorts of benefits will eventually be derived, and what sorts of risks will eventually materialize here. We're speculating largely in the dark. So, we're facing a great amount of uncertainty … What is the proper attitude toward risk when the future is shrouded in uncertainty? That's the big problem, especially when you're dealing with events that are of low probability, but very high impact.[19]

5.2.2 Precaution

There are several legal approaches which have been developed to deal with scientific uncertainty.[20] Of particular relevance for this discussion are 'acknowledgement of scientific uncertainty', 'burden shifting' and 'adaptive management'.[21] Acknowledgement of scientific uncertainty is a legal approach which underpins several regulatory strategies, including that of precaution.[22] Scientific uncertainty is one of the circumstances which can trigger the application of the precautionary principle.[23] This

---

[18] See discussion of 'non-knowing', 'ignorance' and 'unknown unknowns': Ulrich Beck, *World at Risk* (Polity Press, English Edition ed, 2009), 53–4, 115–28; Cass R Sunstein, *Laws of Fear: Beyond the Precautionary Principle* (Cambridge University Press, 2005), 60; Brian Wynne, 'Uncertainty and Environmental Learning: Reconceiving Science and Policy in the Preventive Paradigm' (1992) **2**(2) *Global Environmental Change* 111.

[19] John Arras 'Comments in Commission Deliberations, The Presidential Commission for the Study of Bioethical Issue, Synthetic Biology' 16 November 2010.

[20] Australian environmental lawyer Judith Jones provides a useful typology of these legal approaches to scientific uncertainty in Judith Jones, 'Regulatory Design for Scientific Uncertainty: Acknowledging the Diversity of Approaches in Environmental Regulation and Public Administration' (2007) **19**(3) *Journal of Environmental Law* 347.

[21] Ibid.

[22] Ibid., 353.

[23] Ibid.

principle originated in German domestic law in the 1970s.[24] It has subsequently been adopted in international environmental law as well as being adopted in the domestic law of various jurisdictions including Australia, Canada, and European nations.[25]

The principle guides regulators in responding to scientific uncertainty about potential harm. The principle has been described as a 'better to be safe than sorry' principle.[26] There are many different definitions of the precautionary principle and there are inconsistencies between these.[27] A commonly cited formulation of the precautionary principle is that in the Rio Declaration on Environment and Development: 'Where there are threats of serious or irreversible damage, lack of full scientific certainty shall not be used as a reason for postponing cost-effective measures to prevent environmental degradation.'[28]

The precautionary principle has been reflected in numerous international instruments. It attained legitimacy and wide recognition when it was included in international 'soft law' instruments in the 1990s, among them the Rio Declaration.[29] It has also been incorporated into binding international treaties such as the United Nations Framework Convention on Climate Change, the Convention on Biological Diversity and a protocol to that convention, the Cartagena Protocol on Biosafety.[30] While the principle was originally developed in environmental law, it has also been incorporated into international law beyond the environmental

---

[24] See Jonathan B Wiener, 'The Rhetoric of Precaution' in Jonathan B Wiener et al (eds), *The Reality of Precaution: Comparing Risk Regulation in the United States and Europe* (Routledge, 2011) 3, 9.

[25] Elizabeth Fisher et al, 'Implementing the Precautionary Principle: Perspectives and Prospects' in Elizabeth Fisher, Judith Jones and Rene von Schomberg (eds), *Implementing the Precautionary Principle: Perspectives and Prospects* (Edward Elgar Publishing, 2006) 1, 3.

[26] Tim Lewens, 'Introduction: Risk and Philosophy' in Tim Lewens (ed.), *Risk: Philosophical Perspectives* (Routledge, 2007) 1, 10.

[27] See Sunstein, *Laws of Fear: Beyond the Precautionary Principle* (n 18) 18.

[28] Rio Declaration on Environment and Development Annex 1 UN Doc A/CONF.1.51/26 (Vol. I) (12 August 1992), Principle 15.

[29] Fisher et al, (n 25) 3; ibid.

[30] United Nations Framework Convention on Climate Change opened for signature 9 May 1992 1771 UNTS 107 (entered into force 21 March 1994), art 3(3); Convention on Biological Diversity, opened for signature 5 June 1992 1760 UNTS 79 (entered into force 29 December 1993), Preamble; Cartagena Protocol on Biosafety to the Convention on Biological Diversity, opened for signature 29 January 2000 2226 UNTS 208 (entered into force 11 September 2003), Preamble, arts 1, 10(6), 11(8).

sphere, for example in provisions of the Agreement on the Application of Sanitary and Phytosanitary Measures within the World Trade Organization framework.[31]

The precautionary principle is not triggered either: (a) when the type of harm is known and the probability can be quantified; or (b) when the harm is only 'hypothetical' or an 'imaginary risk'.[32] In the realm between these two scenarios, there are two sets of circumstances where the precautionary principle can be invoked: first, where the type of potential harm is known but causal relationships are unknown or uncertain and so probabilities cannot be estimated; and secondly, where the scope of effects is unknown and their severity can only be qualitatively estimated.[33]

Application of the precautionary principle involves considering the available scientific evidence. Some guidelines on the application of the precautionary principle expressly state that there must first be a scientific examination of the problem before the principle is invoked.[34]

The precautionary principle can be viewed through different lenses, for example as a procedural requirement relating to decision-making processes.[35] It can also be seen as a burden-shifting mechanism in which an applicant wishing to undertake a particular activity or development is required to show that it is safe for health or the environment.[36]

In relation to the way the principle impacts on decision-making, Fisher and Harding stress that the principle is one which is implemented in the administrative decision-making context, noting that 'the principle regulates the reasons for a decision and the process by which a decision is

---

[31] Marrakesh Agreement Establishing the World Trade Organization: Annex 1A Agreement on the Application of Sanitary and Phytosanitary Measures, opened for signature 15 April 1994 (entered into force 1 January 1995), art 5(7).

[32] Schomberg (n 15) 29.

[33] Ibid.

[34] European Commission, Communication from the Commission on the Precautionary Principle (2000) Commission of the European Communities <http://eur-lex.europa.eu/LexUriServ/LexUriServ.do?uri=COM:2000:0001:FIN:EN:PDF>, 3.

[35] Per Sandin, 'Common-sense Precaution and Varieties of the Precautionary Principle' in Tim Lewens (ed.), *Risk: Philosophical Perspectives* (Routledge, 2007) 99. See also Per Sandin, 'A New Virtue Based Understanding of the Precautionary Principle' in Mark A Bedau and Emily C Parke (eds), *The Ethics of Protocells: Moral and Social Implications of Creating Life in the Laboratory* (MIT Press, 2009) 89.

[36] Jones (n 20) 355–6.

made' rather than the outcome of the process.[37] The principle has required a shift away from a previously dominant norm of decision-making which required decisions to be based on fully established facts.[38] It lowers the threshold for level for government action, providing a rationale for action in circumstances where there is the potential for significant adverse impact on the environment or health, but also ongoing scientific uncertainty.[39]

Sunstein, one of the most influential critics of the precautionary principle, argues that if taken too far it can be 'paralysing' because it could be used to argue against all forms of regulatory action as well as inaction.[40] Regulatory action can itself pose risks, for example, there may be health risks associated with genetically modified crops, but there may also be risk of significant loss of life through starvation if these crops are not used.[41] In Sunstein's view both regulation of genetically modified crops and non-regulation fall foul of the precautionary principle if 'taken literally'.[42] While acknowledging the correctness of the principle to the extent that it suggests 'significant investments are worthwhile to prevent speculative harms', Sunstein stressed that what is important is the cost of the measure taken to prevent harm, and the severity of the harm. If action is required in response to all risks of harm, resources would soon be depleted.

Sunstein has argued that the precautionary principle *seems* to provide guidance due to various distortions in the way 'ordinary people' think about risk. People tend not to adequately consider the actual probability of the harm eventuating (particularly where emotions are involved).[43] This 'probability neglect' can be manipulated.[44] It may lead regulators to regulate in response to public fear and demand for intervention.[45]

---

37 Elizabeth Fisher and Ronnie Harding, 'The Precautionary Principle and Administrative Constitutionalism: The Development of Frameworks for Applying the Precautionary Principle' in Elizabeth Fisher, Judith Jones and Rene von Schomberg (eds), *Implementing the Precautionary Principle: Perspectives and Prospects* (Edward Elgar Publishing, 2006) 113, 115.

38 Ibid., 115.

39 Schomberg (n 15) 23, 29.

40 Cass R Sunstein, *Risk and Reason: Safety, Law and the Environment* (Cambridge University Press, 2002), 103–5 and 182–3.

41 Sunstein, *Laws of Fear: Beyond the Precautionary Principle* (n 18) 27–32, 103.

42 Ibid., 27–32.

43 Ibid., 64–5.

44 Ibid., 67–9.

45 Ibid., 69.

Sunstein also argues that people over-estimate some risks because those threats are 'available', which means that an example can be brought easily to mind.[46] Further, people focus on reducing the 'target risk' at the expense of considering the other risks that could be introduced by regulatory action, and they tend to be more concerned about harm caused by loss from the status quo than potential benefits forgone.[47]

Sunstein's critique is based on a 'strong' version of the precautionary principle, which requires regulation 'wherever there is a possible risk to health, safety or the environment, even where the supporting evidence remains speculative and even if the economic costs of regulation are high'.[48] It leaves open the questions of what is required in order to justify regulation, and whether other risks will be considered.[49]

While the principle is often criticised for vagueness or imprecision, there has been extensive scholarly discourse and judicial interpretation regarding its elements and triggers for its application, spanning several decades, which has fleshed out the principle such that extreme interpretations like those underlying the approach of Sunstein are not tenable. Indeed, the history of adoption and implementation of the principle in various jurisdictions shows it is 'both workable and legitimate', as well as being the subject of substantial case law and administrative experience.[50] In Australia, for example, case law in this area has been growing since the 1990s.[51] There is also US case law which has taken a precautionary approach, as will be discussed below. In response to criticism of the precautionary principle as 'incoherent', proponents of the principle have emphasised its nature as a legal principle, with inherent flexibility, and the way in which it is influenced by legal culture.[52]

---

[46] Ibid., 35–9, 81.

[47] Ibid., 40–49.

[48] Ibid., 24.

[49] Ibid., 25.

[50] Fisher et al, 'Implementing the Precautionary Principle: Perspectives and Prospects' (n 25) 2, 4.

[51] See e.g., *Telstra Corporation Ltd v Hornsby Shire Council* (2006) 146 LGERA 10.

[52] Fisher et al, 'Implementing the Precautionary Principle: Perspectives and Prospects' (n 25) 5. However, while defending the legitimacy of the precautionary principle, Fisher, Jones and von Schomberg do acknowledge that there are challenges regarding its implementation and there has not been enough scholarly thought given to when and how it should be implemented; see also Fisher and Harding (n 37) 113.

5.3 CRITIQUE OF PRUDENT VIGILANCE

The Commission's assessment of synthetic biology generally focuses on risk identification and management, and oversight of research in order to minimise risk. The significant level of uncertainty associated with the risks of synthetic biology is often noted in the Commission's report.[53] The Commission established a new principle of 'responsible stewardship' as a basis for its recommendations on risk management. This principle encompasses the concepts of 'prudent vigilance' and 'coordinated regulation'. Prudent vigilance refers to the establishment of processes for assessing the risks and benefits of synthetic biology projects before and after they are undertaken, with the capacity to 'continue to assess safety and security as technologies develop' and to limit their use 'when necessary'.[54] These concepts led the Commission to recommend initial and ongoing review of risk assessment policies. It proposed an interagency discussion of risk assessment activities, including identifying strategies for harmonising these and identifying any gaps in current processes for assessing the risks associated with release of synthetic organisms.[55] The Commission noted the potential for harm to be caused by 'inadvertent environmental release of organisms ... produced by synthetic biology', which requires 'safeguards and monitoring'.[56]

A coalition of non-government organisations (NGOs) including the ETC Group criticised the Commission for 'ignoring' the precautionary principle, stating that 'business as usual has won out over precaution'.[57] This critique of the Commission's approach is valid. The Commission did not need to invent 'prudent vigilance' in favour of adopting precaution.

Insight into why the Commission did not adopt the precautionary principle can be obtained from the transcripts from its public consultations and public deliberations on synthetic biology, as well as its 2010 report. The transcripts show that the Commission received very little evidence about precaution, and the commentary it did receive was entirely negative, apart from a short discussion about the definition of the principle. Further, analysis of these sources reveals that the Commission

---

53 See, e.g., The Presidential Commission for the Study of Bioethical Issues (n 8) 132.

54 Ibid., 4.

55 Ibid., 9.

56 Ibid.

57 ETC Group, *NGOs Blast US Presidential Commission on Bioethics* (16 December 2010) <http://www.etcgroup.org/en/node/5244>.

understood the precautionary principle in an extreme sense; it took a caricatured view of the principle. This meant the Commission saw a need to find a more moderate principle in order to protect scientific progress.

5.3.1 Prudent Vigilance is an Untested Concept

The group of NGOs, including the ETC Group, stridently criticised the Commission's development of a new concept of prudent vigilance: 'it is a completely new concept, apparently invented by the Commission without legal or policy precedent. When dealing with novel synthetic organisms that pose serious risks to the environment and public health, we cannot rely on a new concept with no agreed upon definition, framework, or precedent'.[58]

This critique of prudent vigilance is also valid, because the Commission does not explain the provenance of this principle. It is not clear what drove the Commission to develop the principle of prudent vigilance, apart from its need to create an alternative to its extreme understanding of precaution. The Commission does not explain what theoretical traditions this principle is developed from, or what existing discourses it draws on. Prudent vigilance appears to be drawn from ethics and philosophy, but this needs to be supplemented with regulatory mechanisms developed based on legal analysis. The Commission provided no framework for prudent vigilance's implementation, or guidance as to how it would or should impact on decision-making.

By contrast, the drivers that led to the adoption of the precautionary principle can be identified. It has been described as a response to the failure of previous policies and assumptions to address environmental degradation.[59] It can be seen as a response to 'deep dissatisfaction' with both the decision-making processes and outcomes of the past.[60] Further, the principle's adoption in international instruments and domestic law, along with its interpretation and implementation over several decades,

---

[58] Ibid; Civil Society Organizations, *Letter to Amy Gutmann* (16 December 2010) <http://www.etcgroup.org/sites/www.etcgroup.org/files/publication/pdf_file/Civil%20Society%20Letter%20to%20Presidents%20Commission%20on%20Synthetic%20Biology_0.pdf>.

[59] James Cameron, 'The Precautionary Principle: Core Meaning, Constitutional Framework and Procedures for Implementation' in Ronnie Harding and Elizabeth Fisher (eds), *Perspectives on the Precautionary Principle* (The Federation Press, 1999) 27, 2–3; Fisher et al, 'Implementing the Precautionary Principle: Perspectives and Prospects' (n 25) 3.

[60] Sandin, 'Common-sense Precaution and Varieties of the Precautionary Principle' (n 35) 104.

can be examined.[61] There is jurisprudence regarding its implementation in various domestic jurisdictions, as well as in international fora. This shows that the principle has been put into practice: 'it can be operationalized'.[62] For example, in the European context the European Court of Justice and Court of First Instance developed case law in the 1990s regarding the circumstances which justify precautionary action.[63] In 2000, the European Commission released a Communication from the Commission on the Precautionary Principle. This set out the elements of regulation under the precautionary principle. It states that 'where action is deemed necessary, measures based on the precautionary principle should be':

- *proportional* to the chosen level of protection,
- *non-discriminatory* in their application,
- *consistent* with similar measures already taken,
- *based on an examination of the potential benefits and costs* of action or lack of action (including, where appropriate and feasible, an economic cost/benefit analysis),
- *subject to review*, in the light of new scientific data, and
- *capable of assigning responsibility for producing the scientific evidence* necessary for a more comprehensive risk assessment.[64]

The European Court of Justice and Court of First Instance have subsequently used these guidelines in deciding disputes about precautionary action, and each of these six elements has been interpreted in case law.[65] For example, in *Alpharma Inc v Council of the European Union*, the

---

[61] See Fisher et al, 'Implementing the Precautionary Principle: Perspectives and Prospects' (n 25) 3–5.

[62] Ibid., 4.

[63] Elizabeth Fisher, 'Opening Pandora's Box: Contextualising the Precautionary Principle in the European Union' in Michelle Everson and Ellen Vos (eds), *Uncertain Risks Regulated* (Routledge Cavendish, 2009) 21, 34. The basis for the precautionary principle in EU law is found in Treaty on the Functioning of the European Union, opened for signature 7 February 1992, OJ C 326/47 (entered into force 1 November 1993) [cited as amended], arts 11, 12, 168, 169 and 191.

[64] European Commission (n 34) 4 (emphasis in original).

[65] Mark Cantley and Maurice Lex, 'Genetically Modified Foods and Crops' in Jonathan B Wiener et al (eds), *The Reality of Precaution: Comparing Risk Regulation in the United States and Europe* (Routledge, 2011) 39, 43; Fisher, 'Opening Pandora's Box: Contextualising the Precautionary Principle in the European Union' (n 63) 37.

European Court of Justice examined the application of the precautionary principle in terms of risk assessment, in relation to a regulation banning animal food additives.[66] Further, arguments regarding the proper application of the precautionary principle have been considered in international tribunals including the World Trade Organization's Dispute Resolution Panel and Appellate Body, the International Tribunal on the Law of the Sea, and the International Court of Justice.[67] The Commission has not made the case that this tried and tested principle should be rejected in favour of the new principle of prudent vigilance. Guidance as to the implementation of this new principle is lacking.

5.3.2 Misconception of Precaution by the Presidential Commission

The Commission's report and deliberations on synthetic biology reveal that a strong interpretation was given to the precautionary principle. Precaution was referred to as 'an extreme precautionary approach that blocks technological progress until *all possible risks are known and neutralized*'.[68] While the Commission wrote that *one* interpretation of the precautionary principle mandates such an approach, it is clear that this is the interpretation it worked from because it went on to state that it judged a moratorium to be imprudent.[69] Further, this type of definition was used by Commission members in its deliberations.[70] Precaution, understood in these terms, was seen as too cautious and as failing to appreciate the benefits of synthetic biology research.[71]

---

66 T-70/99, [2002] ECR II-03495, [135]–[146], [152]–[161], [173–176].

67 See e.g., Appellate Body Report, *Measures Affecting the Importation of Apples (Japan v United States)*, DS 245 (26 November 2003), [232]–[238]; Panel Report, *European Communities – Measures Affecting the Approval and Marketing of Biotech Products (United States v European Communities)*, DS291 (29 September 2006) [7.76]–[7.89], [7.1519]–[7.1530], [7.3065]–[7.3069]; *Pulp Mills on the River Uruguay (Argentina v Uruguay) (Judgment)* [2010] ICJ Rep 14 [160]–[164]; *International Tribunal for the Law of the Sea, Responsibilities and Obligations of States Sponsoring Persons and Entities With Respect to Activities in the Area (Advisory Opinion)*, Case No 17 (1 February 2011), [125]–[135], [161]–[163], [236].

68 The Presidential Commission for the Study of Bioethical Issues (n 8) 26 (emphasis added).

69 Ibid., 123–4.

70 Arras (n 19); and Amy Gutmann 'Comments in Commission Deliberations, The Presidential Commission for the Study of Bioethical Issues, Synthetic Biology' 16 November 2010.

71 The Presidential Commission for the Study of Bioethical Issues (n 8) 26.

This approach is problematic because it is based on a misunderstanding of the precautionary principle. It reveals a conflation of the concepts of 'precautionary' and 'preventative' approaches.[72] Unlike precautionary approaches, preventative approaches are those which halt the development of the technology.[73] The regulatory approaches of technology 'promotion', which would increase the spread of the technology, and precaution, are more usefully seen as points on a continuum rather than as a dichotomy.[74] The Commission has assumed the precautionary principle is interpreted and applied in an extreme, black and white fashion. Like Sunstein's 'strong' version of the principle, discussed above, this is problematic because the principle is not intended to mandate regulatory action, whatever the costs, to abrogate *all* potential risks of harm. It does not imply that regulatory action should be taken to prevent harm even when that action is extremely costly or may itself pose a risk of serious health or other consequences.

The precautionary principle's application can be seen as presupposing a scientific risk assessment demonstrating uncertainty, and a decision as to whether action is necessary.[75] This decision relates to the 'chosen level of protection' – the degree of protection from any particular harm that is regarded as appropriate by a particular nation – and is separate from the application of the precautionary principle.[76] It is a 'normative political choice' impacted on by economic and socio-political factors.[77] This means that the question of whether action is necessary in light of scientific information about risk is a political matter.[78]

Once the political decision that action is necessary is made, precaution can come into play.[79] The principle does not always mandate action; depending on the circumstances, implementation of the precautionary principle *justifies* – rather than demands – restrictive measures in

[72] These terms are defined in Kuzma and Tanji (n 1) 106.

[73] Ibid.

[74] Leili Fatehi and Jennifer Kuzma, 'Policy Innovation in Synthetic Biology Governance' (Paper presented at the 21st Century Borders/Synthetic Biology: Focus on Responsibility and Governance, 2013) <http://www.scienceforglobalpolicy.org/LinkClick.aspx?fileticket=VqAc-vBzvMk%3d&tabid=133>, 69; see continuum set out in ibid., 107.

[75] European Commission (n 34) 3, 13–14.

[76] Ibid., 15; Schomberg (n 15) 24.

[77] European Commission, ibid., 3; Schomberg, ibid., 23–4.

[78] Ibid., 15. Similarly, Sunstein notes that the choice of which risks are deemed acceptable is a political one, rather than a technical numerical matter: Sunstein, *Risk and Reason* (n 40) 104.

[79] European Commission (n 34) 12.

situations of uncertainty. There are, however, formulations which require action when certain conditions are met, such as the threat of serious or irreversible harm.[80] Thus the precautionary principle does not necessitate action to address just *any* degree of potential harm; many formulations refer to harm that is 'serious' or 'irreversible', as in the Rio Declaration, while others refer to harm that is 'significant', as in the Convention for the Protection of Biological Diversity.[81] The provisions that refer to 'significant' harm create a right to take precautionary action, while those that refer to 'serious or irreversible harm' create a duty.[82]

Also, precaution does not demand action regardless of how small the likelihood of harm eventuating may be. The degree of proof needed is stated in some international instruments and domestic formulations. The European Commission's Communication refers to '*reasonable grounds for concern* that the potentially dangerous effects on the environment, human, animal or plant health may be inconsistent with the chosen level of protection'.[83] In European Union case law, it has been held that the risk may not be 'purely hypothetical' or 'mere conjecture'.[84]

Implementing the precautionary principle does allow the cost of taking action to be considered. For example, the Rio Declaration definition refers to 'cost effective measures to prevent environmental degradation'.[85] Thus precautionary action does not have to be taken regardless of the costs. Under the European Commission's Communication, cost benefit analysis involves 'comparing the overall cost to the Community of action and lack of action, in both the short and long term'.[86] This shows that other risks, that could arise if the action is taken, can be taken into account, negating the strong interpretation of precaution. Also, non-economic considerations can be taken into account, such as whether options are likely to be acceptable to the public.[87]

---

[80] Arie Trouwborst, *Precautionary Rights and Duties of States* (Martinus Nijhoff Publishers, 2006), 121–4.

[81] Rio Declaration on Environment and Development (n 28) Principle 15; Convention on Biological Diversity (n 30) art. 14. Similarly, the European Commission refers to 'risk … that in the event of non-action may have serious consequences': European Commission (n 34) 15.

[82] Trouwborst (n 80) 121–4.

[83] European Commission (n 34) 9–10.

[84] *Alpharma Inc v Council of the European Union* (n 66) [156]–[158], quoted in subsequent cases also.

[85] Rio Declaration on Environment and Development (n 28) Principle 15.

[86] European Commission (n 34) 17–18.

[87] Ibid., 4.

Further demonstrating the nuances in the implementation of precaution, the European Commission's Communication requires that there be proportionality between the protective measures and the chosen level of protection.[88] Taken together, all these nuances show that the strong interpretation of precaution is not tenable and does not provide a valid reason for preferring the principle of prudent vigilance.

5.3.3 Scientific Progress

The Presidential Commission's assumption that the precautionary principle requires a halt to technological progress is problematic. The principle is implemented in the administrative decision-making context, and relates to how the decision is made and the factors that are taken into account, rather than dictating the outcome.[89] The principle does not dictate which measures are appropriate to protect the environment or public health.[90] Indeed, the precautionary principle does not necessarily require action at all; non-action can be an appropriate response in some circumstances.[91] The 'appropriate measures' are 'influenced by legal and socio-political culture, the specific statutory regime, the nature of particular environmental or public health problems, the availability and ease of implementation of precautionary measures, and more general understandings of legitimate decision-making'.[92] Guidelines as to the implementation of the principle – such as the European Commission's Communication – simply provide rules for use in deciding how to act, rather than stating which action is required in specific circumstances.[93] Thus, a halt to scientific progress is not necessarily required.

It is true that in some cases the precautionary principle has been applied in a way which has been seen to cause damage to technological innovation, such as Europe's effective ban on the import of genetically modified foods.[94] However, in that example, review of the regulatory regimes found that the problem was not the precautionary principle itself, but a range of issues relating to coordination and consistency of regulatory procedures and inadequate review and reform of these.[95]

---

88 Ibid., 3.
89 Fisher and Harding (n 37) 115.
90 Ibid., 116.
91 European Commission (n 34) 3.
92 Fisher and Harding (n 37) 116.
93 European Commission (n 34) 12.
94 Cantley and Lex (n 65) 56–7.
95 Ibid., 59–60.

Europe's former approach to the approval of genetically modified organisms can be seen as more of a *de facto* prohibition or moratorium than a precautionary approach. Indeed it was argued to constitute a *de facto* moratorium by parties in the World Trade Organization litigation.[96]

The Commission thought that the principle of prudent vigilance was a better approach than precaution because it achieved a balance between their extreme understanding of precaution and a 'proactionary' or 'full speed ahead and worry about the risks later' approach.[97] Under such a 'proactionary' approach, synthetic biology would be assumed to be safe and good unless it is demonstrated to be otherwise, placing the burden of proof on those who want to slow down that research.[98] While it was acknowledged that this approach is underpinned by values of scientific freedom and obtaining the benefits of science, it was seen as an 'irresponsibly brazen' approach which does not take adequate account of the risks.[99] Arras explained: 'we're trying to split the difference here between proactive, letting science rip mentality versus a European-style precautionary principle and we framed that in terms of prudent vigilance'.[100]

Similarly, Gutmann commented: 'It won't satisfy either extreme, but I think it is the responsible way of proceeding and advising both private and public parties who are engaged in this field.'[101]

However, the above analysis shows that it was not necessary to come up with a new principle in order to balance the 'proactionary' approach; the precautionary principle could have done this work if understood in a less extreme way. In fact, the precautionary principle *was* developed partly in response to earlier decision-making approaches which required scientific proof before risk-mitigating action would be taken.[102]

The Commission was also concerned to promote scientific progress, and thought that pursuing rather than limiting research was the best way

---

96 Panel Report, *European Communities – Measures Affecting the Approval and Marketing of Biotech Products* (n 67) see e.g., arguments of US at 30.

97 The Presidential Commission for the Study of Bioethical Issues (n 8) 123–4; Arras (n 19); Gutmann 'Comments in Commission Deliberations' (n 70); see also Amy Gutmann, 'The Ethics of Synthetic Biology: Guiding Principles for Emerging Technologies' (2011) **41**(4) *Hastings Center Report* 17.

98 The Presidential Commission for the Study of Bioethical Issues (n 8) 124.

99 Ibid., 26; and Arras (n 19).

100 Arras (n 19).

101 Gutmann 'Comments in Commission Deliberations' (n 70).

102 See Fisher et al, 'Implementing the Precautionary Principle: Perspectives and Prospects' (n 25) 3; Fisher and Harding (n 37) 115.

to promote public beneficence.[103] It was concerned about overregulation 'squelching' scientific creativity and progress.[104] My interpretation is that the Commission concluded that it needed a different principle in order to protect scientific progress from 'extreme' precaution. For example, Gutmann has suggested that both principles 'share a concern for identifying and mitigating risks', but she saw the principle of prudent vigilance as being different from the precautionary principle because it 'advocates continued progress in the *pursuit of potential benefits in tandem* with that ongoing sensitivity to risks and the development of appropriate responses'.[105] Similarly, she noted that failing to save lives or obtain other benefits through developing technology is just as serious as failing to address risks.[106]

While not stated as a reason for adopting prudent vigilance in favour of the precautionary principle, it is fair to say that protection of the US life sciences industry could be a significant factor in this downplaying of the precautionary principle. Indeed, it has been noted that critics of the precautionary principle as insufficiently science-based are often commenting from an industry perspective.[107] There has been huge investment in the biotechnology industry by both government and private sector in the US, particularly in the area of biofuels.[108] This may well have impacted upon the Commission's unwillingness to adopt precaution.

5.3.4 A 'European Idea'

The Commission also saw precaution as a foreign 'European idea', as shown in its public deliberations. Precaution was referred to in these terms several times, for example Commission member John Arras referred to a 'European style precautionary principle' and distinguished 'a precautionary principle, which we've seen adopted in Europe' from the

---

103 See, e.g., James Wagner 'Comments in Commission Deliberations, The Presidential Commission for the Study of Bioethical Issues, Synthetic Biology' 16 November 2010.

104 Nita Farahany 'Comments in Commission Deliberations, The Presidential Commission for the Study of Bioethical Issues, Synthetic Biology' 16 November 2010; ibid.

105 Gutmann, 'The Ethics of Synthetic Biology' (n 97) 22 (emphasis added).

106 Gutmann 'Comments in Commission Deliberations' (n 70).

107 Fisher et al, 'Implementing the Precautionary Principle: Perspectives and Prospects' (n 25) 5.

108 See for example 'National Bioeconomy Blueprint' (The White House, 2012) <http://www.whitehouse.gov/sites/default/files/microsites/ostp/national_bioeconomy_blueprint_april_2012.pdf>.

'proactive approach', implicitly more US in style.[109] The view that precautionary approaches are embraced in Europe, but not in the US, has been described as 'standard' or 'conventional wisdom'.[110] It has been suggested that this is because precaution has been seen as detrimental to innovation and the operation of free markets.[111] The stereotypical view portrays Americans as anti-regulatory and techno-optimistic, while Europeans are afraid of technology, risk-averse and in favour of regulation.[112] However, research shows that the picture is actually more complex.[113] The US was initially quite precautionary in its approach to risk issues in the 1970s.[114] Some scholars propose that Europe and the US have essentially switched so that the US, initially the leader in this area, has now become less inclined to implement precautionary measures, while Europe initially lagged behind and is now more precautionary.[115] Others argue that the US is not inherently less precautionary, but that it prioritises risks differently.[116] While precaution is not expressly adopted as a 'general principle' it is incorporated into some areas of risk regulation.[117] US policy has been more precautionary in relation to some risks such as terrorism, air quality issues, tobacco-related illnesses and risks posed by bovine spongiform encephalopathy.[118]

In particular, statutory requirements for the regulation of some pollutants and contaminants can be seen as precautionary. Various environmental legislation requires the Environmental Protection Agency (EPA) to set standards at a level that allows for a 'margin of safety'. The EPA defines margin of safety to mean: 'Maximum amount of exposure producing no measurable effect in animals (or studied humans) divided

---

109 Arras (n 19).

110 See Sunstein, *Laws of Fear: Beyond the Precautionary Principle* (n 18) 13; and review of literature in Wiener, 'The Rhetoric of Precaution' (n 24) 16.

111 See Cameron (n 59) 12.

112 Wiener, 'The Rhetoric of Precaution' (n 24) 7–8.

113 Ibid., 13.

114 Ibid., 12–23.

115 See ibid.

116 Sunstein, *Laws of Fear: Beyond the Precautionary Principle* (n 18) 13–14, 16.

117 Wiener, 'The Rhetoric of Precaution' (n 24) 10.

118 Sunstein, *Laws of Fear: Beyond the Precautionary Principle* (n 18) 21–2; ibid., 19–20; George Gray et al, 'Beef, Hormones and Mad Cows' in Jonathan B Wiener et al (eds), *The Reality of Precaution: Risk Regulation in the United States and Europe* (Routledge, 2011) 65, 66–7.

by the actual amount of human exposure in a population.'[119] The language of 'margin of safety is used for example in the Clean Water Act, which requires promulgation of toxic and pre-treatment effluent standards that provide an 'ample margin of safety'.[120] There is similar language in the Clean Air Act, which requires that national primary ambient air quality standards allow an 'adequate margin of safety' to protect the public health.[121] For hazardous air pollutants, if Congress has not acted on a recommendation regarding the pollutants, the Administrator must promulgate standards if promulgation is necessary to 'provide an ample margin of safety to protect public health'.[122]

The Clean Air Act also sets out a list of regulated hazardous air pollutants, and provides that a person may petition the Administrator to delete a particular substance from the list.[123] The Administrator can do this where they or the petitioner establish that there is 'adequate data' to determine that the substance 'may not reasonably be anticipated to cause any adverse effects to the human health or adverse environmental effects'.[124] This legislation has been described by some commentators as precautionary.[125]

In addition, there is case law in the US in which courts have required preventative action even in the absence of conclusive scientific evidence of the likelihood of harm. In a significant and much-cited decision in *Reserve Mining Co v Environmental Protection Agency*, the Court of Appeals for the Federal Circuit held that Reserve Mining could be required to cease discharging asbestos contaminants into the air and water, even though it was not established that the contaminants would cause an increase in cancer rates.[126] There was a 'reasonable medical

---

[119] Environmental Protection Agency, *Vocabulary Catalog* Environmental Protection Agency <https://ofmpub.epa.gov/sor_internet/registry/termreg/searchandretrieve/glossariesandkeywordlists/search.do?details=&glossaryName=Exposure%20Factors%20Glossary>.

[120] 33 USC (Office of the Law Revision Counsel, 2017) §1317 (a) (4).

[121] Ibid., 7409 (b) (1).

[122] Ibid., § 7412 (f) (2) (A).

[123] Ibid., § 7412 (b) (3) (A).

[124] 42 USC (Office of the Law Revision Counsel, 2017) § 7412 (b) (3) (C).

[125] See, e.g., Sanford E Gaines, 'Science, Politics, and the Management of Toxic Risks through Law' (1990) **30**(3) *Jurimetrics* 271, 279; Bernard D Goldstein and Russellyn S Carruth, 'Implications of the Precautionary Principle for Environmental Regulation in the United States: Examples from the Control of Hazardous Air Pollutants in the 1990 Clean Air Act Amendments' (2003) **66** *Law and Contemporary Problems* 247.

[126] 514 F 2d (Court of Appeals, Federal Circuit, 1975) 492, 500, 520.

concern for the public health' which was 'of sufficient gravity to be legally cognizable' and to justify an abatement order and injunctive relief.[127] Further, the court held that 'the evidence calls for preventative and precautionary steps'.[128]

All these matters considered, it is overly simplistic for the Commission to reject precaution as a purely 'European' principle. The way in which precautionary regulation of synthetic biology products could be integrated into the US legal system warrants further consideration.

5.3.5 International Leadership

A related issue is whether the Commission's recommendations were intended as a template for international regulation of synthetic biology. The transcript of deliberations reveals that the Commission did see international collaboration in developing regulation as very important, particularly in the context of biosecurity.[129] It saw an opportunity for the US to lead in that collaboration.[130] Similarly, the Commission's report recognised the international nature of synthetic biology research and that 'international coordination is essential for safety and security'.[131] The Commission recommended: 'Oversight and regulatory mechanisms should adopt an analogous approach, so that the United States is involved in regular discussions with other national and transnational organizations, together seeking coordination and consistency when possible.'[132]

However, in rejecting the precautionary principle, the Commission was parochial; it could have given greater consideration to developments in other jurisdictions and in international law. Its goals of international coordination and leadership were poorly served by rejecting the precautionary principle. Reliance on the new principle of prudent vigilance meant that the Commission's approach was out of step with other jurisdictions, and with international efforts to address other major international environmental problems such as climate change through frameworks which incorporate the precautionary principle. This is problematic given the international nature of the field of synthetic biology and the

---

127 Ibid., 500, 520.
128 Ibid., 500.
129 Gutmann 'Comments in Commission Deliberations' (n 70).
130 See, e.g., ibid.
131 The Presidential Commission for the Study of Bioethical Issues (n 8) 10, 132.
132 Ibid., 132.

need to develop international approaches to the regulatory challenges of environmental risk and biosecurity.

5.4 BIOTECHNOLOGY RISK REGULATION IN THE UNITED STATES

5.4.1 Coordinated Framework for the Regulation of Biotechnology

In the US, the Coordinated Framework for the Regulation of Biotechnology (Coordinated Framework) assigns responsibility for regulating biotechnology products to protect health and the environment to three federal agencies, the EPA, the US Food and Drug Administration (FDA), and the US Department of Agriculture (USDA).[133] This Coordinated Framework provides for regulation of biotechnology products based on their specific use.[134]

Under the Coordinated Framework, the EPA is responsible for preventing 'unreasonable adverse effects on the environment', preventing harm from exposure to pesticide residues, and preventing the manufacture and use of chemicals from presenting 'an unreasonable risk of injury to health or the environment'.[135] It regulates various applications of microbial biotechnology, as well as pesticides produced using genetic engineering.[136]

The US FDA is responsible for ensuring the safety of human and animal food and drugs, human medical devices, cosmetics and biological products.[137] This includes human and animal foods that are derived from genetically engineered plants, genetically engineered animals, and human drugs, biological products or medical devices produced using genetic engineering.[138]

The USDA is responsible for protecting agriculture from plant and animal disease risks, and ensuring that certain food products are safe and

---

[133] Environmental Protection Agency et al, 'Modernizing the Regulatory System for Biotechnology Products: Final Version of the 2017 Update to the Coordinated Framework for the Regulation of Biotechnology' (2017) <https://www.epa.gov/sites/production/files/2017-01/documents/2017_coordinated_framework_update.pdf>.

[134] Ibid., 7.

[135] Ibid., 9.

[136] Ibid., 10–14.

[137] Ibid., 9.

[138] Ibid., 15–21.

labelled properly.[139] This includes regulation of various biotechnology products that could cause disease in livestock or plants.[140]

5.4.2 Environmental Risk Regulation for Synthetic Biology Products

In relation to the environmental risk posed by synthetic biology products, the commercial manufacture of certain genetically engineered microorganisms is regulated by the EPA under the Toxic Substances Control Act and associated regulations.[141] Commercial operators may not manufacture a new chemical unless they have first notified the EPA and the EPA has undertaken a risk assessment.[142] This applies to manufacturing a 'new microorganism' for commercial purposes, including commercial research and development.[143] New microorganisms are those that are not listed on the EPA's Inventory.[144] This could include new organisms where genetic material is transferred from one organism, such as yeast, to bacteria for production of a commercial biological product, or other synthetic biology constructs.

Recently this regulatory regime has been made more precautionary by modifying the approach taken to the burden of proof of safety. The EPA is required to prohibit or limit manufacture of the microorganism if it determines that there is an 'unreasonable risk of injury to health or the environment'.[145] A person proposing to manufacture a new chemical (including a microorganism) must provide the EPA with a notice including information that they believe shows that the chemical does not present such a risk.[146] For new microorganisms, this includes information about the characteristics of the new organism, the intended use, controls on release, survival and dissemination, known biological interactions, and any other information that would allow the EPA to evaluate the health and environmental effects.[147]

---

139 Ibid., 9, 22–3.
140 Ibid., 9, 22–7.
141 See 15 USC (Office of the Law Revision Counsel, 2017), §§ 2601–2629; 40 CFR (US Government Publishing Office – e-CFR, 2017), § 725.
142 15 USC (n 141) §§ 2604 (a).
143 Ibid., § 2604 (a); 40 CFR (n 141) § 725.3 and 725.105.
144 40 CFR (n 141) §§ 725.1(a), 725.3.
145 15 USC (n 141) § 2604, 2605.
146 Ibid., § 2604 (b).
147 40 CFR (n 141) §725.155.

Previously, the EPA had discretion to impose prohibitions or limitations where it found a 'reasonable basis to conclude that the manufacture ... will present an unreasonable risk of injury to health or environment', but it did not have an affirmative obligation to perform a risk assessment.[148] The legislation was criticised for placing the burden of proof for establishing a safety problem on the EPA.[149] The EPA is now required to make an assessment of the safety of new chemicals (including microorganisms) before they are manufactured, and determine whether an unreasonable risk is present.[150] That is, new chemicals can only be commercially produced if this assessment has occurred and the risk has been found to be not unreasonable, or, in the case where the EPA determines it does not have enough information to evaluate the risk, manufacture can only occur in accordance with necessary prohibitions or limitations while the required information is being developed.[151] The EPA must require the manufacturer to develop this further information.[152] These new provisions place a greater burden on the manufacturer to establish the safety of their proposed new product.

5.5 PLANNED ADAPTATION

The problem of addressing the regulatory challenges of synthetic biology has been approached not only by leading scientists, sociologists and anthropologists, anti-technology groups and the Commission, but also by political scientists. Political scientists Kenneth A Oye and Lawrence E McCray, and scientist and science policy expert Arthur C Petersen, have proposed that the policy tool of 'planned adaptation' can address the problem of regulating where there is uncertainty.[153]

---

148 15 USC (n 141) §2604.

149 See, e.g., Michael P Wilson and Megan R Schwarzman, 'Toward a New U.S. Chemicals Policy: Rebuilding the Foundation to Advance New Science, Green Chemistry and Environmental Health' (2009) **117**(8) *Environmental Health Perspectives* 1202, 1205.

150 15 USC (n 141) § 2604 (a)(3).

151 Ibid., § 2604 (a) (3), (e), (f).

152 Ibid., § 2603 (a).

153 Lawrence E McCray et al, 'Planned Adaptation in Risk Regulation: An Initial Survey of US Environmental, Health, and Safety Regulation' (2010) **77** *Technological Forecasting & Social Change* 951, 951.

5.5.1 The Concept of Planned Adaptation

Oye has proposed that where there is 'pervasive and irreducible uncertainty', this can mean that 'public policies will be based on incorrect assumptions' and will be wrong because 'the information needed to make right choices is not available up front'.[154] Planned adaptation involves building mechanisms into regulatory regimes for obtaining new knowledge relevant to the policies and their impact, and to allow policies and rules to be changed in order to adapt to this knowledge.[155] Where there is uncertainty, policy actions should be seen as *experimental*.[156] These experiments allow information to be gathered about the environment surrounding regulation in areas of uncertainty, and for the regulation to be updated over time in response to changes in this environment.[157] The changes in the environment may relate to 'political, economic, biological, or technological conditions'.[158] Planned adaptation requires a change in thinking, from regarding policy decisions as 'final' to viewing the making of public policy as 'open ended'.[159] This creates some tensions with administrative law values, as will be seen below.[160]

Much of the literature on adaptive regulation mechanisms is in the environmental regulation sphere, where the policy tool is known as 'adaptive management'. Adaptive management is a form of environmental regulation which involves monitoring and collecting data over time – 'measuring while managing'.[161] In environmental law this involves 'an initial regulatory approval based on initially available information … followed by a period of monitoring and reporting with reassessment of the appropriateness of the initial regulatory action and adjustment or amendment as required in light of new information, and presumably the resolution of previously identified uncertainties'.[162]

In a similar vein, planned adaptation exists as a feature of a regulatory regime where '[a] there is a prior commitment to subject an existing

---

154 Kenneth A Oye, 'Embracing Uncertainty' (2009) **26**(1) *Issues in Science and Technology* 91, 93.

155 McCray et al (n 153) 951.

156 Oye (n 154) 93.

157 Ibid.

158 Ibid.

159 McCray et al (n 153) 958.

160 See Robin Kundis Craig and J B Ruhl, 'Designing Administrative Law for Adaptive Management' (2014) **67**(1) *Vanderbilt Law Review* 1.

161 Jones (n 20) 362.

162 Ibid.

policy to *de novo* re-evaluation and [b] systematic effort is made to mobilize new factual information for use when the re-evaluation takes place'.[163]

McCray and collaborators surveyed US federal regulatory regimes to find cases of planned adaptation. The small number of putative cases led them to query whether the tool was attractive in theory but not in practice.[164] However, they found a number of cases which fit their criteria for planned adaptation, showing that it is indeed 'administratively and politically feasible'.[165] These included air pollution standards, air transportation safety, regulation of pharmaceuticals and nutrition standards. The US Department of Agriculture's Forestry Planning Rule has also been described as an 'adaptive planning framework'.[166] This is a planning regulation made under the National Forest Management Act.[167] The new rule adopted in 2012 is designed to have an improved scope for responding to challenges such as climate change through a process of 'assessment, planning and monitoring'.[168]

In the air pollution example identified by McCray and collaborators, the EPA was required to seek a scientific review of the ambient air standards for particulate matter under the Clean Air Act every five years. The EPA made an 'aggressive effort to advance the relevant science and not just to settle for dealing with advances that happen to turn up'.[169] It was not able to stick to the five-year review cycle but did obtain expert evidence and change the standards accordingly and this in itself provided regulatory learning, suggesting that a longer period between reviews may be appropriate.[170]

Some of the regulatory regimes identified by McCray and collaborators were controversial and described as having 'high visibility'.[171] Therefore, the survey allowed the researchers to propose that:

---

[163] McCray et al (n 153) 952.

[164] Ibid., 952.

[165] Ibid., 954.

[166] United States Department of Agriculture Forestry Service, *FAQs on 2012 Planning Rule* The Forest Planning Rule <https://www.fs.usda.gov/detail/planningrule/faqs#1>.

[167] *National Forest System Land Management Planning*, 36 CFR Part 219, Federal Register 77(68) 21162, 9 April 2012; 16 USC (Office of the Law Revision Counsel, 2017) §1604 (g).

[168] United States Department of Agriculture Forestry Service (n 166).

[169] McCray et al (n 153) 955.

[170] Ibid., 954–5.

[171] Ibid., 954.

> Planned Adaptation is feasible as a decision strategy, even in tough policy areas … varying versions of the approach are being used in policy areas … that involve high stakes, public visibility, and new science – factors that regulatory administrators often find most challenging. It is as if when regulation is most sensitive, the forces of inertia are overcome, and a system of Planned Adaptation is demanded.[172]

The researchers noted some problems associated with building planned adaptation into policy programs, including the cost of obtaining new information and reviewing policies.[173] They pointed out that there are good reasons to value policy stability and that there is a need for agencies to ensure their rules are 'enforceable and credible', which may be seen as more difficult if the rules are changed from time to time.[174] Agencies must provide certainty and consistency in relation to the application of regulatory regimes, and changes to the rules could make their enforcement seem unfair.[175] Further, agencies need to maintain their public reputation, and would not wish to be seen as readily changing their position on policy issues or responding to political pressure.[176]

In addition, some of the features of the administrative law environment in which agencies such as the EPA make decisions can limit the scope for adaptive management.[177] The Administrative Procedure Act sets out rules in relation to public participation in agency decision-making and judicial review of decisions. The public participation requirements include giving notice of proposed rulemaking and giving interested persons the right to petition for the issuance, amendment or repeal of a rule, giving interested parties an opportunity to make submissions in agency hearings, and giving interested persons notice of the denial of their written applications or requests.[178] Legal academics Robin Kundis Craig and J B Ruhl argue that these requirements for public discussion are in tension with adaptive management processes which require decisions to be made and revisited over time.[179] They noted concern among some stakeholders that 'learning while doing' will lead to decision-making which lacks proper public participation.[180]

---

172 Ibid., 958.
173 Ibid., 957.
174 Ibid.
175 Ibid.
176 Ibid.
177 Kundis Craig and Ruhl (n 160) 15–16.
178 5 USC (Office of the Law Revision Counsel, 2017), §§ 553, 554, 555.
179 Kundis Craig and Ruhl (n 160) 30.
180 Ibid.

Judicial review is available for 'final agency action for which there is no adequate remedy in a court'.[181] Adaptive management requires that agencies be able to engage in some degree of 'dial twiddling', and agencies may be unlikely to make adjustments which could trigger judicial review.[182] If the adaptations made by an agency create legal consequences, it is possible that courts will consider them to be final actions.[183]

Craig and Ruhl argue that these features of administrative law push agencies to channel their investigation and public participation efforts into the 'front-end' of decision-making, and make a final decision that is defensible when judicially reviewed, rather than reconsidering and adjusting decisions over time.[184] The administrative law 'drivers towards finality' limit the scope for ongoing discussion, experimentation and adaptation.[185] However, Craig and Ruhl argue that there is scope to accommodate public participation and judicial review in a way that allows for a true adaptive management.[186] This would involve reconsidering the way in which agency actions trigger public participation and judicial review requirements, and the timing of participation and review.[187] It would also require acknowledging that adaptive management is not suited to all decision-making contexts.[188]

McCray and collaborators noted that in the examples they identified, the planned adaptation was imposed not by the agencies themselves but from the outside, for example by Congress.[189] They argue that this demonstrates that planned adaptation may be seen as 'destabilizing' by both agency decision-makers and interest groups. It requires a change in thinking, from regarding policy decisions as 'final' to viewing the making of public policy as 'open ended'.[190]

One of the advantages of planned adaptation is that it creates incentives for obtaining new knowledge and information which are not present when policy decisions are seen as permanent.[191] In those cases resources

---

181 5 USC (n 178) § 704.
182 Kundis Craig and Ruhl (n 160) 33–8.
183 Ibid., 33.
184 Ibid., 4–6, 36–7.
185 Ibid., 36.
186 Ibid., 42–7.
187 Ibid., 42–4.
188 Ibid., 59.
189 McCray et al (n 153) 958.
190 Ibid.
191 Ibid.

are more likely to be directed to acquiring the knowledge required to make *future* policy decisions.[192] McCray and colleagues point to a need to improve incentives for generating better knowledge, relevant to existing regulations, and exposing underlying false assumptions. Planned adaptation also provides a way to minimise the stymieing effect of uncertainty on policy making, which can delay action at a cost.[193]

Oye, McCray and Petersen stopped short of advocating the broad introduction of planned adaptation, posing the question of whether planned adaptation 'should be emulated across US regulatory programs, and perhaps beyond'.[194] They point out that although it is feasible, regulatory regimes utilising planned adaptation are 'relatively rare'.[195]

In addition to managing scientific uncertainty, adaptive mechanisms have also been proposed to assess the effectiveness of the regulatory regime. For example, assessment of whether the regulatory regime has performed well, and achieved its objectives, is part of the 'Really Responsive Regulation' approach to regulatory theory.[196] That theory proposes that it is necessary for the regulatory body to be sensitive to change, and to modify the regulatory tools in response to changes in the regulatory environment.[197]

In relation to emerging technologies specifically, Gregory N Mandel has proposed that an adaptive mechanism be used to both manage changing scientific knowledge and technology development, as well as to review the effectiveness of the regulatory tools used.[198] Thus according to Mandel, governance should follow an adaptive and iterative model. This can occur through a 'continuing cycle' of regulation, data gathering on the risks associated with the technology, evaluation and modification of the regime.[199]

---

192 Ibid.

193 Oye (n 154) 93.

194 McCray et al (n 153) 959.

195 Ibid., 957.

196 Robert Baldwin and Julia Black, 'Really Responsive Regulation' (2008) **71**(1) *The Modern Law Review* 59, 72–3.

197 Ibid 72–5, 87–91.

198 Gregory N Mandel, 'Regulating Emerging Technologies' (2009) **1** *Law, Innovation & Technology* 75, 83–9.

199 Ibid., 89.

5.5.2 Strengths and Weaknesses of Planned Adaptation for Synthetic Biology

Oye and colleagues, political scientist Gautam Mukunda and chemist Scott Mohr, have argued that the examples of planned adaptation they identified could provide a model for 'risk governance' in synthetic biology.[200] They made specific recommendations about risk management in synthetic biology and explained that each of these recommendations should be regarded as a 'policy experiment': 'intended to accentuate benefits and minimize risks while also eliciting information on the nature of security and safety risks, the extent of economic, environmental, security and medical benefits, and the costs and the benefits of suggested policies'.[201]

As McCray, Oye and Petersen note, planned adaptation can even work in 'high visibility' policy areas.[202] Synthetic biology is now a highly visible regulatory area due to the media attention to the synthetic cell announcement and the subsequent Commission review.

The importance of adaptability of regulatory regimes for synthetic biology is also implicit in the sociological perspective on this issue presented by Zhang, Rose and Marris.[203] Their paper has connections with the ideas of planned adaptation, for example in the statement that synthetic biology regulation should not be a 'rigid regulatory regime' but 'a flexible and evolving "art of governance"'.[204]

The main strength of the planned adaptation approach in synthetic biology is that it provides a way to get started, to cope with policy making under pervasive uncertainty about the risks of synthetic biology. Adaptive regulation also has 'logical appeal' because it allows regulatory action to be modified in response to information about whether it has been successful.[205] This means that regulators can design measures, or assess existing regulatory measures, taking into account the best available

---

200 Gautam Mukunda et al, 'What Rough Beast? Synthetic Biology, Uncertainty, and the Future of Biosecurity' (2009) **28**(2) *Politics and the Life Sciences* 2, 21.

201 Ibid., 21.

202 McCray et al (n 153) 954.

203 Joy Y Zhang et al, 'The Transnational Governance of Synthetic Biology: Scientific Uncertainty, Cross-borderness and the "Art" of Governance' (BIOS Working Paper No: 4, BIOS (Centre for the Study of Bioscience, Biomedicine, Biotechnology and Society), May 2011) <http://www2.lse.ac.uk/BIOS/publications/Working%20Papers/WP4TransnationalGovernance.pdf>.

204 Ibid., 3.

205 Jones (n 20) 361, 364–5.

information about magnitude and severity of risks of synthetic biology, the state of scientific knowledge, technological development, practices of research scientists and industry, as well as other relevant factors. But these starting points are not permanent; regulators can build in flexibility to review these policies at set points in future. Regulation can be provisional; regulators can do their best and then reflect on how it went.

This means that for the particular issues raised by synthetic biology, regulation does not have to be delayed until a later time when uncertainty is reduced (which may not occur). Regulators can begin to address gaps *now*. They do not have to wait for a crisis to arise, a criticism levelled at the current approach to technology regulation.[206] Thus planned adaptation promotes both regulatory effectiveness and regulatory connection to scientific development. Regulatory effectiveness refers to whether a regulatory intervention is 'making a difference' in terms of achieving its purpose.[207] Regulatory 'connection' refers to maintaining an appropriate connection between the stage of technological development and the regulation.[208] An adaptive review process will help the regulatory strategy remain connected to the state of scientific development, and assist in identifying and managing disconnection. In turn, a regime that is adjusted after proactive review both of the regulatory mechanism and the changing state of scientific development and other changes in the social and scientific context is more likely to work as intended. Thus connection and effectiveness are both enhanced.

Further, adaptation allows compromise between the conflicting norms of law and science. Scientists want to find the truth of things, while regulators want to make things safe.[209] While scientific enquiry may demand ongoing investigation, law requires timely decisions and information seeking cannot continue forever.[210] Adaptive management allows regulators to make decisions, while scientists can continue the search for knowledge and truth.

Some of Sunstein's critique of the precautionary principle can be viewed as points in favour of adaptive management. For example, Sunstein argued that the precautionary principle cannot be defended by reference to situations of uncertainty or catastrophe, because things will

---

206 Fatehi and Kuzma (n 74) 69.

207 Brownsword and Morag Goodwin (n 12) 277.

208 See ibid 372–97; Brownsword and Han Somsen (n 12) 26–31.

209 Lewens (n 26) 14.

210 Judith Jones, 'Certainty as Illusion: The Nature and Purpose of Uncertainty in the Law' in Gabriele Bammer and Michael Smithson (eds), *Uncertainty and Risk: Multidisciplinary Perspectives* (Earthscan, 2008), 278.

often go from being in the category of 'uncertainty' – where probabilities of bad outcomes cannot be assigned – to the category of 'risk' – where probability can be estimated – due to improved knowledge gained over time.[211] This move from uncertainty to risk could be seen as one of the beneficial outcomes of adaptive management.

Sunstein has also argued that 'acquiring information is far better than responding to the worst-case scenario, at least when that response itself creates dangers in the realm of both uncertainty and risk'.[212] This suggests that a process of information seeking about uncertain risks is important. Further, Sunstein argues that if we assume that we will gain more knowledge about the probability of an uncertain harm over time, steps can be taken on an 'Act, then learn' basis.[213] Regulatory steps could be designed that allow a change in approach if harm becomes known to be likely to occur.[214] Such an adaptive approach enables regulating in the face of uncertainty, building in periodic review of decisions or legislation so that improved scientific information can be reflected. Sunstein concluded that 'On this view, an understanding of what we do not know means not that regulators should do little, but that they should act in stages over time'.[215]

In defence of the Commission's adoption of prudent vigilance in preference to precaution, Gutmann distinguished the two principles by arguing that prudent vigilance facilitates adaptation to changing circumstances.[216] However, it is also possible to incorporate adaptive management as an element of a precautionary approach under conditions of scientific uncertainty. For example, under the European Commission's Communication, measures taken under the precautionary principle must be 'subject to review, in the light of new data'.[217] This means that the measures are provisional, they must be 're-examined and if necessary modified depending on the results of the scientific research and the follow-up of the impact'.[218] Adaptive management has 'dovetailed' with precaution in risk management in the Australian regulatory regime for

---

211 Sunstein, *Laws of Fear: Beyond the Precautionary Principle* (n 18) 61.
212 Ibid.
213 Ibid., 117.
214 Ibid., 117–18.
215 Ibid., 118.
216 Gutmann, 'The Ethics of Synthetic Biology' (n 97) 21–2.
217 European Commission (n 34) 3.
218 Cantley and Lex (n 65) 44; ibid., 20.

gene technology.[219] The legislation incorporates the precautionary principle and provides for the issuing of licences in some circumstances on the condition that the applicant is required to obtain further information during their activities.[220] Similarly, the international Agreement on Sanitary and Phytosanitary Measures provides for provisional measures to be adopted 'where scientific information is insufficient' and requires member states to seek to obtain 'the additional information necessary for a more objective assessment of risk and review the sanitary or phytosanitary measure accordingly within a reasonable period of time'.[221]

Planned adaptation creates an incentive for the acquiring of 'new policy-relevant knowledge'.[222] In the rapidly changing field of synthetic biology, this is crucial. Effective regulation for synthetic biology can *only* be developed in a situation where new knowledge both about the science and how existing regulatory mechanisms are working is proactively sought.

Of course, in applying the policy tool of 'planned adaptation' to synthetic biology, there is devil in the detail. For example, how often should planned review occur? Synthetic biology research is moving quickly and a review undertaken now could be very different from one undertaken in three or five years' time. However, overly frequent review would become resource intensive. There are other issues such as the process which should be undertaken for acquiring and assessing new knowledge, the sorts of professionals who should be tasked with managing this process, the factors on which assessment of this knowledge should be based, the time allocated for this process and the mechanism for consultation of stakeholders on proposed adaptations to policies.

In addition, the issues raised above regarding the way in which administrative law requirements can hamper adaptive decision-making require greater consideration in the synthetic biology context. In particular, the way in which decision-making could be adaptive while preserving proper public participation warrants further investigation.

Importantly, using adaptive approaches to the regulatory issues raised by synthetic biology may not address all problems arising from uncertainty. Zhang and collaborators have analysed synthetic biology regulation and concluded that scientific uncertainty is not a temporary state of

---

219 Jones (n 20) 362; see *Gene Technology Act 2000* (Cth) s 4.

220 Jones, ibid., 353, 362–3; see *Gene Technology Act*, ibid., ss 4, 65.

221 Marrakesh Agreement Establishing the World Trade Organization: Annex 1A Agreement on the Application of Sanitary and Phytosanitary Measures (n 31) art. 5(7).

222 McCray et al (n 153) 958.

affairs that can be remedied by obtaining knowledge.[223] In fact, in their view 'increasing knowledge often leads to increasing uncertainty', so that there is a need to manage 'the implications of uncertainty'.[224] Zhang, Marris and Rose argue that it is important to 'directly incorporate non-knowing' into governance and adopt 'regulatory practices [that] do not rely on scientific knowledge but build social resilience to not knowing through continuous provocation of reflexiveness among all stakeholders'.[225] However, it is not clear how a regulatory program could 'give explicit attention to both knowledge and non-knowing' as proposed by Zhang, Marris and Rose.[226] It would be preferable to use precaution, in combination with adaptive regulation, to develop a regulatory approach that is reviewed from time to time to address the uncertainties as far as possible. This would allow mitigation of some risk and incorporation of feedback about the effectiveness of the regulation and its impact on scientific research. These are valuable ends even though scientific uncertainty may not be eliminated.

5.6 APPLICATION OF PRECAUTION AND ADAPTATION TO ENVIRONMENTAL RISK

As discussed, risk of environmental harm is one of the key regulatory challenges raised by synthetic biology. There is a risk of accidental harm to the environment or human health from the release or escape of organisms produced using synthetic biology. These may be completely synthetic organisms (although these have not yet been developed) or organisms which have been modified using the techniques of synthetic biology outlined in Chapter 2. In her evidence to the Commission in 2010, ecologist Allison Snow explained that:

> [we] do need to be very careful whenever self-replicating organisms are released in the environment, especially if it's an intentional release but also for unintentional releases. Many of them will do no harm whatsoever. But important exceptions could occur, especially if the genetically engineered organism can multiply and become more abundant out in the environment.[227]

---

[223] Zhang et al, (n 203) 4.
[224] Ibid., 4, 9.
[225] Ibid., 28.
[226] Ibid., 4.
[227] Snow, 'Evidence to The Presidential Commission for the Study of Bioethical Issues, Synthetic Biology' (n 17).

A key issue in this context is whether existing risk assessment policies, developed for assessing genetically modified organisms, are capable of dealing with synthetic biology.[228]

5.6.1 Presidential Commission's Approach

The Commission recommended a review of risk assessment practices across government agencies, in order to identify gaps in risk assessment that would apply to release of synthetic organisms and look at the differences between practices and opportunities for harmonisation.[229] This was part of the general recommendation of taking a coordinated approach to the regulation of synthetic biology. The Commission acknowledged the need for particular care in relation to 'deliberate release' of synthetic organisms, especially where 'the probability or magnitude of risks are high or highly uncertain'.[230] It also proposed 'ongoing review of the ability of synthetic organisms to multiply in their natural environment and identify, as needed, reliable containment and control mechanisms'.[231] Further, the Commission recommended 'reasonable risk assessment' prior to field release of synthetic organisms.[232]

5.6.2 Critique of the Commission's Recommendations

The recommendations were cautious in that they recommended further review by various bodies in order to identify gaps, and did not propose any significant changes to the regulatory status quo. The major critique of the Commission's environmental recommendations is that it should have adopted the precautionary principle rather than the new principle of prudent vigilance. Notably, the Commission shied away from using the

---

228 This issue has been raised from the early days of synthetic biology, with the 2009 report *Ethical Aspects of Synthetic Biology* of the European Group on Ethics in Science and New Technologies questioning whether synthetic biology is different because it potentially involves altering multiple traits, and might present unknown risks: European Group on Ethics in Science and New Technologies to the European Commission, 'Ethical Aspects of Synthetic Biology: Proceedings of the Roundtable Debate' (European Commission, 19 May 2009) <http://ec.europa.eu/bepa/european-group-ethics/docs/publications/round_table_ethical_aspects_of_synthetic_biology.pdf>, 26–7.

229 The Presidential Commission for the Study of Bioethical Issues (n 8) 8.

230 Ibid., 10.

231 Ibid., 9.

232 Ibid., 131.

words 'caution' or 'precaution' in its recommendations about risk assessment and management. The report occasionally referred to the need for 'caution' or 'taking precautions' in specific circumstances, but did not recommend a precautionary approach to risk. Instead the Commission recommended 'careful risk analysis' – seen as an essential aspect of responsible stewardship – 'reasonable risk assessment' prior to field release, and noted the need for '*adequate consideration* of risk'.[233] These formulations are problematic because it is not clear what they mean, or how they would be implemented. What kind of risk assessment procedure would be 'reasonable', 'adequate' or sufficiently 'careful'? The choice of the word 'reasonable' implies that the Commission wanted to counter the demands of 'strong precaution', which is unnecessary as established above. The lack of clarity regarding these terms is compounded by the lack of clarity as to how to implement the principles of responsible stewardship and prudent vigilance.

The Commission discussed technical scientific measures, such as built-in features of synthetic organisms, for 'reliable containment and control' of synthetic organisms and stated that 'Surveillance or containment of synthetic organisms is a concrete way to embrace responsible stewardship'.[234] Technical safety measures may include techniques such as designing organisms which are unable to reproduce, or with minimal genomes which only provide basic function, so that the organism's survival is highly dependent on being in a laboratory environment.[235] There are also strategies such as engineering microbes to be dependent on an artificially supplied nutrient, so that they would die off if they were released into nature.[236] Cells can also be programmed to self-destruct when certain conditions are induced.[237] These techniques are known as 'suicide genes' and 'kill switches' respectively. In the future it may also be possible to design organisms with novel chemical structures incorporated into essential molecules, for example, using different chemical units

---

233 Ibid., 10 (emphasis added).

234 Ibid., 130.

235 Markus Schmidt, 'Do I Understand What I Can Create? Biosafety Issues in Synthetic Biology' in Markus Schmidt et al (eds), *Synthetic Biology: The Technoscience and its Societal Consequences* (Springer Academic Publishing, 2009) 81, 91–4.

236 See Gerd HG Moe-Behrens et al, 'Preparing Synthetic Biology for the World' (25 January 2013) **4** *Frontiers in Microbiology* 5 <http://www.ncbi.nlm.nih.gov/pmc/articles/PMC3554958/pdf/fmicb-04-00005.pdf>.

237 Jarred M Callura et al, 'Tracking, Tuning, and Terminating Microbial Physiology Using Synthetic Riboregulators' (2010) **107**(36) *Proceedings of the National Academy of Sciences* 15898; ibid.

to create a nucleotide that is similar to DNA, but is not found in nature.[238] This would limit the ability of the engineered organisms to transfer genetic material to the natural environment.[239]

However, in focusing on technical measures the Commission was overly reassured by the possibility of using such measures to biologically 'contain' synthetic biology organisms. It noted that synthetic organisms could potentially mutate and evolve in the natural environment, and referred to the scope for using technology to 'build in safeguards' through mechanisms such as incorporating 'suicide genes' and nutritional dependencies.[240] The Commission was techno-optimistic in this regard. While these possibilities are often raised by scientists such as Church and Venter, they should be only part of the approach to risk management in this area and their limitations need exploration and consideration. While the Commission did not advocate risk management solely using technological safeguards, this focus on technological measures appears to have shut down discussion of other ways of thinking about managing risk.[241] The technological measures – as the Commission itself acknowledged – are still evolving. As Snow explained, new organisms can behave in unexpected ways once released into the environment, and even those that seem 'innocuous' can change and evolve in the environment. Once released, they might survive better than expected, and they cannot later be removed.[242]

Instead of relying on prudent vigilance and technical measures, the Commission should have adopted the precautionary principle due to the high level of uncertainty created by the scale at which the techniques of

---

238 Markus Schmidt, 'Safeguarding the Genetic Firewall with Xenobiology' (Paper presented at the 21st Century Borders/Synthetic Biology: Focus on Responsibility and Governance, 2013) <http://www.scienceforglobalpolicy.org/LinkClick.aspx?fileticket=VqAc-vBzvMk%3d&tabid=133>; and Schmidt, 'Do I Understand What I Can Create? Biosafety Issues in Synthetic Biology' (n 235) 94–5.

239 Schmidt, 'Safeguarding the Genetic Firewall with Xenobiology', ibid; and Schmidt, 'Do I Understand What I Can Create?' ibid.

240 The Presidential Commission for the Study of Bioethical Issues (n 8) 9–10, 129–31.

241 Social science researchers Zhang et al have identified a tendency for a 'closing down' of discussion about synthetic biology. Discussion tends to revolve around questions about specific risks, and is 'closed down' by saying that all risks are covered by existing regulations. This means that indeterminate, unknown aspects are not analysed: Zhang et al, (n 203) 16–17.

242 Snow, 'Evidence to The Presidential Commission for the Study of Bioethical Issues, Synthetic Biology' (n 17).

synthetic biology operate in relation to the genome of a cell. The Commission pinpointed this issue when it stated: ‘In synthetic biology, a major concern is whether the scale of manipulation, using de novo chemical synthesis instead of conventional recombinant DNA techniques, raises sufficiently new levels of uncertainty about products, such as their characteristics or safety profile, to warrant new levels or forms of oversight.’[243]

The Commission stated that the concerns raised by synthetic biology in the context of food, agriculture and environmental applications are ‘broadly similar’ to those previously raised by genetically modified technologies.[244] However, this approach marginalises the way in which synthetic biology *differs* from the preceding classical recombinant DNA technology. Traditional genetic engineering aimed to manipulate the traits of organisms by making changes at the level of one or several genes, for example, inserting one gene into the genome of a bacterium to confer a new property such as green fluorescence. One branch of synthetic biology involves large-scale genetic engineering such as altering a whole metabolic pathway involving many genes, as was done by Jay Keasling and colleagues.[245] Another branch of synthetic biology involves genetic engineering at the level of the whole genome and trying to generate completely novel organisms using synthetic DNA. This goes beyond recombinant DNA technology by attempting to build new organisms with functions that do not exist in nature.[246] Thus organisms produced using the techniques of synthetic biology will have a high degree of ‘engineered complexity’ as compared to traditional genetically modified organisms.[247]

The theme of a key difference being the scale of the technology is reflected in scientists’ opinions. For example, Venter explained the difference between synthetic biology, and his favoured term ‘synthetic genomics’ as one of scale. He explained that although the terms are sometimes used interchangeably: ‘synthetic genomics differs from the more widely known synthetic biology in the scale of changes that can be made and in the kinds of experiments that it enables … Synthetic

---

243 The Presidential Commission for the Study of Bioethical Issues (n 8) 83.

244 Ibid., 70.

245 Jay D Keasling, ‘Synthetic Biology for Synthetic Chemistry’ (2008) **3**(1) *ACS Chemical Biology* 64.

246 Joachim Boldt and Oliver Müller, ‘Newtons of the Leaves of Grass: Certain Ethical Implications of Synthetic Biology Research Go Beyond Those of Genetic Engineering’ (2008) **26**(4) *Nature Biotechnology* 387, 387.

247 Kuzma and Tanji (n 1) 96.

genomics technologies ... provide the capability to build whole genomes and can examine how best to organize them'.[248] He identified a key feature of synthetic genomics capability in his research team as ability to rapidly assemble DNA molecules of large size (up to millions of base pairs).[249] Similarly, George Church explained:

> we should not be reassured that biology is not capable of engineering and there's no difference between what we're doing and what I did as a graduate student.
>
> Genetic engineering was one or two genes. Genome engineering as it's commonly used ... is also a couple of genes it's just done in the chromosome rather than on a plasmid ... Metabolic engineering you might do a pathway or a small network—30 genes or less. But genetic code offers us multi-virus resistance and safety measures and some use of new amino acids and this is truly genome-wide.[250]

Church also emphasised the rapid pace of change and development in synthetic biology, stating 'my major take-home for all this is that we are going much faster than it appears'.[251] In relation to the rapid improvement in DNA synthesis capability, he told the Commission: 'What's happened since 1999 is this exponential curve has gotten steeper. And I think that's something you can't ignore. So I would say that it's time to go beyond having more discussions.'[252]

This rapid pace of developments in synthetic biology, as compared to recombinant DNA technology, amplifies uncertainty about the environmental implications. This suggests that a precautionary approach is appropriate.

The large scale of synthetic biology techniques is relevant to environmental risk because the traits of organisms will be changed more drastically than with traditional genetic engineering, making their environmental impacts less predictable. It is not clear how predictable their

---

248 Prepared Statement of J Craig Venter, 'Evidence to United States House of Representatives Committee on Energy and Commerce, Effects of Developments in Synthetic Genomics and Implications for Energy and Health' (27 May 2010) Washington DC, 8.

249 Ibid., 9.

250 George Church, 'Evidence to The Presidential Commission for the Study of Bioethical Issues, Synthetic Biology' (8 July 2010) Washington DC.

251 Ibid.

252 Ibid.

impacts upon natural organisms will be.[253] They may be more difficult to manage and monitor.[254]

Some take the view that synthetic biology is *not* actually so unfamiliar at present, that highly engineered organisms could actually be *more* safe and have reduced ability to survive in nature, be *less* difficult to manage than traditional gene technology and that there is sufficient relevant knowledge to conduct safety assessments.[255] It could be argued that a heavily modified organism might be less likely to survive in the natural environment.

However, there is still debate and more information is needed about potential environmental impacts. This information gathering was recommended by the Conference of the Parties to the Convention on Biological Diversity. Considering the precautionary approach, it noted 'the need to consider the potential positive and negative impacts of components, organisms and products resulting from synthetic biology techniques on the conservation and sustainable use of biodiversity'.[256] It requested that information on these organisms and their impacts be sought from parties, governments, communities and other stakeholders and synthesised for further consideration and peer review. Discussion of synthetic biology's impacts within the framework of the Convention is ongoing.

While the Commission acknowledged uncertainty about the impact of releasing synthetic organisms into the environment, it glossed over the issue of highly modified organisms generated using the techniques of synthetic biology, but which are not wholly synthetic. Consideration of the environmental effects of these organisms is also required. In her evidence to the Commission, Snow explained the uncertainty in both areas:

> we don't have much experience with cultivating micro-algae or bacteria outdoors, let alone new life forms that might be entirely synthetic. New types

---

253 Dana et al (n 2).

254 Ibid.

255 See Tania Bubela et al, 'Synthetic Biology Confronts Publics and Policy Makers: Challenges for Communication, Regulation and Commercialization' (2012) **30**(3) *Trends in Biotechnology* 132, 134; Victor de Lorenzo, 'Environmental Biosafety in the Age of Synthetic Biology: Do we Really Need a Radical New Approach?' (2010) **32**(11) *BioEssays* 926; and Joyce Tait and David Castle, 'Artificial Microbes: Balanced Regulation of Synthetic Biology' (2012) **484**(7392) *Nature* 37.

256 Decisions Adopted by the Conference of the Parties to the Convention on Biological Diversity at its Eleventh Meeting, Annex I UN Doc UNEP/CBD/COP/11/35 (8–19 October 2012), 141.

> of really different bacteria? No experience. This brings up the question of whether regulatory agencies will be able to monitor and evaluate new types of synthetic or partially synthetic organisms that are proposed for release.[257]

As an example of a scenario in which synthetic biology could create risks of environmental harm, photosynthetic bacteria which have been engineered using synthetic biology to produce high levels of sugar (a real technology developed by the Silver lab at Harvard) could potentially escape into the environment.[258] The engineered bacteria would be contained within a bioreactor, but an extreme event such as a hurricane could cause their release into the environment.[259] In that case, questions would arise such as how the engineered genes affect 'fitness' – that is whether the modified organisms would quickly die out when released into nature, or persist, if they survive, whether they would they produce toxins, how quickly they might mutate, and how they would interact with natural organisms.[260] A lack of knowledge of natural cyanobacteria would impede safety analyses.[261] Regulatory agencies can certainly draw upon existing knowledge and experience obtained through the oversight of recombinant DNA technology, but simply equating the two fields is not sufficient as an approach to environmental risk.

Emphasising the similarities between synthetic biology and traditional genetic engineering, as in the Commission's approach, also does not

---

[257] Snow, 'Evidence to The Presidential Commission for the Study of Bioethical Issues, Synthetic Biology' (n 17).

[258] The Woodrow Wilson International Center for Scholars: Science Technology & Innovation Program, *Comprehensive Environmental Assessment and Synthetic Biology Applications Workshop: Background Materials* (n 7).

[259] The Woodrow Wilson International Center for Scholars: Science Technology & Innovation Program, *Comprehensive Environmental Assessment and Synthetic Biology Applications Workshop: Conference Notes Version 1.0* (n 6) 6.

[260] The Woodrow Wilson International Center for Scholars: Science Technology & Innovation Program, *Comprehensive Environmental Assessment and Synthetic Biology Applications Workshop: Background Materials* (n 7); The Woodrow Wilson International Center for Scholars: Science Technology & Innovation Program, *Comprehensive Environmental Assessment and Synthetic Biology Applications Workshop: Conference Notes Version 1.0* (n 6) 5–8.

[261] The Woodrow Wilson International Center for Scholars: Science Technology & Innovation Program, *Comprehensive Environmental Assessment and Synthetic Biology Applications Workshop: Background Materials* (n 7); The Woodrow Wilson International Center for Scholars: Science Technology & Innovation Program, *Comprehensive Environmental Assessment and Synthetic Biology Applications Workshop: Conference Notes Version 1.0* (n 6), 6.

consider the inadequacies associated with that regime of regulation.[262] It has been argued that the regulatory mechanisms are 'challenged' even by traditional genetic modification technology.[263] Thus a problem of regulatory connection is already in existence. It has been proposed that a more 'dynamic' approach to governance of synthetic biology is required.[264]

The uncertainty surrounding the environmental impacts of synthetic biology, lack of scientific information, and the rapid pace of scientific change suggest that a process of adaptation of regulatory instruments is crucial in promoting regulatory effectiveness and connection in the area of environmental risk. Indeed, the major positive aspect of the Commission's recommendations is that they reflect the importance of ongoing assessment of risk and evaluation and updating of regulatory policies.[265] The Commission understood the problems created by the high level of uncertainty in relation to environmental risk and the need to adapt policies over time. It saw this adaptability as part of prudent vigilance. For example, in relation to technical methods for risk assessment, Commission member Stephen Hauser stated: 'Prudent vigilance is required to assure that the strategy of how we compare new to old organisms, when the old organisms exist, remains effective as the field matures'.[266] As the Commission noted, synthetic biology is likely to produce organisms that are more and more removed from natural organisms as the field advances. This means that a program of planned adaptation of environmental risk management policies in the light of scientific developments will be crucial.

5.7 CONCLUSION

This chapter has considered the regulatory challenges posed by the uncertainty surrounding synthetic biology and the potential for synthetic biology to cause environmental harm. The Commission should have advocated use of the precautionary principle rather than the new, amorphous concept of responsible stewardship. Precaution is necessary

---

[262] Zhang et al, (n 203) 16–17.

[263] Fatehi and Kuzma (n 74) 67.

[264] Ibid., 69.

[265] See The Presidential Commission for the Study of Bioethical Issues (n 8) 8, 12, 27.

[266] Stephen Hauser 'Comments in Commission Deliberations, The Presidential Commission for the Study of Bioethical Issues, Synthetic Biology' 16 November 2010; see also ibid., 131.

because of the uncertainty associated with the environmental risks arising from key differences between synthetic biology and earlier gene technology. Precaution would provide a better approach to managing the risks raised by synthetic biology as well as facilitating the international cooperation and dialogue that is crucial to regulating the international field.

Regulators should adopt adaptive approaches and invest in improving knowledge over time, not only about the science but also about the impact of regulatory mechanisms. Such mechanisms allow regulation to be introduced, perhaps to mitigate the most significant risks, even though there is uncertainty and the science is moving rapidly. Combined with precaution, this is the best way to pursue the key role of regulation in managing the environmental risks posed by synthetic biology.

6. Synthetic biology, biosafety and 'biopunks'

6.1 INTRODUCTION

Interest in experimenting with synthetic biology, and biotechnology generally, has spread outside research institutions and biotech companies. Various labels are used to refer to those biologists who conduct experiments outside traditional research settings, including 'amateurs', 'citizen scientists', 'do-it-yourselfers', the 'biocurious' and 'outsider biologists'.[1]

There has been significant concern about biosafety issues and biosecurity problems posed by synthetic biology, including the existence of do-it-yourselfers. This chapter focuses on biosafety and the next chapter addresses biosecurity and the dual-use nature of synthetic biology. Addressing the safety and security risks posed by synthetic biology are important aspects of the role of regulation in relation to this field.

This chapter analyses the existing biosafety oversight mechanisms and guidelines which apply to academic biological research and biotechnology companies. This discussion will show that there is an area of risk in relation to the new breed of 'do-it-yourself' biologists, which overlaps with and mirrors aspects of the synthetic biology community. This chapter investigates what these biologists are doing and why, with a view to illuminating the problems posed. It argues that the main regulatory challenge posed is one of biosafety regulation, and that risks are currently identifiable. It proposes that it is not appropriate to ignore these, and that there needs to be careful monitoring of this field.

In this chapter, I have drawn on material that is publicly available. That is, what those involved in the community have been willing to write and say about themselves in a public way.

---

[1] This last term is used in Marcus Wohlsen, *Biopunk: DIY Scientists Hack the Software of Life* (Current, 2011).

6.2 BIOSAFETY, BIOSECURITY AND DUAL-USE RESEARCH

A problem in discussions of the regulatory issues raised by synthetic biology is that there is overlap and confusion in how the terms 'biosafety' and 'biosecurity' are used in the literature and discourse. Sometimes the terms are used interchangeably.[2] They are also used by different actors to mean different things.[3] To further complicate matters, 'biosafety' is used in some countries to describe the concepts described by 'biosecurity' in other countries.[4] This section defines the way in which these terms will be used in this chapter and subsequently.

Biosafety refers to taking care in laboratory research by ensuring that staff are properly trained to handle dangerous biological materials and that organisms used in research are contained. The World Health Organization's (WHO) guidance document Laboratory Biosafety Manual defines laboratory biosafety as 'the containment principles, technologies and practices that are implemented to prevent *unintentional* exposure to pathogens and toxins, or their accidental release'.[5]

The WHO's biosecurity guidance defines biosecurity as 'the protection, control and accountability for valuable biological materials ... within laboratories, in order to prevent their unauthorized access, loss, theft, misuse, diversion or intentional release'.[6]

The manual Biosafety in Microbiological and Biomedical Laboratories published by the US Department of Health and Human Services defines biosafety and biosecurity in a similar way to the WHO. It explains that:

---

[2] See, e.g., The Editors, 'How to Kill Synthetic Biology' (2006) **294**(6) *Scientific American* 4, where 'making synthetic biology secure' is discussed under a heading relating to 'biosafety'.

[3] For example, a selection of scientific experts and biosecurity professionals' conceptions of biosecurity are set out in Editorial, 'What is Biosecurity?' (2007) **63**(2) *Bulletin of the Atomic Scientists* 18.

[4] US Department of Health and Human Services et al, 'Biosafety in Microbiological and Biomedical Laboratories 5th Edition' (US Department of Health and Human Services, Public Health Service, Centers for Disease Control and Prevention, National Institutes of Health, 2009) <http://www.cdc.gov/biosafety/publications/bmbl5/BMBL.pdf> 106.

[5] WHO, 'Laboratory Biosafety Manual 3rd Edition' (WHO, 2004) <http://www.who.int/csr/resources/publications/biosafety/Biosafety7.pdf>, 47 (emphasis added).

[6] WHO, 'Biorisk Management: Laboratory Biosecurity Guidance' (WHO, 2006) <http://www.who.int/csr/resources/publications/biosafety/WHO_CDS_EPR_2006_6.pdf>, iv.

> Biosafety programs reduce or eliminate exposure of individuals and the environment to potentially hazardous biological agents. Biosafety is achieved by implementing various degrees of laboratory control and containment, through laboratory design and access restrictions, personnel expertise and training, use of containment equipment, and safe methods of managing infectious materials.[7]

Biosecurity is defined as the 'discipline addressing the security of microbiological agents and toxins and the threats posed to human and animal health, the environment, and the economy by deliberate misuse or release'.[8] The misuse of valuable biological materials is defined as 'inappropriate or illegitimate use, despite existing and subscribed agreements, treaties and conventions'.[9] This manual refers to protecting research information as well as material.[10]

The manual explains that in some research, biosafety practices such as controlling access to materials and ensuring appropriate training would be sufficient to also address security issues. However, where there are biological agents which have high public health concern or commercial value associated with them, additional measures may be needed to promote biosecurity.

I use a combination of the WHO and Biosafety in Microbiological and Biomedical Laboratories definitions of biosafety and biosecurity. For clarity, I use the element of intention to distinguish these concepts. Thus I use 'biosafety' to refer to preventing accidental events which pose a risk to those working in the laboratory (or do-it-yourself settings) and the risk of accidental release of dangerous organisms posing a threat to human health or the environment. I use the term 'biosecurity' to refer to intentional theft or misuse of biological tools, materials or information, and intentional release of dangerous biological material.

A further relevant concept is 'dual-use' research. Synthetic biology is a dual-use research area, meaning that 'one and the same piece of scientific research has the potential to be used for bad as well as good purposes'.[11] This creates a 'dual-use dilemma'. This is generally described as an ethical dilemma 'since it is about promoting good in the context of the potential for also causing harm, for example, the promotion of health in

---

7 US Department of Health and Human Services (n 4) 105.

8 Ibid.

9 Ibid., iv.

10 Ibid., 105.

11 Seumas Miller and Michael J Selgelid, 'Ethical and Philosophical Consideration of the Dual-use Dilemma in the Biological Sciences' (2007) **13** *Science and Engineering Ethics* 523, 534.

the context of providing the wherewithal for the killing of innocents'.[12] This is a dilemma faced by individual researchers and research institutions.[13] It also creates a regulatory challenge for governments: how should the research be regulated so that scientific progress is not impeded, while also minimising the risks?

6.3 BIOSAFETY IN SYNTHETIC BIOLOGY

While the Presidential Commission report on synthetic biology focused on biosafety in the context of commercial environmental release of organisms produced using synthetic biology, there was only a short discussion of laboratory biosafety. This section examines the regulatory mechanisms that apply to laboratory biosafety, and shows that the main areas of risk are in do-it-yourself biology.

In the US, there is a biosafety framework that applies to research in institutions such as universities. At the federal level, the National Institutes of Health (NIH) has promulgated the NIH Guidelines for Research Involving Recombinant or Synthetic Nucleic Acid Molecules (NIH Guidelines).[14] These have the legal status of regulations. They were first introduced in the 1980s, and undergo review and amendment as advised by the Recombinant DNA Advisory Committee. Amendments to incorporate synthetic DNA – as well as recombinant DNA – became effective in 2013.

Research involving recombinant or synthetic DNA must comply with the NIH Guidelines where it is funded by the NIH, or where any recombinant or synthetic DNA work at that institution is funded by the NIH.[15] Under these Guidelines, non-compliance can result in loss of funding. Other federal agencies can also require compliance with the NIH Guidelines as a condition of funding award. The NIH Guidelines require institutions to set up Institutional Biosafety Committees which review and give approval to laboratory work with recombinant DNA. Different categories of experiment require different levels of approval, ranging from notifying the local Institutional Biosafety Committee to

---

[12] See ibid., 524.

[13] Ibid., 527.

[14] *NIH Guidelines for Research Involving Recombinant or Synthetic Nucleic Acid Molecules*, Federal Register **81**(73) 22286, 22 March 2016.

[15] Ibid., s 1-D.

obtaining approval from the Director of the NIH and the NIH's Recombinant DNA Advisory Committee. For different categories of research, certain biosafety containment levels are set out by the NIH Guidelines.

Researchers in university settings in the US have to undergo biosafety training as part of their apprenticeship in the life sciences. This training is the responsibility of the research institution under the NIH Guidelines. Some training must be updated periodically. For example, at Stanford University where synthetic biology leader Drew Endy is based, all who work with biological agents must be trained in biosafety and certain further training must be repeated annually or every two years.[16]

Under the NIH Guidelines, research at biotechnology companies must comply with these guidelines where any recombinant DNA work at that company is funded by the NIH,[17] or where the conditions of other federal funding incorporate this requirement. Where they do not receive federal funding, commercial entities participating in biotechnology research still have a commercial imperative to comply with the NIH Guidelines. The Presidential Commission pointed out that biotechnology companies are generally staffed by scientists who have trained in academia and are familiar with the biosafety regulations.[18] Further, the NIH Guidelines have been widely adopted across life sciences research and, according to the Commission, have now become the 'gold standard'.[19] Therefore, large companies with a public profile such as J Craig Venter's Synthetic Genomics would need to comply with these NIH Guidelines in order to demonstrate that they are good corporate citizens. There may be less public relations pressure for small biotechnology firms, although it would be positive to be able to demonstrate compliance when seeking funding from both private and public sources.

As discussed in the previous chapter, the EPA has responsibility for regulating commercial research and development and commercial manufacture of certain genetically engineered microorganisms under the

---

16 Stanford University, *Biosafety Training Requirements* Stanford University: Environmental Health and Safety <https://ehs.stanford.edu/reference/biosafety-training-requirements>.

17 *NIH Guidelines for Research Involving Recombinant or Synthetic Nucleic Acid Molecules* (n 14) 1-C-1-a-(1).

18 The Presidential Commission for the Study of Bioethical Issues, 'New Directions: The Ethics of Synthetic Biology and Emerging Technologies' (The Presidential Commission for the Study of Bioethical Issues, December 2010) <http://bioethics.gov/sites/default/files/PCSBI-Synthetic-Biology-Report-12.16.10_0.pdf>, 146.

19 Ibid., 89.

Coordinated Framework for the Regulation of Biotechnology. Commercial operators may not manufacture a new microorganism chemical unless they have first notified the EPA, which must undertake a risk assessment.[20] The manufacturer must provide information about the characteristics of the new organism, the intended use, controls on release, survival and dissemination, known biological interactions, and any other information that would allow the EPA to evaluate the health and environmental effects.[21] The EPA will assess whether any restrictions are needed. These requirements do not apply to certain research and development activities that are covered by other schemes such as the NIH Guidelines.[22]

While biotechnology companies would be subject to this regulatory scheme, it would not generally apply to do-it-yourself work.[23] This is not normally for commercial advantage or commercial distribution and therefore would not meet the definition of 'Manufacture, import, or process for commercial purposes' in the regulations:

(1) To import, produce, manufacture, or process with the purpose of obtaining an immediate or eventual commercial advantage for the manufacturer, importer, or processor, and includes, among other things, 'manufacture' or 'processing' of any amount of a microorganism or microbial mixture:
  (i) For commercial distribution, including for test marketing.
  (ii) For use by the manufacturer, including use for product research and development or as an intermediate.[24]

Commercial research and development conducted for 'commercial advantage' is further defined in these regulations as including activities wholly or partly funded by a commercial entity and research that is not commercially funded but where the researcher intends to obtain a commercial advantage, such as research to develop a commercial application. Therefore, if a do-it-yourselfer sought to commercialise their activities or seek funding from the commercial sector, they would come under this regulatory scheme.

---

[20] 15 USC (Office of the Law Revision Counsel, 2017), § 2604 (a); 40 CFR (US Government Publishing Office – e-CFR, 2017) § 725.3 and 725.105.
[21] 40 CFR (n 20) §725.155.
[22] See ibid., §§ 725.232.
[23] The Presidential Commission for the Study of Bioethical Issues (n 18) 94.
[24] 40 CFR (n 20) § 725.3.

Depending on the state, there are either federal or state occupational safety and health requirements which apply to biological research.[25] For example in California – where key synthetic biology company Amyris is based – industrial relations regulations set out requirements for laboratory safety. There are General Industry Safety Orders which include procedures for the 'use, handling and storage' of hazardous substances such as blood-borne pathogens and airborne diseases, as well as ventilation requirements for biological safety cabinets and fume hoods.[26] These requirements apply to all places of employment, and therefore would apply to both academic research institutions and biotechnology companies.[27] However, they would not apply to do-it-yourselfers where these individuals do not have employees.

Both universities and industry researchers are regulated by the rules controlling access to and use of certain disease-causing organisms under the Select Agents and Toxins system.[28] The system applies to a prescribed list of particularly dangerous infectious organisms (such as plague and Ebola), and toxins (such as ricin) which pose a severe public health risk. It also applies recombinant and synthetic nucleic acids that can produce infectious forms of the listed viruses or functional forms of the listed toxins. It restricts the possession, use and transfer of these prescribed biological agents and requires registration of these activities. By restricting who can access these agents and what can be done with them, the select agent system is targeted to both accidental and intentional events; that is, both biosafety and biosecurity risks. For approval to possess a select agent, both biosafety and security plans are required under this system.

There has been concern about do-it-yourself biologists potentially creating dangerous disease-causing organisms.[29] The Select Agents and Toxins regulatory regime applies to possession, transfer or use of the prescribed by any 'individual or entity'.[30] Thus individual do-it-yourselfers would come within the application of this scheme. Under the

---

25 The Presidential Commission for the Study of Bioethical Issues (n 18) 91.

26 8 Cal [Industrial Relations] Code (Barclays Official California Code of Regulations, 2016), see e.g., §§ 5154.1, 5154.2, 5193, 5199.

27 Ibid., § 3202.

28 See 42 USC (Office of the Law Revision Counsel, 2016) 262a and the regulations made under this, 42 CFR (US Government Publishing Office – e-CFR, 2016) Part 73.

29 See e.g., Howard Wolinsky, 'Kitchen Biology' (2009) **10**(7) *EMBO Reports* 683.

30 42 CFR (n 28) § 73.7(a).

regulations, select agents or toxins can only be possessed, used or transferred where the relevant individual, or in the case of an 'entity', the entity's designated Responsible Official, has registered this activity. Registration requires a security risk assessment of the relevant person or entity. The organisms can only be kept in the specific location for which approval has been given and those who have approved access cannot give access to other individuals unless they too have undergone security assessment.

Further, there are criminal penalties under this regime for transferring a select agent to an unregistered person and for unregistered possession of a select agent.[31] This means that any person or entity that held select agents could not sell or give them to a do-it-yourselfer without seeking an amendment to their regulatory approval. This would involve giving information about the purpose of accessing the agent and the do-it-yourselfer would have to undergo security assessment. Thus, it would be difficult for do-it-yourselfers to work with any of these organisms or toxins. However, this system only regulates those agents which generate the most severe public health concern; there are still other less dangerous pathogens.

Given that scientists working in institutions have training in managing biosafety problems and their work is subject to institutional review, and that commercial biotechnology firms have reasons to comply with the relevant guidelines, problematic accidents in synthetic biology are more likely to occur in the do-it-yourself community. While there may be scope to review and update legal mechanisms for oversight of research in institutions and corporations and rules for risk assessment and management, a pressing gap in discussion and oversight is presented by the do-it-yourself area. Thus the rest of this chapter focuses on the do-it-yourself community.

6.4 DO-IT-YOURSELF BIOLOGY

The group 'DIYbio' has become a focal point for do-it-yourselfers in biology to meet each other, form local groups, discuss issues and experiments and share information and ideas. I use the term 'DIYbio' to refer to members or activities of that specific group, and the terms 'do-it-yourselfer' or 'do-it-yourself biology' to refer to non-institutional biologists more broadly.

---

[31] 18 USC (Office of the Law Revision Counsel, 2016) § 175b.

DIYbio was founded by two individuals with science training, Jason Bobe and Mac Cowell, during a meeting in Boston in 2008. The group operates through the central website DIYbio.org, with local groups which meet to discuss and run experiments. DIYbio has now grown from the original 25 attendees at the first meeting to a network with local groups in cities in North America, Europe and Asia. These groups conduct social activities, workshops, and experiments. By the time of the Presidential Commission's consultations on synthetic biology, the group's mailing list had grown to about 2000 members (though a proportion of these were people working in the biosecurity area who wished to understand what DIYbio was doing).[32]

In addition to individuals experimenting in biology at home and connecting through DIYbio.org, the development of community biology labs is an emerging trend. These are labs where people can share resources such as equipment, and teach each other skills in biology. As Bobe explained:

> A community lab is taking a page out of the book of clay potters and woodworkers pooling their resources to buy expensive equipment that would otherwise be beyond the scope of an individual to put something in their garage. And this is also places where people have classes. They get access to expertise … .[33]

There are community labs in New York, Boston and around the San Francisco Bay area. Local DIYbio groups have set up BosLab, the Boston Open Science Laboratory and GenSpace in New York. Some of these groups are using crowd funding.[34] The BioBricks Foundation is also using this funding model, asking for donations from its mailing list members to keep the BioBricks community going.

The links between DIYbio and synthetic biology are apparent from the beginning of the DIYbio story. There are connections in terms of the key figures involved, the open, sharing culture of the DIYbio and BioBricks communities, and the inspiration taken from computer hacking and open source movements. Both of the co-founders of DIYbio have been involved in synthetic biology. Bobe had previously worked on Open Wetware, an online space for sharing of information about biological engineering run by the BioBricks Foundation, and on the Personal

---

[32] Jason Bobe, 'Evidence to The Presidential Commission for the Study of Bioethical Issues, Synthetic Biology' (14 September 2010) Philadelphia Penn.

[33] Ibid.

[34] Ibid.

Genome Project under the stewardship of synthetic biology leader George Church.[35] He is now an Associate Professor at the Mount Sinai School of Medicine. Cowell had previously worked on iGEM with Randy Rettberg who is also an instrumental figure in the field of synthetic biology.[36] The first DIYbio meeting was also attended by Tom Knight, one of the early founders of synthetic biology.[37]

At this meeting, Cowell posed big questions linking do-it-yourself biology to synthetic biology: 'Can molecular biology or biotechnology be a hobby? Will advancements in synthetic biology be the tipping point that enables DIYers and garagistas to make meaningful contributions to the biological sciences, outside of traditional institutions?'[38]

There are strong similarities in culture between the two fields. Like the BioBricks community, the do-it-yourselfers show a preference for openness, sharing, and positive engagement with ethical and regulatory issues both at the level of community grass roots discussion and engagement with the Commission. Also like the BioBricks community, they emphasise the importance of making biotechnology easier to participate in. There is also similarity in culture as many of the do-it-yourselfers are computer programmers.[39] They are inspired by open source software and collaboration among computer hackers. For example, at DIYbio's first meeting, Cowell asked 'Can DIYbio.org be the Homebrew Computer Club of biology?'[40] The Homebrew Computer Club was an early venue for computer tinkerers to get together and share ideas in the 1970s and 80s, with members including Apple Computer Inc. co-founders Stephen Wozniak and Steve Jobs.

There is also overlap in how individuals who are members of the synthetic biology and do-it-yourself biology communities identify themselves. For example, there were attendees from the world of do-it-yourself and citizen science attending the international Synthetic Biology 5.0 meeting. These included Mac Cowell, Ellen Jorgensen of community

---

35 Jason Bobe, *About* Jason Bobe <http://www.jasonbobe.net/about.html>.

36 Greg Boustead, 'The Biohacking Hobbyist' (11 December 2008, 2008) *Seed Magazine* <http://seedmagazine.com/content/article/the_biohacking_hobbyist/>.

37 Jason Bobe, 'Don't Phage Me, Bro' (24 May, 2008) *DIYbio Blog* <http://diybio.org/2008/05/>.

38 Ibid.

39 Prashant Nair, 'Straight talk with … Mac Cowell and Jason Bobe' (2009) **15**(3) *Nature Medicine* 230, 231.

40 Bobe 'Evidence to The Presidential Commission for the Study of Bioethical Issues, Synthetic Biology' (n 32).

lab Genspace and Tito Jankowski of Open PCR and DIYBio. Jorgensen explained that synthetic biology and genetic engineering are occurring at Genspace and applauded synthetic biology's goal of making biology more accessible.[41] Cowell spoke about building low-cost tools for use in synthetic biology, as is being done in the do-it-yourself space.[42] Jankowski presented a poster on 'OpenPCR', a low-cost, open source tool for running the polymerase chain reaction (PCR).[43] Further, DIYbio has tried to become involved with synthetic biology's flagship education program, iGEM. There was considerable enthusiasm early on within the DIYbio community for entering an iGEM team.[44] However, the organisers of iGEM decided to disallow this in 2009 in part due to the lack of institutional safety frameworks which would apply to do-it-yourself biologists.[45]

Some of the key tools of synthetic biology are also enabling the development of do-it-yourself biology. DNA synthesis allows time-consuming lab work to be outsourced, so that a do-it-yourself biologist could order synthetic DNA for experiments online.[46] The parts agenda, with its emphasis on modularity, predictability and de-skilling, is also attractive to do-it-yourself biologists and likely to increase the scope of what it is possible for non-institutional scientists to achieve.

The Presidential Commission noted the connection between do-it-yourself biology and synthetic biology in its deliberations, highlighting that do-it-yourself biology is something which distinguishes synthetic biology. For example, Commission member Alexander Garza commented that: 'The thing that makes syn-bio a little unique is that it's been embraced by those outside of academia and outside of these private

[41] Ellen Jorgensen, 'Isn't That Cute? Little Johnny's First New Life Forms' (Speech delivered at Synthetic Biology Slam, Stanford University, Ca, June 14 2011) <http://www.youtube.com/watch?v=LqHdK_yYbiM&list=UUFHpWUDcjnjNAXrhGSb9cqw&index=7&feature=plcp>.

[42] Mac Cowell, 'Low-Throughput Lab Automation' (Speech delivered at Synbio Slam, Stanford University, Ca, 14 June 2011) <http://www.youtube.com/watch?v=WoE2DnIDikw>.

[43] See *OpenPCR* (2011) OpenPCR <http://openpcr.org/>.

[44] See, e.g., DIYbio Blog entries: Jason Bobe, 'DIYbio Meetup in Boston on Thursday, October 16' (11 October, 2008) *DIYbio Blog* <http://diybio.org/2008/10/>; Tito Jankowski, 'This week in DIYbio' (17 February, 2009) *DIYbio Blog* <http://diybio.org/2009/02/17/this-week-in-diybio-2/>.

[45] Mac Cowell, 'iGEM Closes Doors to Amateurs' (10 April, 2009) *DIYbio Blog* <http://diybio.org/2009/04/10/igem-closes-doors-to-amateurs/>.

[46] Wohlsen (n 1) 22.

companies, these so-called DIY communities'.[47] The Commission's report also stated that synthetic biology poses 'some unusual potential risks' due to the work of those outside 'traditional research settings'.[48]

6.4.1 What are Do-it-yourself Biologists Actually Doing?

It is difficult to get an idea of what do-it-yourself biologists are doing because the usual methods of obtaining an overview of research activity do not apply. While do-it-youselfers may attend some mainstream science conferences, and indeed hold their own conferences, do-it-yourself biologists do not generally apply for grants from funding bodies or publish their results in scientific journals. For these reasons, it is challenging to work out how much synthetic biology and do-it-yourself biology overlap and intersect in terms of the actual biological activities undertaken. In the Presidential Commission's consultations on synthetic biology, terms such as 'do-it-yourself' and 'DIYbio' were used frequently, but little substantive evidence was received on do-it-yourself biology.

In his evidence to the Commission, Bobe explained that there are broadly two types of activities being undertaken by do-it-yourselfers. The first he called 'exploratory biology', which 'tends to be focused on discovery and the application of technologies like DNA sequencing to exploring the environment and oneself'.[49] This involves activities such as looking into one's own genetic code. For example, Kay Aull constructed a home laboratory in order to test her own DNA for a gene that would put her at risk for the genetic disease hemochromatosis.[50] Another example is Bioweathermap, a project in which do-it-yourselfers engage in a 'global, grassroots, distributed environmental sensing effort aimed at answering some very basic questions about the geographic and temporal distribution patterns of microbial life'.[51] In one project they have sampled

---

47 Alexander Garza 'Comments in Commission Deliberations, The Presidential Commission for the Study of Bioethical Issues, Synthetic Biology' 16 November 2010.

48 The Presidential Commission for the Study of Bioethical Issues (n 18) 125.

49 Bobe 'Evidence to The Presidential Commission for the Study of Bioethical Issues, Synthetic Biology' (n 32).

50 See ibid; Wohlsen (n 1) 12–16.

51 Bobe 'Evidence to The Presidential Commission for the Study of Bioethical Issues, Synthetic Biology' (n 32); BioWeatherMap, *The BioWeatherMap Initiative* BioWeatherMap <http://bioweathermap.org/>.

the microorganisms present on paper money in various locations and used DNA sequencing to identify them.[52]

Bobe identified a second area called 'constructive biology', which is engineering focused and involves refashioning the 'tools and toys' used in molecular biology laboratories.[53] It was not clear whether Bobe was talking about the techniques of molecular biology or the equipment at this point – probably both. There are several examples of do-it-yourself biology entrepreneurs who have made cheaper versions of essential lab equipment, such as the LavaAmp, an inexpensive, portable thermocycling machine for conducting PCR. PCR is a ubiquitous process in DNA technology, used to create copies of fragments of DNA for use in experiments or medical diagnostic tests.[54] Experiments involving genetic engineering are also occurring. For example, at BasLab do-it-yourselfers have run classes in basic recombinant DNA techniques.[55]

In reviewing the literature about DIYbio and the movement's own website and materials, I have found that what enthusiastic individuals are able to achieve outside institutional research settings is surprisingly sophisticated. For example, Eri Gentry and John Schloendorm conducted home experiments in which they extracted their own white blood cells and observed whether these could attack various types of cancer cells which they had been able to obtain.[56] They were later able to attract enough investment to start a company to pursue this work.[57]

Many of the do-it-yourselfers are entrepreneurs and inventors, others simply want to understand how things work and learn more about processes like DNA sequencing; and others want to find ways to work around the expense of traditional biology equipment and build cheaper alternatives.[58] In this context, Bobe emphasised that do-it-yourself biology is not exactly the same as the kind of work that is done in research institutions:

---

52 Bruce Upbin, 'The Bioweathermap Sees The Germs Around Us' (13 July, 2011) *Forbes* <http://www.forbes.com/sites/bruceupbin/2011/07/13/the-bioweathermap-sees-the-germs-around-us/>; BioWeatherMap (n 51).

53 Bobe 'Evidence to The Presidential Commission for the Study of Bioethical Issues, Synthetic Biology' (n 32).

54 LavaAmp, *Pocket PCR for Pennies* (4 June 2010) <http://lava-amp.com/>.

55 Boston Open Source Science Lab, *Classes* BOSSLAB <http://bosslab.org/classes>.

56 Wohlsen (n 1) 99–111.

57 Ibid.

58 Bobe, 'Evidence to The Presidential Commission for the Study of Bioethical Issues, Synthetic Biology' (n 32).

> I expect that one of the things we'll see happen is that you have to take existing science and look at it with a new lens. When you are forced to do things in a low cost, low waste environment, you actually come up with something that looks quite a bit different.[59]

6.4.2 Why Do-it-yourself?

As do-it-yourself biology is an informal movement, there are no official accounts of the group's philosophy which are endorsed by its members. There are self-described thought leaders[60] in certain groups such as DIYbio. However, non-traditional biology is not a coherent movement represented by a clearly defined, unified group.[61] There is no broadly adopted philosophy or common goals. Reasons for being interested in this type of research may vary widely across the community. This section provides an overview of the reasons of those do-it-yourselfers who have engaged with the media, and those who have engaged with other do-it-yourselfers at publicly accessible conferences or through DIYbio.org.

The reasons for doing biotechnology research outside the usual venues are important because they can be used to work out whether there are problems with mainstream ways of doing things, such as whether the intellectual property system is having negative consequences for innovation. Investigating motivations also sheds further light on what research do-it-yourselfers are actually doing, and the types of risks this presents. This assists in analysing the unique regulatory challenges presented by do-it-yourself biology.

The reasons for engaging in do-it-yourself biology include passionate interest in biological problems, wanting to make biotechnology more accessible, interest in designing cheaper equipment for biotechnology, the ability to explore topics that may not be possible to explore in institutions due to funding priorities and other limitations, and the desire to look into one's own genetics and disease risks. It is underpinned by a mentality of 'hacking', in which allowing many individuals to look at a problem is seen to increase the likelihood of finding a solution.[62] For some, working around intellectual property constraints is also a factor.

---

59 Ibid.

60 See Nair (n 39).

61 Rob Carlson, 'Garage Biology in Silicon Valley' (8 March, 2010) *Synthesis* <http://www.synthesis.cc/2010/03/garage-biology-in-silicon-valley.html>.

62 This understanding of 'hacking' is drawn from Wohlsen (n 1) 5, where Wohlsen explains that this is different from negative connotations sometimes given to this term.

Do-it-yourself biologists all have a desire to explore biology. However, some lack the training or opportunity to do this through traditional routes. Do-it-yourself biologists are finding a way for people who are interested in biology, but do not have the necessary post-graduate qualifications, to be involved.[63] They are trying to make working with biology more widely accessible. Indeed, Bobe has commented that developing exciting biotechnology is 'one part inspiration and one part accessibility to these technologies'.[64] In the early days of DIYbio Bobe outlined his vision for 'science without scientists':

> As molecular tools get cheaper, and the know-how for using them more widely distributed, I think we're going to see a renaissance in science. The peculiar feature of this renaissance is that it's going to take place outside of 'science proper', away from the universities which dominate now, and funded out-of-pocket by enthusiasts without PhDs.[65]

Thus for some do-it-yourselfers, the goal is democratisation of biological science. This is seen as enabling 'more people to do their own science: make hypotheses, design experiments, collect large data sets, and apply a mixture of reasoning and cloud computing to make discoveries'.[66] Specifically, the goal is democratisation of the ability to *participate* in science, to run biotechnology experiments. In her *Biopunk Manifesto* DIYbio member Meredith Patterson emphasised the importance of 'scientific literacy':

> A person educated in science can understand science; a scientifically literate person can *do* science. Scientific literacy empowers everyone who possesses it to be active contributors to their own health care, the quality of their food, water, and air, their very interactions with their own bodies and the complex world around them.[67]

This Manifesto explains one person's philosophy in relation to do-it-yourself biology.

Do-it-yourself biology is about accessibility not only of scientific information but also the ability to explore scientific questions. DIYbio's home page explains, 'DIYbio.org is an organization dedicated to making

---

[63] See, e.g., Mac Cowell in Boustead (n 36); Wohlsen (n 1) 6, 161.

[64] Wolinsky (n 29) 684.

[65] Jason Bobe, 'Science without Scientists' (22 August, 2008) *DIYbio Blog* <http://diybio.org/2008/08/>.

[66] Ibid.

[67] Meredith Patterson, *A Biopunk Manifesto* (30 January 2010) LiveJournal <http://maradydd.livejournal.com/496085.html>.

biology an *accessible* pursuit for citizen scientists, amateur biologists and biological engineers who value openness and safety'.[68] Bobe has pointed out that there is a 'very, very low barrier to entry' into the DIYbio field, which is regarded as a positive thing.[69] The Biopunk Manifesto goes further, not limiting the accessibility to those who value openness and safety but stating that 'We the biopunks are dedicated to putting the tools of scientific investigation into the hands of anyone who wants them.'[70]

The Biopunk Manifesto goes so far as to consider this access a *right*:

> the right to arrive independently at an understanding of the world around oneself is a fundamental human right ... the opportunity to satisfy that curiosity all too often turns on economic opportunity, and we aim to break down that barrier ... we assert that the right of freedom of inquiry, to do research and pursue understanding under one's own direction, is as fundamental a right as that of free speech or freedom of religion.[71]

In addition to its goals of democratising access to biological science, DIYbio is also a response to frustration with the limitations of the traditional way of doing science in research institutions. Do-it-yourself biologists can look at different problems, or do different types of work, from what institutional scientists would focus on (which must usually be work that is likely to be publishable).[72] Cowell argues that do-it-yourselfers have abilities such as conducting work that is important but not exciting enough to be published in academic journals, or which is 'whimsical', but could yield useful results.[73] Patterson, who has done do-it-yourself experiments to test baby formula for toxic melamine, argues that 'if we have to wait for university labs to do this, we're going to be waiting forever'.[74] For some, the focus on career advancement and institutional politics in academic science is seen as limiting the pace of scientific progress.[75] Non-institutional scientists can by contrast think in

---

[68] DIYbio, *An Institution for the Do-It-Yourself Biologist* DIYbio <http://diybio.org/>, (emphasis added).

[69] Bobe, 'Evidence to The Presidential Commission for the Study of Bioethical Issues, Synthetic Biology' (n 32).

[70] Patterson (n 67).

[71] Ibid.

[72] See, e.g., Boustead (n 36).

[73] See Nair (n 39) 231.

[74] Wohlsen (n 1) 38–40.

[75] Ibid.

a more creative, less 'establishment' or traditional way.[76] Do-it-yourselfer Hugh Reinhoff has explained that as an outsider biologist, he is free to speculate about hypotheses, and 'the dots can be further apart'.[77] He explained that as he is not working within an institution, he takes no professional risk in doing this.

This frustration with institutional science and response by taking it into their own hands has similarities to work by patients with poorly understood diseases and their families to advocate for awareness of their condition and to find and share information about causes and treatments. For some this extends to seeking to push the science itself forward.[78]

For some do-it-yourselfers, accessibility of biotechnology is important because the particular problem they are concerned with is not being addressed by traditional science. This might relate to a neglected disease or an issue that affects a population in the developing world. In an example of studying neglected disease, Hugh Reinhoff has explored his own daughter's DNA in the hope of understanding her illness.[79] He has had to do this research himself because her genetic condition is so rare that it is a low priority for traditional researchers.[80] In an example of aiming to assist developing countries, Joseph Jackson and Guido Nunez-Mujica have developed the LavaAmp, a cheap PCR machine for use in molecular biotechnology and diagnostics.[81] This would usually be an expensive item, but they have designed an affordable, portable version for use in the developing world, such as diagnosing the Chagas disease in

---

[76] Laura H Kahn, 'DIY Biology' (18 June 2012) *Bulletin of the Atomic Scientists* <http://www.thebulletin.org/web-edition/columnists/laura-h-kahn/diy-biology>; Wohlsen (n 1) 39–40, 85; Hugh Reinhoff, 'Comments in Panel Discussion' (Speech delivered at Outlaw Biology Symposium, Los Angeles, 29–31 January 2010) <http://artsci.ucla.edu/videos/outlaw_biology/OB9_Panel_Discussion1/>.

[77] Reinhoff, ibid.

[78] For example, Josh Sommer, a young man diagnosed with the fatal illness chordoma, responded to what he saw as the slow progress in traditional research into this disease by co-founding a foundation to enhance sharing of results and collaboration amongst scientists in the field. See Chordoma Foundation, *About Us* (2012) Chordoma Foundation <http://www.chordomafoundation.org/about-us/>.

[79] Wohlsen (n 1) 46.

[80] Reinhoff (n 76).

[81] Justin Bourke, *Empowering Citizen Scientists to Do It Themselves* (22 July 2011) Better Bio: The Life of Science <http://www.betterbio.org/empowering_citizen_scientists_to_do_it_themselves>; Wohlsen (n 1) 54–9.

Venezuela.[82] Such do-it-yourselfers are trying to make both the *doing* of biotechnology innovation and the *outcomes* of this innovation accessible.[83]

In relation to intellectual property, there is a strong preference for openness, sharing and community in do-it-yourself biology. There is a sense among do-it-yourselfers that science will yield better results when it is done by more people in a distributed way.[84] This reflects the philosophy underpinning the standard biological parts agenda. Further, one of the main benefits of do-it-yourself biology is seen as 'cross-pollination' of expertise, where individuals with various fields of expertise can work on problems.[85] Do-it-yourselfers are in favour of open source approaches to innovation and have used strategies such as developing a new piece of biotechnology equipment and not only selling it for a small price but freely revealing the designs so that others might improve them.[86]

Cowell has stated that he is excited about being able to use the BioBricks Public Agreements.[87] As will be seen in subsequent chapters, these agreements allow standard biological parts to be donated to a commons and accessed by other synthetic biologists under standard intellectual property terms.[88] The focus on openness and sharing in both the BioBricks commons and do-it-yourself biology community comes from the philosophy that people will learn and solve problems more effectively if they are sharing their knowledge and ideas. This philosophy is similar to that underpinning innovation communities in the computer programming world.[89] Open source innovation in that sphere has inspired leaders in both the do-it-yourself and BioBricks communities.[90]

---

82 Ibid.

83 Wohlsen (n 1) 27.

84 Nair (n 39) 231; Wohlsen (n 1).

85 Boustead (n 36); Nair (n 39).

86 See Wohlsen (n 1) 61–8.

87 Nair (n 39); see also The BioBricks Foundation, *The BioBrick™ Public Agreement* (2013) The BioBricks Foundation <http://biobricks.org/bpa/>.

88 See Chapters 8 and 9. See also James Boyle, *The Public Domain: Enclosing the Commons of the Mind* (Yale University Press, 2008), 176–8; Sapna Kumar and Arti Rai, 'Synthetic Biology: The Intellectual Property Puzzle' (2007) **85** *Texas Law Review* 1745, 1763–7; Arti Rai and James Boyle, 'Synthetic Biology: Caught Between Property Rights, the Public Domain and the Commons' (2007) **5**(3) *PLoS Biology* 0389, 0391–2.

89 See theory relating to innovation communities in Eric von Hippel, *Democratizing Innovation* (The MIT Press, 2005) and discussion in Chapter 9.

90 See Nair (n 39) and Chapter 9.

In the BioBricks context, the sharing mechanism is also expressly used to get around problems caused by the intellectual property system. By contrast, avoiding intellectual property rights does not seem to be the major motivator in do-it-yourself biology. The focus is more on accessibility of the chance to participate in biological research. However, intellectual property is a factor in some do-it-yourselfers' work. For example, Eri Gentry has stated that 'until very recently, commercial biology was something done in universities or big companies. But those organizations generally own the majority of the work, which stifles innovation'.[91] She also stated that scientists' attempts to protect their intellectual property can lead to 'creativity malaise'.[92]

Restrictive licence terms are relevant for some of those do-it-yourselfers who strive to develop their own, cheaper versions of the equipment used in biotechnology research. For example, Norm Wang, who has developed improved versions of some standard biotechnology equipment, was frustrated by licences which did not allow him to modify purchased equipment.[93] The high cost of biotechnology equipment such as PCR machines, seen as a problem by do-it-yourselfers who want to address problems in the developing world as in the LavaAmp example above, could also be related to intellectual property rights over that equipment.

While do-it-yourself biology is not always an attempt to avoid particular intellectual property rights, some of this work arises from a rejection of the very *idea* of intellectual property. Journalist Marcus Wohlsen's impression after interviewing do-it-yourselfers was that 'in their eyes, the traditional model of intellectual property keeps knowledge barricaded behind the castle wall … knowledge walled up will only ever benefit the few, while knowledge set free will empower everyone'.[94] While getting around intellectual property rights may not be the main motivator for do-it-yourselfers, some have a dim view of these rights. For example, Cowell has said that he himself is not interested in owning intellectual property, and has held a DIYbio workshop in which he invited attendees to experiment on a patented gene.[95]

---

91 Zain Jaffer, 'Eri Gentry's Garage Biotech Revolution' (15 July 2010, 2010) *Wired UK* <http://www.wired.co.uk/news/archive/2010-07/15/eri-gentry-garage-biotech-revolution>.

92 Wohlsen (n 1) 103.

93 Ibid., 62–5.

94 Ibid., 80.

95 Nair (n 39) 231; Khadijah Britton, *Do It Your Damn Self* Better Bio: The Life of Science <http://www.betterbio.org/do_it_your_damn_self>.

6.5 WHAT PROBLEMS ARE POSED BY DO-IT-YOURSELF BIOLOGY?

The do-it-yourself biology movement in itself is not a symptom of major problems with the intellectual property system. Due to the small numbers of these biologists, the variety of types of experiments they are working on and a variety of motivations, it is unlikely that there is a crisis looming in terms of large amounts of patent infringement litigation against do-it-yourselfers. The major regulatory challenge posed by do-it-yourself biology is one of biosafety regulation.

DIYbio raises safety issues because the activities and people involved are not subject to institutional safety frameworks and review, and may not have a comprehensive understanding of safety risks and requirements. Leading synthetic biologists such as Drew Endy have highlighted this issue. Endy explained that iGEM uses existing biosafety frameworks based on recombinant DNA technology to educate students about biosafety issues and require them to engage with their university's institutional biosafety framework.[96] However, he identified the problem of biosafety in do-it-yourself biology:

> Where we run into puzzles on this one particular topic of governance is when the technology platforms begin to move beyond an institutional oversight framework ...
>
> So, what are the challenges in biosafety governance? ... It's going to be to figure out how to deal with an increased scope of genetic engineering work, an increased pace of the work ... And that's dealt with practically already at institutional levels. But we now need to figure out how to promulgate that or to disallow outright such work beyond, you know, relatively rich institutional bounds that can afford to have an established and well-functioning biosafety committee.[97]

In her evidence to the Commission, chemical engineer and synthetic biologist Kristala Prather also raised the need to think about a way of dealing with risk in both institutional and non-institutional settings that is constructive, but also incorporates a degree of 'surveillance' or 'monitoring'.[98] The Federal Bureau of Investigation (FBI) has also identified

---

[96] Drew Endy, 'Evidence to The Presidential Commission for the Study of Bioethical Issues, Synthetic Biology' (8 July 2010) Washington DC.

[97] Ibid.

[98] Kristala L J Prather, 'Evidence to The Presidential Commission for the Study of Bioethical Issues, Synthetic Biology' (8 July 2010) Washington DC.

do-it-yourself biology as a potential risk area and has undertaken a range of educational engagement activities directed at the do-it-yourself community under its Weapons of Mass Destruction Directorate Bioterrorism Prevention Program.[99]

6.5.1 DIYbio Community Engagement with Safety Concerns

Due to the openness of the community, its engagement with safety issues and focus on achieving positive applications, the most likely risk of harm arising from DIYbio is from an accidental event. That is, there is greater biosafety risk than biosecurity risk. DIYbio's founders are aware of the concerns about biosafety and biosecurity surrounding their work. In his evidence to the Presidential Commission, Bobe noted that the press has repeatedly connected do-it-yourself biology with weapons of mass destruction and biosecurity.[100] Not only are they aware of these concerns, but in establishing the organisation of DIYbio they had the goal of finding ways of addressing them, to 'really get ahead of widespread amateur activity and try to establish a framework for best practices global wide'.[101]

As early as July 2008, two months after the very first DIYbio meeting, Bobe raised the issue of needing to demonstrate that biosafety issues were being managed.[102] DIYbio's website now includes valuing openness and safety as part of the identity of a do-it-yourself biologist:

> DIYbio.org is an organization dedicated to making biology an accessible pursuit for citizen scientists, amateur biologists and biological engineers *who value openness and safety*. This will require mechanisms for amateurs to increase their knowledge and skills, access to a community of experts, the development of a code of ethics, responsible oversight, and leadership on issues that are unique to doing biology outside of traditional professional settings.[103]

A key issue, however, is that some people working in do-it-yourself biology do not necessarily have awareness of biosafety issues or training

---

[99] Edward H You, 'Evidence to The Presidential Commission for the Study of Bioethical Issues, Synthetic Biology' (9 July 2010) Washington DC.

[100] Bobe, 'Evidence to The Presidential Commission for the Study of Bioethical Issues, Synthetic Biology' (n 32).

[101] Ibid.

[102] Jason Bobe, 'Environmentally Benign DIYbio' (15 July, 2008) *DIYbio Blog* <http://diybio.org/2008/07/> .

[103] DIYbio (n 58).

in biology.[104] Indeed, DIYbio's cross-disciplinary nature is part of what the founders see as creating a uniquely fertile ground for innovation.[105] The community includes people who may have low awareness of issues of biosafety as well as some who are professional scientists.[106] Some amateur biologists may not have the ability to implement requirements imposed by regulatory mechanisms like the NIH Guidelines for experiments involving recombinant DNA.[107]

Bobe has made the point that dealing with safety issues cannot be done by amateurs individually, and it is important for there to be an organisation such as DIYbio to form a focal point for addressing these issues.[108] Otherwise, each individual would have to address questions such as how to dispose of something safely as though for the first time.[109] Therefore, Bobe sees his role as leading discussion not only about ways to innovate but also about ways to manage risks.[110] Both of the co-founders of DIYbio have proposed use of specific safety models designed for this community, Cowell proposing that this be in the form of easy to understand measures that explain the rules and Bobe suggesting that there should be a new model of safety regulations for individuals who are outside institutions.[111]

In the Commission's consultations, Bobe said that there was no community consensus or norms that had developed in relation to biosafety or ethics, as the field was then very young.[112] But he did see scope for developing incentives for fostering ethical conduct. He saw the developing of norms as very important. He thought doing this early was important because he expected the phenomenon of do-it-yourself biology to grow globally, and saw an opportunity to start setting a positive pattern in relation to biosafety.

---

[104] Bobe, 'Evidence to The Presidential Commission for the Study of Bioethical Issues, Synthetic Biology' (n 32).

[105] Nair (n 39) 231.

[106] Bobe, 'Evidence to The Presidential Commission for the Study of Bioethical Issues, Synthetic Biology' (n 32).

[107] Ibid.

[108] Nair (n 39) 231; Bobe, 'Evidence to The Presidential Commission for the Study of Bioethical Issues, Synthetic Biology' (n 32).

[109] Bobe, 'Evidence to The Presidential Commission for the Study of Bioethical Issues, Synthetic Biology' (n 32).

[110] Nair (n 39) 231.

[111] Ibid.

[112] Bobe, 'Evidence to The Presidential Commission for the Study of Bioethical Issues, Synthetic Biology' (n 32).

DIYbio has since been involved in a project with the Woodrow Wilson Center to develop 'norms of practice and biosafety resources for amateurs worldwide'.[113] In 2011 Bobe convened discussions about developing a code of ethics for DIYbio in the US and Europe, resulting in draft codes which are still in draft form. These emphasise transparency, open access, safety, use of biotechnology for peaceful purposes, education and respect for the environment.[114] In addition, the European draft includes principles of community, responsibility and accountability.[115]

DIYbio has also set up a biosafety advisory board to allow its members to consult experts on specific safety questions. Further demonstrating its willingness to deal with safety (and security) issues, the do-it-yourself community has engaged with the FBI's outreach program.[116] Indeed, it even invited the FBI to attend a do-it-yourself biology conference to discuss responsible research.[117] According to the FBI, its outreach activities – such as convening a conference to talk about risk management in the area – have been well received and the community is taking concerns about safety and security seriously.[118]

Given this evidence that do-it-yourselfers are highly engaged in discussion about biosafety, have shown awareness of the need to ensure safe practices and the desire to do this, are participating in development of norms and have set up a biosafety advisory board, it is unlikely that anyone who self-identifies as a member of this community would be deliberately trying to develop something dangerous in their experiments. Further, the discussion of their activities shows that they are trying to develop positive advances in biological science. Also, the recognition of transparency and openness in their ethical principles suggests that they are not seeking to engage in dangerous or illegal activities. For these reasons, the main area of risk is unintentional biosafety problems rather than intentional misuse of biological tools and processes.

However, those who *are* interested in intentional harm would not be likely to participate in the public discussions I have drawn on here. While the founders of DIYBio are engaging in discussions about safety, the

---

113 Ibid.

114 *Draft DIYbio Code of Ethics: As Agreed by US Delegates* (July 2011) DIYbiology <http://diybio.org/codes/>.

115 *DIYbio Code of Ethics: Draft from the European Delegation* (7 September 2011) DIYbio <http://diybio.org/codes/>.

116 Bobe, 'Evidence to The Presidential Commission for the Study of Bioethical Issues, Synthetic Biology' (n 32).

117 You (n 99).

118 Ibid.

regulatory problems in relation to biosafety are further exacerbated by the fact that not all regulatees are identifiable or amenable to discussions about regulation. It is difficult to apply the teachings of empirical research in regulation and theories such as 'responsive regulation' and 'really responsive regulation'[119] to the do-it-yourself biology context, because those theories tend to assume that the regulatees are identifiable 'firms'.

6.6 PRESIDENTIAL COMMISSION

In its report, the Presidential Commission viewed do-it-yourself biology as bolstering the notions of citizenship and community which underpin a democracy: 'science and its oversight do not belong exclusively to experts, highly trained professionals, or government officials. Science is a shared resource, affecting and belonging to all citizens'.[120] The Commission focused on the risk of do-it-yourself biologists having capacity to construct a 'completely novel organism' and concluded that there was 'no serious risk' at that time.[121] While not referenced, this conclusion seems to be drawn from the evidence of scientists Kristala Prather and Robert Carlson, who emphasised the cost and difficulty of such a task.

Further, Bobe was asked whether 'people can build pathogens at home', and replied:

> I think the capabilities of individuals at this time are really, really low. … It's difficult to know what everybody is doing. But based on the discussions that are happening online, people don't in fact yet have these new desktop DNA sequencing devices. They are just too expensive. And I don't know of anybody having a DNA synthesis device at home that's functional.[122]

The Presidential Commission also pointed to the expense and difficulty associated with the Venter Institute's synthetic cell development, which

---

[119] See Ian Ayres and John Braithwaite, *Responsive Regulation: Transcending the Deregulation Debate* (Oxford University Press, 1992); Robert Baldwin and Julia Black, 'Really Responsive Regulation' (2008) **71**(1) *The Modern Law Review* 59.

[120] The Presidential Commission for the Study of Bioethical Issues (n 18) 160.

[121] For example, ibid., 13, 148.

[122] Bobe, 'Evidence to The Presidential Commission for the Study of Bioethical Issues, Synthetic Biology' (n 32).

took 15 years and 40 million dollars in funding.[123] The Commission has taken significant comfort from this. For example, Gutmann stated 'what we've learned is that there's no cause for panic today'.[124] Therefore, the Commission did not see a need to impose unique restrictions on do-it-yourself biologists.[125]

In focusing on the present inability of do-it-yourself biologists to construct novel pathogens, the Commission has glossed over risks of unexpected or unintended outcomes of other activities such as the construction of organisms with *some* novel or modified traits, or experiments involving *re*construction of some of the genes of natural organisms using synthetic DNA. The Commission did note that there are some current risks associated with do-it-yourself biology such as growth of potentially dangerous organisms using conventional means, and inadequate waste disposal.[126] However, it did not discuss how these risks could be managed.

A positive aspect of the Commission's approach is that it recognised the need for assessment of risk in this area to be revisited in the future. It noted the need to monitor the field of do-it-yourself biology as it changes over time, both in terms of the growth of the community and its capabilities.[127] This was part of a broader recommendation to continue to assess safety and security risks posed by synthetic biology in both institutional and non-institutional settings. The Presidential Commission recommended that this review be undertaken by the Executive Office of the President with the FBI and others, and periodically updated. This recommendation was itself part of the broader recommendation to develop a coordinated approach to regulating synthetic biology.

This adaptive approach will promote regulatory connection in relation to the unpredictable development of the do-it-yourself biology community. Adaptation is important in the do-it-yourself biology context, as the Commission received evidence that both the capability of do-it-yourself biologists and the general biotechnology landscape *will* change. In relation to do-it-yourself biology, Carlson speculated that building

[123] The Presidential Commission for the Study of Bioethical Issues (n 18) 148.

[124] Amy Gutmann 'Comments in Commission Deliberations, The Presidential Commission for the Study of Bioethical Issues, Synthetic Biology' 16 November 2010.

[125] The Presidential Commission for the Study of Bioethical Issues (n 18) 148.

[126] Ibid.

[127] Ibid., 146.

synthetic viruses is likely to get easier over the medium to long term.[128] Bobe stated that while DNA synthesis capability is currently too expensive for do-it-yourselfers, 'I'm certain in the future there will be people who have DNA synthesizers and DNA sequencing machines in their home workshops or in community labs'.[129] In relation to the life sciences landscape generally, infectious disease specialist and member of the National Advisory Board for Biosecurity member David Relman told the Commission 'I think the only thing that we can be confident about today is that we should not be confident about what this landscape will look like in a year's time or certainly five years' time.'[130]

The Presidential Commission recommended that if significant 'unmanaged risks' are identified in the review it proposed, consideration be given to making certain existing measures mandatory both within and outside institutional research, independent of the source of funding.[131] In the meantime, the Commission chose to focus on efforts to engage and educate rather than restrain the do-it-yourself biology community. These efforts were seen as an important implementation of the principles of 'responsible stewardship' and 'intellectual freedom and responsibility' espoused by the Commission. Rather than imposing restrictions from *outside* the community, the Commission wanted to foster the norm of responsible stewardship *within* that community: 'As these communities grow, organized efforts to engage this community in discussions of safety and security and to foster a commitment to responsible stewardship will be increasingly important.'[132]

While the Presidential Commission conceded that this is more difficult than in institutional settings, it thought that where responsibility is exercised by individuals rather than institutions, this requires increased vigilance but not necessarily increased limitations on research.[133] The Commission stressed the need to find ways to foster a culture of responsibility in non-institutional research. It identified a need to ensure

---

128 Robert Carlson, 'Evidence to The Presidential Commission for the Study of Bioethical Issues, Synthetic Biology' (8 July 2010) Washington DC.

129 Bobe, 'Evidence to The Presidential Commission for the Study of Bioethical Issues, Synthetic Biology' (n 32).

130 David Relman, 'Evidence to The Presidential Commission for the Study of Bioethical Issues, Synthetic Biology' (14 September 2010) Philadelphia Penn.

131 The Presidential Commission for the Study of Bioethical Issues (n 18) 150.

132 Ibid., 129.

133 Ibid., 146.

do-it-yourself scientists understand safety and security issues, and suggested that options for enculturation could include education programs and grants for training.

The Presidential Commission put the principle of intellectual freedom at the fore of its discussion of do-it-yourself biology by stating that 'Intellectual freedom lies at the heart of America's scientific enterprise' and highlighting examples in which 'unfettered creativity' has led to developments which are to the public good.[134] It explained intellectual freedom as including not only the academic freedom that applies to researchers in universities, but also the 'right of *all* individuals to freedom of enquiry'.[135] This freedom is limited by the other side of the coin, responsibility. Responsibility is maintained through 'certain regulatory and norm-based constraints on academic and intellectual freedom'.[136]

A confused message is given in the report about the extent to which this applies in industry settings. The report states that the regulatory and norm-based constraints apply both in academic and other settings, and that 'industry research programs employ extensive quality assurance and control processes that satisfy both external mandates and internal needs'.[137] It notes that scientists trained in academia working in biotechnology companies are used to complying with various requirements. It also states that 'recognition and acceptance of these processes are less common' in non-academic settings such as some of the do-it-yourself community.[138] However, the report also states that the do-it-yourself research community and 'other private researchers' are exercising intellectual freedom without the structures and norms which create responsibility. It is not clear whether this refers to private individuals or private sector research, or both. Overall, it seems that the Commission was arguing that there are norms and procedures which maintain responsibility both within academia and biotechnology companies, which can be traced back to the Asilomar Conference, but these are not intrinsic to do-it-yourself communities.

The Commission came to the conclusion that the principle of intellectual freedom and responsibility requires that 'restrictions on research, whether by regulation among scientists or by government intervention, should limit the free pursuit of knowledge only when the perceived risk

[134] Ibid., 141–2.
[135] Ibid., 142 (emphasis added).
[136] Ibid., 143.
[137] Ibid.
[138] Ibid.

is too great to proceed without limit. Restrictions can prevent research harms but also can impede innovation and progress that may itself reduce harms'.[139]

6.7 WHAT SHOULD BE DONE ABOUT THE RISKS?

There are two schools of thought about making do-it-yourself biology safer. One is that oversight measures should be applied to do-it-yourself biologists, whose activities should be monitored. The other is that education, openness and engagement are more likely to be effective in managing the safety (and security) risks. In this view, the more distributed the study of biology, the better for all. Thus there is a tension between the values of openness and access on the one hand and oversight and control on the other.

George Church has been emphasising the importance of community responsibility and monitoring of compliance with laws since as early as 2005.[140] In his evidence to the Commission, he stressed the importance of having not only rules about biotechnology activities, but also 'government-mandated surveillance' and enforcement of those rules.[141] He told the Commission: 'You can do licensing as we do driver's license but you have to do surveillance to make sure people are obeying the laws.'[142] He pointed out that implementation of the NIH Guidelines is dependent on scientists having federal funding or 'in some other way being a responsible citizen', and there is a lack of oversight ensuring that the rules are followed by all, rather than only 'the standard members of society'.[143] He pointed out that there is more to the synthetic biology landscape than federal grantees, and took the view that licensing should be required across 'the entire ecosystem in synthetic biology'.[144] While he did not specifically mention do-it-yourselfers here, they appear to be part of his concern in his reference to those outside the 'standard members of society'. He also pointed out that licensing and oversight is

---

139 Ibid., 144.

140 George Church, 'Let Us Go Forth and Safely Multiply' (2005) **438** *Nature* 423.

141 George Church, 'Evidence to The Presidential Commission for the Study of Bioethical Issues, Synthetic Biology' (8 July 2010) Washington DC.

142 Ibid.

143 Ibid; George Church, 'Let Us Go Forth and Safely Multiply' (n 140).

144 Church, 'Evidence to The Presidential Commission for the Study of Bioethical Issues, Synthetic Biology' (n 141).

preferable to driving do-it-yourself biology underground by alienating the do-it-yourselfers and destroying their openness: 'it's going to happen anyway; you can make it go underground or you can try to shape it'.[145] Venter has suggested that there should be 'reasonable restraints' on do-it-yourself biology, but without 'destroying this wonderful creativity'.[146]

In contrast, Carlson's concern is that garage biology not be stifled, as it has great potential to contribute to innovation. In his evidence to the Presidential Commission, he stressed the importance of innovative 'garage' work to progress in areas such as aviation.[147] He proposed that distributed innovation is the best way to respond to emerging problems such as novel infections: 'The only way we're going to get there, in my opinion, is to have innovation everywhere and anywhere as fast as we can go.'[148] Carlson did not think limiting access to the technology would be a practical solution and compared it with regulation which created a black market for illicit drug production.

Instead, Carlson thought that encouraging openness was the only way in which do-it-yourself biology could be made safer:

> I think that's our only tool in fact. I think that trying to keep track of what people are doing, trying to have people volunteer and do it together, that's great …
>
> And so, going forward, if you are worried about threats, if you are worried about mistakes, I think the only way we can deal with this issue is to try to make sure everyone is open about what they are up to.[149]

He thought that all that could really be done to promote safety in do-it-yourself biology is to create a framework for do-it-yourselfers to ask questions about safety issues. DIYbio established a mechanism for doing this, but it appears that it is not currently functioning.

The Biopunk Manifesto's assertion that the opportunity to experiment in the biological sciences is a right is problematic. The ability to use laboratory equipment and conduct experiments is more akin to a privilege, and this privilege is earned through training, experience and the

---

145 Phil McKenna, 'Rise of the Garage Genome Hackers' (2009) **2689** *New Scientist* 20, 21.

146 J Craig Venter, 'Evidence to The Presidential Commission for the Study of Bioethical Issues, Synthetic Biology' (8 July 2010) Washington DC.

147 Carlson, 'Evidence to The Presidential Commission for the Study of Bioethical Issues, Synthetic Biology' (n 128).

148 Ibid.

149 Ibid.

demonstrated ability to do this work safely. As discussed above, researchers in university settings not only have to undergo training in biosafety, but must undertake courses in this area repeatedly throughout their careers. The privilege carries with it individual *responsibility* to comply with all safety and legal requirements relating to the work. As stated by the Presidential Commission, science is a shared resource 'affecting and belonging to all citizens'.[150] However, this should not extend in an unlimited way to the doing of biotechnology experiments.

Venter has explained efficacy of standard safety arrangements and guidelines in arguing for light-handed regulation of synthetic biology:

> There's a common sense level to all this work that I think most labs—it's a reason with millions of experiments in molecular biology, there's been no fundamental problems or accidents in part, the *ethics of the scientists doing this and how they train scientists and the simple guidelines that are out there.* I think they work very effectively.[151]

However, this enculturation in ethics, training in safety and adherence to the existing guidelines does not necessarily apply to do-it-yourself scientists who have not undergone the long apprenticeship in the life sciences represented by high level study or technical training.

While the more radical elements of DIYbio want to use science as a 'liberator',[152] there is actually potential for do-it-yourself experimentation to begin to infringe on liberal principles. It is a different nature of activity from computer building or experimentation in radio and aviation, often cited as analogies. While computer hardware and homemade radios or aircraft are unlikely to physically harm people, and cannot accidentally spread to others from their designer, biological hazards are different in that there is potential for others beyond the tinkerer to be harmed. This is particularly the case where the genes of living organisms are tinkered with. If other people who have not agreed to be involved in the experimentation were accidentally harmed, this would infringe the liberal principles the Biopunk Manifesto propounds.

If do-it-yourself biologists want access to the ability to experiment, they must accept the flip side of this access, responsibility for managing safety risks and the intrusion of some degree of regulation into their work. Some argue that DIYbio has potential precisely *because* it does not

---

150 See The Presidential Commission for the Study of Bioethical Issues (n 19) 160.

151 Venter, 'Evidence to The Presidential Commission for the Study of Bioethical Issues, Synthetic Biology' (n 146) (emphasis added).

152 See discussion of motivations in Wohlsen (n 1) 44, 81.

have to fit into the frameworks of institutional science, therefore it can progress more rapidly and creatively. However, institutional scientists do not have liberty to do whatever they like in their laboratories. If do-it-yourselfers want access to the same opportunity that institutional scientists have, then it is the opportunity to conduct regulated, limited and reviewed experimentation that they are seeking access to. If do-it-yourselfers do not want to be regulated, they are seeking access to a different kind of opportunity which is not afforded to institutional scientists. While there are fewer controls on biological research conducted in companies, there are still regulations which apply, for example to environmental release. It would be counterintuitive to give untrained hackers greater liberty and latitude. This would be against the interests of public safety.

Even in the regulated world of institutional synthetic biology, scientists are still concerned about risk. For example, Prather told the Commission:

> The potential benefits I think are enormous … At the same time, the risks are real. Because there is this information gap between what we really understand about biology and what our capabilities are, it's impossible for us to really predict what's going to happen in every single experiment. And so I do think it's very worthwhile to think about being as careful as possible as we do this to minimize those risks.[153]

Indeed, debate at a 2012 conference on regulation of synthetic biology showed significant concern and apprehensiveness about do-it-yourself biology among the professional scientists, other academics and regulators participating in the debate.[154] It was noted that not all do-it-yourselfers are amenable to discussing biosafety concerns.

For these reasons, synthetic biology should not be completely democratised and decentralised. The attempts at democratising innovation[155] in the BioBricks and standard biological parts communities – which involve community building, sharing and setting up contractual arrangements –

---

153 Prather, 'Evidence to The Presidential Commission for the Study of Bioethical Issues, Synthetic Biology' (n 98).

154 Comments of conference attendees during debate on paper: Todd Kuiken, 'Do-It-Yourself Biology: Reality and the Path Toward Innovation' (Paper presented at the 21st Century Borders/Synthetic Biology: Focus on Responsibility and Governance, 2013) <http://www.scienceforglobalpolicy.org/LinkClick.aspx?fileticket=VqAc-vBzvMk%3d&tabid=133>, 80–86.

155 See the discussion in Chapter 9, which draws on the theory of democratising innovation set out in Eric von Hippel, *Democratizing Innovation* (The MIT Press, 2005).

are democratising in an appropriate way. However, biotechnology should not necessarily be democratised to the extent that it can be done by anyone, without training or oversight.

Thus, the Commission's approach is not sufficiently cautious. It recommended that if significant 'unmanaged risks' are identified in the review it proposed, consideration be given to making certain oversight measures mandatory.[156] However, unmanaged risk can already be seen in this area, which suggests steps should be taken to impose certain restrictions.

Applying a more cautious approach would assist in developing this regulation. Here I use the term 'caution' in the sense used by philosopher and scholar of bioethics and environmental ethics, Per Sandin. Sandin has proposed that in addressing the risks posed by synthetic organisms, a useful principle may be 'perform those, and only those, actions that a cautious agent would perform in the circumstances'.[157] A cautious agent would gather as much information as possible, would favour 'small, reversible steps in introducing new technologies' and would expect laboratory research to be subject to 'democratic supervision and accountability'.[158] In the area of biosafety in do-it-yourself biology, there is uncertainty regarding the degree of risk, and there is also 'ignorance' in that the precise nature of potential harm is not fully known.[159] That is, regulators do not know what the elements of do-it-yourself biology who do not publicly discuss their work are doing. For this reason, caution should be exercised.

A major regulatory challenge in the area of do-it-yourself biology is that there is already concern in the do-it-yourself community about the potential for this work to arouse the suspicions of law enforcement authorities and lead to penalties. This causes some do-it-yourselfers to

---

[156] The Presidential Commission for the Study of Bioethical Issues (n 18) 150.

[157] Per Sandin, 'A New Virtue Based Understanding of the Precautionary Principle' in Mark A Bedau and Emily C Parke (eds), *The Ethics of Protocells: Moral and Social Implications of Creating Life in the Laboratory* (MIT Press, 2009) 89, 98.

[158] Ibid., 101–2.

[159] See discussion of 'non-knowing', 'ignorance' and 'unknown unknowns': Ulrich Beck, *World at Risk* (Polity Press, English Edition ed, 2009), 53–4, 115–28; Cass R Sunstein, *Laws of Fear: Beyond the Precautionary Principle* (Cambridge University Press, 2005), 60; Brian Wynne, 'Uncertainty and Environmental Learning: Reconceiving Science and Policy in the Preventive Paradigm' (1992) **2**(2) *Global Environmental Change* 111.

want to keep their work and identity secret.[160] Provisions of the system of Select Agents and Toxins regulation have arguably also been applied to some of the wrong people, and fostered unease amongst scientists.[161] Any criminal provisions would need to be carefully crafted to avoid misapplication to non-dangerous activities of backyard tinkerers and there would need to be clear educative guidance available to do-it-yourself biologists about how the provisions work. Even if this were achieved, the fear of criminal sanctions could cause do-it-yourself biologists to be less open about what they are doing. This is problematic because it reduces the open discussion of risks or mistakes that arise.[162]

It may be that civil approaches are the best way to balance the goals of being able to have open discussion of do-it-yourself biology and continue the engagement work by the FBI, while also putting some measures in place to promote safety. It is certainly difficult to delineate what people should be allowed to experiment with in their garage, and what should be restricted to 'professionals'. While the Presidential Commission did not see a need to impose unique restrictions on this group, some guidelines could be *extended* to apply to their work. For example, Bobe's proposal of a version of the NIH Guidelines designed to be understandable to non-professional scientists could be implemented.

Engagement with the do-it-yourself biologist regulatees is very important in order to promote compliance with any regulation that is introduced. In general, if regulatees do not view the regulation as legitimate, they may not comply with it.[163] The regulation would need to be attuned to the motivations and attitudes of the do-it-yourself biologists.[164] The maintenance of the positive relationships between do-it-yourself firms and law enforcement agencies such as the FBI will be critical in achieving this.

Civil penalties could be combined with a regulatory option discussed in the Presidential Commission's deliberations, that of having a registry of experiments in synthetic biology and requiring do-it-yourselfers to

---

160 Carlson, 'Garage Biology in Silicon Valley' (n 61).

161 See Christian Enemark, 'Law in the Time of Anthrax: Biosecurity Lessons from the United States' (2010) **17** *Journal of Law and Medicine* 748, 752–5.

162 Carlson, 'Garage Biology in Silicon Valley' (n 61); Kahn (n 76).

163 Roger Brownsword and Han Somsen, 'Law, Innovation and Technology: Before We Fast Forward – A Forum for Debate' (2009) **1** *Law, Innovation and Technology* 111.

164 Drawing on the 'really responsive regulation' theory proposed in Baldwin and Black (n 119).

register. This did not get much traction in the deliberations, and did not appear in the final report. Indeed, Gutman thought that the do-it-yourself community would be 'up in arms and rightly so'.[165] However, it is fair to expect a degree of responsibility in do-it-yourself work, and these biologists should not expect to have fewer requirements imposed upon them than companies. The registration requirement could be implemented at the level of community labs and combined with a mechanism for seeking advice on safety management. Companies could be required to register also, unless they can demonstrate compliance with the NIH Guidelines. This would reduce any sense that do-it-yourselfers are being singled out for unique limitations, and would address the gap posed by very small firms which may not have academically-trained scientists, federal funding or commercial reasons for complying with the NIH Guidelines.

In order to promote regulatory connection, the ability to adapt regulation over time is important in the do-it-yourself biology context. Adaptation is not limited to managing *scientific* uncertainty. Information about what kind of capability the do-it-yourself community has, who is part of this community and the efficacy of engagement and oversight strategies can be used to adapt the regulatory mechanism over time. Oversight measures introduced for do-it-yourself biology should have this adaptive feature, so that regulatory connection can be maintained.

There are further measures relating to screening orders for synthetic DNA, which could be implemented to identify non-traditional science being conducted using DNA from pathogenic organisms. These are discussed in the following chapter.

6.8 CONCLUSION

This chapter has revealed a picture of perceived limitations of the usual ways of doing scientific research, and how individuals are finding ways to work around these. This discussion shows that the emergence of synthetic biology, and do-it-yourself biology, creates a need to re-think and build upon the existing regulatory environment for biosafety. Addressing this biosafety risk is a key role of regulation of synthetic biology.

---

[165] Gutmann (n 124).

Some of the existing regulatory mechanisms do not currently apply to the do-it-yourself community and it may not be possible for do-it-yourselfers to voluntarily apply these systems to themselves without modification or guidance. In this context, biosafety requirements which can be understood and implemented by do-it-yourself biologists should be developed. Consideration should be given to registration requirements for do-it-yourself biology, particularly at the community lab level. Penalties imposed for failure to comply with these requirements should be civil. In order to maximise cooperation within the do-it-yourself community and regulatory effectiveness, engagement and education activities should continue. In order to maintain regulatory connection, any regulation which is introduced should be reviewed and adapted in future.

7. Biosecurity: potential for deliberate misuse of synthetic biology

7.1 INTRODUCTION

The announcement of the creation of a 'synthetic cell' by the Venter Institute in 2010 prompted a significant panic in the United States about the ethical and regulatory implications of synthetic biology, particularly in the area of bioterrorism and biosecurity. Three events in the weeks following the announcement demonstrate its impact on the policy environment surrounding synthetic biology in that jurisdiction. On the day of the announcement, President Obama asked the Presidential Commission to investigate the 'implications of this scientific milestone'.[1] He asked them to consider 'the potential medical, environmental, security, and other benefits of this field of research, as well as any potential health, security or other risks'.[2] At the Commission's first consultation meeting the key issues raised were: What is the significance of the development of a synthetic cell? What is the likely significance of the field of synthetic biology, in terms of how much it will change our world and benefit us? What is it likely to be used for? How afraid should we be of the materialisation of various risks of harm? What should the government's role be?

A week after the synthetic cell announcement, the House Committee held hearings about the implications of the development.[3] They invited

[1] The Presidential Commission for the Study of Bioethical Issues, 'New Directions: The Ethics of Synthetic Biology and Emerging Technologies' (The Presidential Commission for the Study of Bioethical Issues, December 2010) <http://bioethics.gov/sites/default/files/PCSBI-Synthetic-Biology-Report-12.16.10_0.pdf>, vi.

[2] Ibid., vi.

[3] Committee on Energy and Commerce, 'Preliminary Transcript: Hearing on Effects of Developments in Synthetic Genomics and Implications for Energy and Health' (United States House of Representatives, 27 May 2010) <http://democrats.energycommerce.house.gov/documents/20100527/transcript.05.27.2010.fc.pdf>.

synthetic biology leaders Venter, Endy and Keasling to provide testimony. Matters of safety, security and regulation of this research were key themes in the hearing.

Three weeks after the announcement, new legislation on weapons of mass destruction – containing for the first time a section highlighting dual-use concern about synthetic biology – was introduced into the United States House of Representatives.[4] While this legislation has not been passed, together with the hearings and reference to the Presidential Commission, it shows the level of concern generated by the synthetic cell.

This chapter focuses on the biosecurity issues raised by synthetic biology and the regulatory challenges they present. The key regulatory question is whether synthetic biology raises new issues beyond those already raised by recombinant DNA technology, and whether it can be properly managed using the existing regulatory environment. Regulating to address biosecurity risk is an aspect of the role of regulation of emerging technologies identified in the Introduction: to address issues of risk so as to minimise harm, and to promote the development of synthetic biology in a way that improves 'the basic conditions of human social existence'.[5] Thus far, the major mechanism under discussion to respond to the biosecurity challenges is screening of orders for synthetic DNA. The gene synthesis industry has developed self-regulatory approaches to this, which are examined in this chapter.

This chapter discusses the regulatory challenges posed by the emergence of biosecurity risks and the biosecurity issues raised specifically by synthetic biology. It examines the existing approaches to biosecurity regulation for synthetic biology and scrutinises these in terms of the regulatory criteria of legitimacy, connection and effectiveness. I argue that these approaches are insufficient both in terms of regulatory effectiveness and legitimacy. An enhanced, enforceable regulatory approach to the biosecurity risks of synthetic biology is needed. This should be based on the rule of law, and should encompass adaptation to the changing

---

[4] Originally Weapons of Mass Destruction Prevention and Preparedness Act 2010, HR 5498, 111th Congress (2010), now Weapons of Mass Destruction Prevention and Preparedness Act 2012, HR 2356, 112th Congress (2012) s. 304.

[5] See Roger Brownsword and Han Somsen, 'Law, Innovation and Technology: Before We Fast Forward – A Forum for Debate' (2009) **1** *Law, Innovation and Technology* 1, 2–3; see also Roger Brownsword and Morag Goodwin, *Law and the Technologies of the Twenty-First Century: Text and Materials* (Cambridge University Press, 2012), 46.

stage of scientific development and changes in the general biosecurity environment.

7.2 BIOSECURITY IS A REGULATORY CHALLENGE

As discussed in the previous chapter, synthetic biology is a dual-use research area. When dual-use research is appropriated for dangerous purposes, the malicious user could be either the scientist who originally undertook the research (an insider) or a secondary outside user who might appropriate it for their own purposes.[6] An example of the insider threat is Bruce Ivins, a government scientist suspected of orchestrating the 2001 anthrax attacks in the US.[7]

The use of biotechnology for bioterrorism might be seen as relatively unlikely, as a level of expertise and technical capability would be required. However, it has been argued that – in general – the danger of a bioterrorism event is increasing.[8] This is due to the expansion of the biotechnology industry, improved knowledge of disease-causing organisms and dissemination of this knowledge, and a decreasing cost and skill threshold for biological engineering.[9] Advances in biological science mean that the scientific difficulties associated with creating a biological weapon have decreased, and this trend can be expected to continue.[10] As judge, legal theorist and economist Richard A Posner points out, 'What is certain is that the technology usable for bioterrorism will improve.'[11] The spread of this biological knowledge internationally means that the dangerous possibilities are changing rapidly.[12] In the synthetic biology context, the spread of gene synthesis services across jurisdictions is also relevant.

The rapid changes in biological science should not be ignored.[13] However, it is not possible to estimate the probability of a bioterrorism

---

[6] Christian Enemark, 'Law in the Time of Anthrax: Biosecurity Lessons from the United States' (2010) **17** *Journal of Law and Medicine* 748, 748.

[7] Ibid., 750, 755–6.

[8] Richard A Posner, *Catastrophe: Risk and Response* (Oxford University Press Inc, 2004), 83.

[9] David P Fidler and Lawrence O Gostin, *Biosecurity in the Global Age: Biological Weapons, Public Health and the Rule of Law*, Stanford Law and Politics (Stanford University Press, 2008), 30–32; Posner (n 8) 83–4.

[10] Fidler and Gostin, ibid., 30–31.

[11] Posner (n 8) 83.

[12] Fidler and Gostin (n 9) 41.

[13] Ibid.

event because it is not possible to obtain all the intelligence information necessary for this exercise.[14] There is ongoing, strongly contested debate about the likelihood of an attack.[15]

While the likelihood of a bioterrorism event may be low, the consequences are potentially catastrophic.[16] For this reason, it has been proposed that greater preventative efforts should be considered.[17] The problem is how to design regulation when the events may be low probability, but the consequences could be severe.

Addressing biosecurity concerns requires, among other things, regulatory supervision of biological research.[18] However, there are competing considerations at play which make biosecurity a challenging area in which to develop regulation.[19] On the one hand, there is a desire to give scientists freedom in research and innovation.[20] Society wishes to obtain the benefits of scientific research in terms of improved understanding of diseases, new medicines and other developments.[21] Facilitating this beneficial innovation is another aspect of the role of the regulation of emerging technologies.[22] In addition, publishing and discussing research is an important element in the progress of scientific research and discourse.[23] These values can be seen as part of scientific research culture. In the context of public health, it has traditionally been assumed that 'scientific freedom, dissemination and innovation are unquestionably public goods'.[24] However, each of these values is in tension with regulation for the purposes of promoting biosecurity.[25] Further, regulators need to avoid imposing so much regulation that it actually creates an incentive for scientists to move away from the fields of research where biosecurity oversight is in place.[26] This would be a negative unintended consequence of that regulation.

---

14 Posner (n 8) 174.

15 Fidler and Gostin (n 9) 25.

16 Posner (n 8) 197–8.

17 Ibid.

18 Fidler and Gostin (n 9) 8.

19 Ibid., 11–13.

20 Ibid., 12.

21 Lawrence O Gostin and David P Fidler, 'Biosecurity under the Rule of Law' (2007) **38** *Case Western Journal of International Law* 438, 447.

22 See Brownsword and Somsen (n 5) 2.

23 Fidler and Gostin (n 9) 12; Gostin and Fidler (n 21) 447–8.

24 Fidler and Gostin (n 9) 13.

25 Ibid., 12–13.

26 Enemark (n 6), 757.

There are additional reasons why research in biology may be more difficult to regulate than the field which previously attracted major security concern, nuclear science. It is difficult to control quantities of biological materials, because a sample of living microorganisms can be cultured to produce a larger quantity.[27] By contrast, quantities of nuclear material are fixed; a sample of nuclear material cannot be used to generate a larger quantity.[28] Further, there are many more scientists working in microbiology than were engaged in nuclear research, and they are working in many legitimate places such as companies and universities rather than mainly in military research programs.[29]

Thus there are many regulatory challenges in relation to biosecurity regulation of life sciences research. Due to these complexities, it has been suggested that biosecurity regulation cannot be approached using the 'command and control' system applied to nuclear technology; a different approach is required.[30] In the past, the primary method of regulating biological weapons has been an arms control approach at the international level under the Convention on the Prohibition of the Development, Production and Stockpiling of Bacteriological (Biological) and Toxin Weapons and on their Destruction (the Biological Weapons Convention).[31] State parties have undertaken not to develop or acquire biological agents that have no justification for protective or peaceful purposes, or weapons for using such agents for hostile purposes or armed conflict.[32] They are required to take measures to prohibit and prevent the development, production, stockpiling, acquisition or retention of such biological agents, toxins, weapons, and equipment within their territory or jurisdiction.[33]

However, this arms control approach is no longer the major regulatory strategy in relation to biological weapons.[34] There has been a trend towards criminalisation of activities relating to biological weapons in

---

27 Ibid., 749.
28 Ibid.
29 Ibid.
30 Fidler and Gostin (n 9) 13.
31 Convention on the Prohibition of the Development, Production and Stockpiling of Bacteriological (Biological) and Toxin Weapons and on their Destruction opened for signature 10 April 1972 (entered into force 26 March 1975).
32 Ibid., art. I.
33 Ibid., art. IV.
34 Fidler and Gostin (n 9) 59.

international and domestic law.[35] For example, there are provisions in US legislation criminalising the development, production or possession of 'any biological agent, toxin or delivery system for use as a weapon'.[36] These measures are more suited to the current reality in which non-state actors rather than governments might engage in these activities.[37] Rogue states are also a security concern.

7.3 SYNTHETIC BIOLOGY POSES BIOSECURITY ISSUES

7.3.1 New Capabilities Offered by Synthetic Biology

The new capabilities provided by synthetic biology are relevant to its biosecurity risks. Synthetic biology represents 'a new set of tools starting from a new vantage point ... it changes a lot of the rules'.[38] Some areas of synthetic biology involve large-scale synthesis of DNA for use in experiments. This technology – combined with synthetic biology's research into modifying whole systems within natural organisms and constructing novel organisms – has the potential to be used for creating dangerous microorganisms. The concerns are three-fold: that synthetic biology will be used to re-create existing dangerous pathogens such as smallpox through DNA synthesis, that it will be used to engineer naturally occurring microorganisms to be more dangerous, or that it will be used to create completely new pathogens.[39]

Biosecurity has been raised as an issue in the synthetic biology context as early as 1999, when Cho and collaborators were asked by the Venter Institute to conduct an ethics review of the minimal genome research.[40]

---

35 Ibid., 60–73.

36 18 USC (Office of the Law Revision Counsel, 2017), § 175.

37 Fidler and Gostin (n 9) 64–5.

38 J Craig Venter, 'Evidence to The Presidential Commission for the Study of Bioethical Issues, Synthetic Biology' (8 July 2010) Washington DC.

39 David Relman, 'Evidence to The Presidential Commission for the Study of Bioethical Issues, Synthetic Biology' (14 September 2010) Philadelphia Penn; Jonathon B Tucker and Raymond A Zilinskas, 'The Promise and Perils of Synthetic Biology' (2006) Spring 2006 *The New Atlantis* 25.

40 Mildred K Cho et al, 'Ethical Considerations in Synthesizing a Minimal Genome' (1999) **286**(5447) *Science* 2087.

Since then, various writers have commented on these issues, including scientists, political scientists, lawyers and risk experts.[41]

Leading scientists in synthetic biology are concerned about the security issues. George Church has been highlighting the issues since 2005, when he wrote a proposal for 'non-proliferation' in synthetic biology based on the idea of checking orders for synthetic DNA for sequences from dangerous organisms.[42] This proposal appears to have been drawn upon in development of the US Government's voluntary guidelines for oversight of gene synthesis activities, to be discussed below.

Drew Endy described the problem in this way:

> It also challenges safety and security frameworks where if you want to limit access to a gene encoding a toxin by only having experts working with that, what if you make that sequence available on the Internet and now somebody could print the gene or a pathogen, and so on. Being able to go from material to information and back to material can stress and change current practices and relationships. It's part of the reason this capacity to construct and reinstall and initiate replication of a genome is a big deal.[43]

In 2007, the Venter Institute and others published a report on these issues: Synthetic Genomics: Options for Governance.[44] The research commissioned for that study concluded that in 2007, constructing a virus *de novo* was more difficult than obtaining a natural virus.[45] However, it was judged that this situation could be reversed within ten years (by

---

[41] See, e.g., Hans Bügl et al, 'DNA Synthesis and Biological Security' (2007) **25**(6) *Nature Biotechnology* 627; Robert Carlson, 'The Changing Economics of DNA Synthesis' (2009) **27**(12) *Nature Biotechnology* 1091; Stephen M Maurer and Laurie Zoloth, 'Synthesising Biosecurity' (2007) **63**(6) *Bulletin of the Atomic Scientists* 16; Gautam Mukunda et al, 'What Rough Beast? Synthetic Biology, Uncertainty, and the Future of Biosecurity' (2009) **28**(2) *Politics and the Life Sciences* 2. Concern about underestimation of security concerns damaging the field was raised in The Editors, 'How to Kill Synthetic Biology' (2006) **294**(6) *Scientific American* 4.

[42] George Church, *A Synthetic Biohazard Non-proliferation Proposal* (21 May 2005) <http://arep.med.harvard.edu/SBP/Church_Biohazard04c.htm>; George Church, 'Let Us Go Forth and Safely Multiply' (2005) **438** *Nature* 423.

[43] Drew Endy, 'Evidence to The Presidential Commission for the Study of Bioethical Issues, Synthetic Biology' (8 July 2010) Washington DC.

[44] M S Garfinkel et al, 'Synthetic Genomics: Options for Governance' (The J Craig Venter Institute, Massachusetts Institute of Technology, Center for Strategic and International Studies, October 2007) <http://www.jcvi.org/cms/fileadmin/site/research/projects/synthetic-genomics-report/synthetic-genomics-report.pdf>.

[45] Ibid., 16.

2017).[46] As with bioterrorism events in general, there is debate about how feasible this is. Some regard the chance of a rogue actor constructing a synthetic pathogen as very small, due to the technical difficulties of that task.[47] In this vein, the NSABB noted in 2010 that there were 'significant limitations to our current ability to custom design and produce novel organisms, especially pathogens, either by *de novo* synthesis or by engineering extant organisms in predictable ways'.[48] The Venter Institute's 2015 report on DNA synthesis and biosecurity stated that at that time, it was feasible for a well-trained scientist in a well-equipped laboratory to build a viral genome, but that bacterial genomes were still too difficult to assemble.[49] The report stated that only the researchers at the Venter Institute have been able to do that, in the work that led to the synthetic cell announcement in 2010.[50]

Nonetheless, the US Government is concerned about these issues. In 2012, the first US National Strategy for Biosurveillance stated that: 'Innovative developments mark the advent of threats of a new kind. Specifically, recent breakthroughs in synthetic biology offer the promise but also potentially the peril of technological progress, where the field is advancing at a staggering pace.'[51]

---

[46] Ibid.

[47] See, e.g., Robert Carlson, 'Evidence to The Presidential Commission for the Study of Bioethical Issues, Synthetic Biology' (8 July 2010) Washington DC; Gerald L Epstein, 'The Challenges of Developing Synthetic Pathogens' (19 May 2008) *Bulletin of the Atomic Scientists* <http://www.thebulletin.org/web-edition/features/the-challenges-of-developing-synthetic-pathogens>; Tucker and Zilinskas (n 39) 38–9. Marris and Rose argue that much discussion of risks posed by synthetic biology – including biosecurity risk – tends to assume that the ambitious goals of synthetic biology have already been achieved, for example that biological weapons are already easy to build: Claire Marris and Nikolas Rose, 'Let's Get Real on Synthetic Biology' (2012) **214**(2868) *New Scientist* 28.

[48] National Science Advisory Board for Biosecurity, 'Addressing Biosecurity Concerns Relating to Synthetic Biology' (National Institutes of Health and Department of Health and Human Services, 2010) <http://oba.od.nih.gov/biosecurity/pdf/NSABB%20SynBio%20DRAFT%20Report-FINAL%20(2)_6-7-10.pdf>, 7.

[49] Sarah R Carter and Robert M Friedman, 'DNA Synthesis and Biosecurity: Lessons Learned and Options for the Future' (J Craig Venter Institute, 2015), 5.

[50] Ibid., 5.

[51] 'National Strategy for Biosurveillance' (The White House, 2012) <http://www.whitehouse.gov/sites/default/files/National_Strategy_for_Biosurveillance_July_2012.pdf>, 3.

The NSABB has reported on the specific challenges presented by synthetic biology, and recommended that oversight of DNA synthesis activities should be introduced and existing legislative provisions relating to genetic sequences should be clarified.[52] It subsequently also recommended institutional review of the biosecurity risks of synthetic biology, oversight of dual-use research not only in academia but also in the private and 'voluntary' sectors, and 'outreach and education strategies'.[53] Since these recommendations there have been moves to educate do-it-yourself biologists about dual-use risks.[54]

The FBI also has a program encompassing investigation, threat assessment, outreach and education relating to synthetic biology within its Weapons of Mass Destruction Directorate.[55]

(a) Scale of capacity to alter genetic material

The techniques of synthetic biology enable large-scale changes to natural genomes and the design and construction of synthetic genomes. Venter explained that 'Synthetic genomics technologies ... provide the capability to build whole genomes and can examine how best to organize them'.[56] For example, the researchers who developed the 'synthetic cell' noted that in their experiment the synthetic genome used was similar to a natural genome, but that the synthesis and transplant of more novel genomes is likely to be possible in future.[57] This means that humans are likely to become more able to design the genetic features of organisms to

---

[52] National Science Advisory Board for Biosecurity (NSABB), 'Addressing Biosecurity Concerns Related to the Synthesis of Select Agents' (National Institutes of Health and Department of Health and Human Services, 2006) <http://oba.od.nih.gov/biosecurity/pdf/Final_NSABB_Report_on_Synthetic_Genomics.pdf>, 10–11.

[53] See NSABB, 'Addressing Biosecurity Concerns Relating to Synthetic Biology' (n 48) 13–14.

[54] NSABB, 'Strategies to Educate Amateur Biologists and Scientists in Non-life Science Disciplines About Dual Use Research in the Life Sciences' (National Institutes of Health, June 2011) <https://osp.od.nih.gov/wp-content/uploads/2013/06/FinalNSABBReport-AmateurBiologist-NonlifeScientists_June-2011_0.pdf>.

[55] See, e.g., Edward H You, 'Evidence to The Presidential Commission for the Study of Bioethical Issues, Synthetic Biology' (9 July 2010) Washington DC.

[56] Prepared Statement of J Craig Venter, 'Evidence to United States House of Representatives Committee on Energy and Commerce, Effects of Developments in Synthetic Genomics and Implications for Energy and Health' (27 May 2010) Washington DC, 8.

[57] Daniel G Gibson et al, 'Creation of a Bacterial Cell Controlled by a Chemically Synthesized Genome' (2010) **329**(5987) *Science* 52.

carry out functions of our choosing. These developing capabilities increase the risks associated with synthetic biology.

(b) DNA synthesis and assembly technology

The development of DNA synthesis technology is a key reason why synthetic biology poses new security risks. There are regulatory regimes in place to control access to dangerous pathogens and toxins as well as their transport, storage, sale and use. In the US, these are regulated under the Select Agents and Toxins system.[58] However, the ability to synthesise DNA means that physical access to these tightly controlled organisms may not be needed in order to use their genes for nefarious purposes; they could potentially be constructed in a laboratory. For example, scientists have already demonstrated that it is possible to construct, from short pieces of DNA, the genomes of pathogens such as polio virus and the 1918 Spanish influenza.[59] In 2002 scientists in the US produced the poliovirus genome by assembling and manipulating mail-ordered pieces of DNA.[60] They were then able to create an infectious polio virus particle. The virus was reduced to a 'chemical entity' which can be synthesised using information that is in the public domain, forming part of 'a new reality of synthetic viruses'.[61] The publishing of this work caused controversy regarding the ethics of conducting this experiment and publishing the methods, whether the sequences of such dangerous entities should be freely available, and how access to DNA sequences originating from dangerous organisms might be controlled. The sequences of many organisms are now freely available online.

These viruses were built in the laboratory using short pieces of *natural* DNA and chemically joining them together. Scientists can now not only convert a physical, biological piece of DNA into a sequence (information), but can also take sequence information and build from it the

---

[58] 42 CFR (US Government Publishing Office – e-CFR, 2016) Part 73.

[59] See Jeronimo Cello et al, 'Chemical Synthesis of Poliovirus cDNA: Generation of Infectious Virus in the Absence of Natural Template' (2002) **297**(5583) *Science* 1016; Terrence M Tumpey et al, 'Characterization of the Reconstructed 1918 Spanish Influenza Pandemic Virus' (2005) **310**(5745) *Science* 77; Eckard Wimmer, 'The Test-Tube Synthesis of a Chemical Called Poliovirus: The Simple Synthesis of a Virus Has Far-Reaching Societal Implications' (2006) **7**(SI) *EMBO Rep* S3.

[60] Wimmer (n 59) S3.

[61] Ibid., S5.

matching physical piece of DNA.[62] Venter has identified two key features of synthetic genomics capability as the technology to rapidly assemble DNA molecules of large size, and the ability to assemble fragments of DNA in different arrangements.[63] The ability to make longer pieces of synthetic DNA, and assemble these into large constructs is significant because very short pieces of DNA would not have function or utility on their own, but with longer stretches of DNA whole microbial genes and even genomes can be synthesised, increasing the range of possible experiments.

Synthetic DNA and DNA assembly technology now make it possible for scientists to actually design genomes, rather than cutting and pasting sections of naturally occurring genomes. This alters what can be done in biological engineering. DNA synthesis is a critical tool because now 'the ability to specify genetic instructions by writing DNA from scratch provides, in principle, the opportunity to specify the behaviour of biological systems'.[64] Venter pointed out that:

> One of the major advantages of synthetic genomics over classical biotechnology techniques – such as recombinant DNA – is that there is no need to have access to a physical supply of a particular DNA sequence. Sequence fragments are simply created de novo by chemical synthesis and assembled into entire chromosomes and organisms. This ability to synthesize (write) DNA and use it in the construction of new cells can catalyze a major change in what organisms can be engineered to do.[65]

Scientists and other individuals can now order increasingly long stretches of human designed and built DNA from gene synthesis companies in the US and other countries. Bioengineer and expert in biotechnology trends Robert Carlson explained that 'You send them a sequence and a credit card and they FedEx you the DNA two weeks later.'[66] This ability to build synthetic DNA is increasing exponentially, while the costs are

62 Venter, 'Evidence to The Presidential Commission for the Study of Bioethical Issues, Synthetic Biology' (n 38).

63 Venter, Prepared Statement 'Evidence to United States House of Representatives Committee on Energy and Commerce, Effects of Developments in Synthetic Genomics and Implications for Energy and Health' (n 56) 9.

64 Rob Carlson, *Biology is Technology: The Promise, Peril, and New Business of Engineering Life* (Harvard University Press, 2010), 73.

65 Venter, Prepared Statement 'Evidence to United States House of Representatives Committee on Energy and Commerce, Effects of Developments in Synthetic Genomics and Implications for Energy and Health' (n 56) 8.

66 Carlson, 'Evidence to The Presidential Commission for the Study of Bioethical Issues, Synthetic Biology' (n 47).

falling.[67] Carlson states that 'The rate at which DNA synthesis capacity is changing is thus a measure of the improvement in our ability to manipulate biological systems and biological information.'[68]

A related issue is the public availability of the genetic sequences of pathogenic organisms. Venter has explained that before gene synthesis technology: 'Scientists sort of controlled who got what … Now, any virus sequence that's in the public databases can be pretty readily remade … All you need is the digital information and a DNA synthesizer'.[69]

In synthetic biology, a debate is occurring regarding regulatory mechanisms for controlling access to sequences or screening research projects for use of sequences derived from pathogenic organisms. One way to do this is to require DNA synthesis companies to check orders for synthetic DNA, both in relation to the customer and the sequence requested. In order to do this, companies would need both information about genetic sequences related to disease-causing organisms (public health information) and access to lists of suspicious persons (security information).[70] The idea is that companies would look for a similarity between the requested sequence and sequences from dangerous organisms.

In 2007, Church, Endy and scientists at various gene synthesis companies and officers of the FBI proposed a model for addressing security issues relating to DNA synthesis.[71] The authors identified the potential for DNA synthesis technology to be 'purposefully misapplied' and argued that oversight at that time was not sufficient. They proposed that customers of DNA synthesis companies should be required to provide information about themselves, their institutions and their project to gene synthesis companies. The synthesis companies would use software tools to check the requested sequence against agreed lists of dangerous organisms or DNA sequences derived from these. If any issues or concerns were identified, these would be communicated to the relevant government agencies. In this way, researchers and institutional biosafety review bodies would also have a degree of responsibility. This model

---

67 Carlson, *Biology is Technology: The Promise, Peril, and New Business of Engineering Life* (n 64) 66–7.

68 Robert Carlson, 'The Pace and Proliferation of Biological Technologies' (2003) **1** *Biosecurity and Bioterrorism: Biodefense Strategy, Practice, and Science* 203.

69 Venter, 'Evidence to The Presidential Commission for the Study of Bioethical Issues, Synthetic Biology' (n 38).

70 Ralf Wagner, 'Evidence to The Presidential Commission for the Study of Bioethical Issues, Synthetic Biology' (14 September 2010) Philadelphia, Penn.

71 Bügl et al (n 41).

proposed collaboration and exchange of information between industry and government. For example, it recommended that DNA synthesis companies work together with government to improve the technology used to screen orders for synthetic DNA. The model was also designed to have international application and promote international cooperation.

A key benefit of this model is that it would apply to a broader scope of DNA synthesis customers than the NIH guidelines, which only apply to certain federally funded research or research at institutions which receive Federal funding. As noted by the proposers of the model, those guidelines have been successful partly because recombinant DNA research was being carried out in organisations which 'endorse and support an oversight-based safety process'.[72] But the social and scientific context of synthetic biology is different, for example in relation to the emergence of do-it-yourself biology. Though these individuals are operating outside organisations with oversight frameworks, they can order synthetic DNA from gene synthesis companies. This proposed model would apply to their orders as well as those of scientists in universities and biotechnology companies. Thus it is more effective than a security oversight model based on the NIH guidelines.

Indeed, this is essentially the model which has been adopted by the US Government. As will be seen below, the government model extends on this proposal by providing more detailed customer screening principles, such as checking the customer against certain lists of persons, and setting out procedures for reporting suspicious orders to law enforcement agencies.

While they need to be able to identify possible misuse of dangerous sequences of DNA, companies also need to be able to provide synthetic genes from disease-causing organisms for legitimate purposes.[73] There are many legitimate research reasons for requiring access to synthetic DNA, such as vaccine development.[74] Therefore, there is concern that regulatory control of DNA synthesis technology will be detrimental for scientific progress. While he has pointed out security issues, Endy has also highlighted that access to DNA synthesis technology and sequence information is very important:

> an attempt to ban or limit access to DNA synthesis technology … would only be guaranteed to cripple biological engineering research and biomedical research … our future biological security will depend not on limits to

---

[72] Ibid., 628.
[73] Wagner (n 70).
[74] Ibid.

> technology and research, but rather on the successful development of general, agile capabilities for detecting, understanding and responding to biological risks.[75]

In their 2007 proposal, Endy, Church and collaborators argued that improved technologies for DNA synthesis were crucial to advancement in biological research capability, and that therefore 'open commercial development of synthesis technology' should not be inhibited.[76] This shows the way in which the regulatory challenge identified earlier – that of balancing the need for promoting scientific freedom and creativity in order to obtain beneficial applications of research with the need for biosecurity oversight – plays out in the synthetic biology context.

7.3.2 Sharing and Openness in Synthetic Biology

In the synthetic biology environment, there is a culture of sharing of research results and information. This is particularly the case in the parts agenda promoted by the BioBricks Foundation and Registry of Standard Biological Parts. This open, sharing approach is beneficial for the progress of research, and the promotion of socially desirable innovation. However, it is also in conflict with the traditional goals of biosecurity policy. Generally, biosecurity policy would aim to control scientific research and restrict dissemination of information.[77] This illustrates the tension between openness and publication of research and limiting of information for security reasons, identified above. This is another of the regulatory challenges which arise when synthetic biology research is to be regulated for biosecurity purposes. By contrast, there are also those who regard openness and sharing as the best way to *promote* security, because the more minds working on scientific problems with relevance to biosecurity, the more likely it is that adequate responses to a biological threat can be generated.[78]

---

[75] Drew Endy, 'Foundations for Engineering Biology' (2005) **438**(7067) *Nature* 449, 452–3.

[76] Bügl et al (n 41) 628.

[77] For general discussion of this challenge, Fidler and Gostin (n 9) 12–13, 78–9.

[78] Carlson, 'Evidence to The Presidential Commission for the Study of Bioethical Issues, Synthetic Biology' (n 47).

7.3.3 De-skilling of Biology

Synthetic biology research also raises security concerns due to its emphasis on 'de-skilling' biology.[79] Synthetic biologists expressly aim to make biological engineering techniques more accessible, reducing the requirement for large amounts of 'tacit knowledge' and skill gained through years of experience in laboratories.[80] For example, one of the goals of the parts agenda is to make standardised biological parts that can be easily assembled to make small machines that work in predictable ways. Synthetic biologists are trying to achieve this de-skilling by applying the principles of engineering, such as standardisation and modularity, to biological systems.[81]

This de-skilling agenda has security implications.[82] In general, technical and cost barriers to genetic engineering are decreasing, as are the technical difficulties of biological weapons development.[83] While synthetic biology's focus on 'making biology easier to engineer' is designed to generate beneficial applications, it could have the unintended effect of accelerating this trend towards less difficult weapons development.

Related to this de-skilling, security concerns are also raised by the emergence of the do-it-yourself biology community, which encompasses work on synthetic biology. No transparency or accountability obligations are applicable to this 'backyard biology'. Unlike research institutions, there is no oversight of biosafety standards, or ethical review of such experimentation. It is also difficult for regulators to monitor these activities, because not all of the practitioners of do-it-yourself biology are readily identifiable.

7.3.4 Military and Security Research in Synthetic Biology

Significant amounts of funding are being provided for synthetic biology research by military agencies, as described in Chapter 2. The defensive

---

[79] Mukunda et al (n 41) 14.

[80] Ibid., 14.

[81] Ibid.

[82] Ibid.; the 'safety' implications are noted in M Schmidt, 'Diffusion of Synthetic Biology: A Challenge to Biosafety' (2008) **2**(1–2) *Systems and Synthetic Biology* 1.

[83] Fidler and Gostin (n 9) 30–32; Posner (n 8) 83–4.

military applications of synthetic biology include biosensors for explosives, contaminants or pathogens, and improved vaccines.[84] Of course, not all of these projects will directly result in military applications, but dual-use technology may be generated.

7.4 EXISTING APPROACHES TO BIOSECURITY REGULATION OF SYNTHETIC BIOLOGY

7.4.1 Self-regulation by Synthetic Biologists

Security issues have been discussed in the synthetic biology community since early in the development of the new field.[85] At the international conference Synthetic Biology 2.0 in 2006, scientists, bioethicists, lawyers, economists, policy analysts and others discussed scientific advances as well as biological safety and security.[86] They agreed on a Draft Declaration of the Second International Meeting on Synthetic Biology which was opened for public comment.[87] At the time, they noted that DNA synthesis was 'relatively cheap', that not all synthesis companies screened orders for sequences from hazardous organisms, and that DNA synthesis provides a 'plausible path for circumventing current best practice in biological safety, including institutional, peer, or government review of genetic research'.[88] The Draft Declaration expressed support for a working group to establish software tools for checking orders for synthetic DNA, 'adoption of best practice sequence checking technology' by all DNA synthesis companies, ongoing international discussions in the science and engineering communities to develop solutions and ongoing discussion with 'all stakeholders' to develop governance options – including self-governance – for the consideration of policymakers 'such

---

[84] See, e.g., Mauricio S Antunes et al, 'Programmable Ligand Detection System in Plants through a Synthetic Signal Transduction Pathway' (2011) **6**(1) *PLoS ONE* e16292, e16292; Mukunda et al (n 41) 12.

[85] See e.g., Erika Check, 'Synthetic Biologists Face Up to Security Issues' (2005) **436** *Nature* 895.

[86] Second International Meeting on Synthetic Biology, *Declaration of the Second International Meeting on Synthetic Biology* (2006) syntheticbiology.org <http://syntheticbiology.org/SB2Declaration.html>.

[87] Ibid.

[88] Ibid.

that the development and application of biological technology remains overwhelmingly constructive'.[89]

These self-regulation efforts were not successful. There has been no change in relation to the draft declaration since 2006. It remains in draft form and does not appear to have been further discussed. An interpretation of these events is that the BioBricks Foundation wanted to be, and be seen as being, proactive regarding safety and security risks. It wanted to show leadership and engagement in these areas, in line with the norms I have identified of engagement and positive contribution. The BioBricks Foundation's emphasis on promoting biosafety and biosecurity in synthetic biology research is part of this desire to make a positive contribution.

However, this discussion was abandoned, at least in part, due to objections from the ETC group.[90] In 2007, the ETC Group called on the synthetic biology community to 'abandon their proposals for self-governance and engage in an inclusive process of global societal debate'.[91] It alleged that this action by scientists was designed to prevent government action to address risks.[92]

Despite these objections, the synthetic biology community's efforts to develop and promote self-regulation and the leadership of the BioBricks community are constructive for the safe, secure and responsible development of the field of synthetic biology. Given the norms of positive contribution and engagement identified in earlier chapters, these activities are unlikely to be intended to completely exclude government action in the area. Nonetheless, this example shows that efforts to promulgate an agreed set of principles for a scientific community's work can be derailed by objections from civil society organisations.

The ETC Group also claimed that the biosecurity management proposal published in 2007 by scientists including Church, Endy and collaborators is not valid because some of the scientists have competing

---

89 Ibid.

90 Joy Y Zhang et al, 'The Transnational Governance of Synthetic Biology: Scientific Uncertainty, Cross-borderness and the "Art" of Governance' (BIOS Working Paper No: 4, BIOS (Centre for the Study of Bioscience, Biomedicine, Biotechnology and Society, May 2011) <http://www2.lse.ac.uk/BIOS/publications/Working%20Papers/WP4TransnationalGovernance.pdf>, 19.

91 ETC Group, 'Global Coalition Sounds the Alarm on Synthetic Biology, Demands Oversight and Societal Debate' (News Release, 19 May 2006).

92 Ibid.

financial interests.[93] In relation to the 2007 report Synthetic Genomics: Options for Governance, the ETC Group argued that the authors were 'synthusiasts', 'unabashed synthetic biology boosters'.[94] It described this study as an 'insular process' which failed to address wider societal concerns and issues of ownership and monopoly.[95] Overall, the ETC Group has argued that 'it is not for scientists to either control public discourse or determine regulatory frameworks'.[96]

The synthetic biology community also aims to self-regulate using the education of young scientists in the synthetic biology field through iGEM. The university students consider safety issues in their project, and the FBI goes to iGEM to educate the budding scientists about safety and security issues.[97]

7.4.2 Existing Regulation of Synthetic Biology in the United States

(a) Historical background: regulating recombinant DNA research

Regulation of the use of recombinant DNA has developed alongside the technology from the 1970s. The seminal paper announcing recombinant

---

[93] ETC Group, 'The G(e)nomes of Zurich: Civil Society Calls for Urgent Controls on Synthetic Life' (News Release, 25 June 2007).

[94] ETC Group, 'As Extreme Genetic Engineers Gather in Hong Kong, Critics Warn of Corporate Grab on Plant Life: SynBio 4.0 = SynBio-4-profit' (News Release, 9 October 2008).

[95] ETC Group, 'Syns of Omission: Civil Society Organizations Respond to Report on Synthetic Biology Governance from the J. Craig Venter Institute and Alfred P. Sloan Foundation' (News Release, 17 October 2007).

[96] ETC Group, 'Extreme Genetic Engineering: An Introduction to Synthetic Biology' (ETC Group, January 2007) <http://www.etcgroup.org/upload/publication/602/01/synbioreportweb.pdf>, 50.

[97] For example, George Church told the Presidential Commission:

> So I teach a course on responsible conduct of science, which is based on some of the NIH literature. In addition, in every other standard course that I teach, I inject some, both lectures and exercises, where they have to seriously understand it in order to do well in the course. I would say in addition, iGEM and SynBERC as part of our training includes what's called a human practices component where really they are judged in their competition as to how well they conquer that particular knowledge base as well. So I think there's certainly room for improvement, but there's some hope as well.

George Church, 'Evidence to The Presidential Commission for the Study of Bioethical Issues, Synthetic Biology' (8 July 2010) Washington DC.

DNA technology was published in 1973.[98] Subsequently a group of scientists published a letter in *Science* warning of the potential biohazards of this technology and recommended a moratorium until these risks were better evaluated.[99] They also recommended that the NIH establish an advisory committee to oversee recombinant DNA experiments. The NIH established the Recombinant DNA Advisory Committee (RAC) in 1974.[100] The RAC advises the NIH about 'the conduct and oversight of research involving recombinant DNA'.[101]

In 1975, the Asilomar Conference on the risks of recombinant DNA technology took place and the scientists – along with civil servants, lawyers and others – agreed on recommendations for biosafety management in experiments with recombinant DNA.[102] The RAC then developed the NIH Guidelines, which set out guidelines for the safe conduct of recombinant DNA research.[103] These are now the primary mechanism for regulating biological engineering experiments. The guidelines must be followed where research involving recombinant DNA or synthetic DNA is funded by the NIH, or where any recombinant DNA or synthetic DNA work at that institution (or company) is funded by the NIH.[104] Amendments to the NIH Guidelines to incorporate synthetic DNA – as well as recombinant DNA – into the scope of the guidelines became effective in 2013.[105] Other Federal agencies can also make compliance a condition of funding.

---

[98] Stanley M Cohen and Annie CY Chang, 'Recircularisation and Autonomous Replication of a Sheared R-Factor DNA Segment in Escherichia Coli Transformants' (1973) **70**(5) *Proceedings of the National Academy of Sciences* 1293.

[99] Paul Berg et al, 'Potential Biohazards of Recombinant DNA Molecules' (1974) **185**(4148) *Science* 303.

[100] Office of Biotechnology Activities, *About Recombinant DNA Advisory Committee (RAC)* Office of Science Policy – National Institutes Of Health <http://oba.od.nih.gov/rdna_rac/rac_about.html>.

[101] Department of Health and Human Services, *Charter: Recombinant DNA Advisory Committee* (2011) Office of Science Policy – National Institutes of Health <http://oba.od.nih.gov/oba/RAC/Signed_RAC_Charter_2011.pdf>.

[102] P Berg et al, 'Summary Statement of the Asilomar Conference on Recombinant DNA Molecules' (1975) **72**(6) *Proceedings of the National Academy of Sciences* 1981.

[103] *NIH Guidelines for Research Involving Recombinant or Synthetic Nucleic Acid Molecules*, Federal Register **81**(73) 22286, 22 March 2016, first issued in 1976.

[104] Ibid., s. 1-D.

[105] Ibid.

(b) Select Agent Regulations

As discussed, in the United States, access to and use of dangerous disease-causing organisms is tightly controlled under the Select Agents and Toxins system.[106] The system applies to a prescribed list of particularly dangerous infectious organisms and toxins, which pose a severe public health risk. These can only be possessed, used or transferred where the relevant individual has registered this activity. Registration requires a security risk assessment. For approval to possess a select agent, a security plan is required. The organisms can only be kept in the specific location for which approval has been given. Those who have approved access cannot give access to other individuals unless they have also undergone security assessment. Special approval is required for 'restricted experiments' which raise particular concern, such as transferring a drug resistance trait to a select agent, or forming synthetic or recombinant DNA containing the genes to produce a lethal select toxin.

The legislation provides for civil money penalties for breach of the requirements regarding possession and use.[107] There are criminal penalties for transferring a select agent to an unregistered person and for unregistered possession of a select agent. In addition, certain 'restricted persons' may not receive, transport or possess select agents and toxins in 'interstate or foreign commerce'.

The Select Agents and Toxins List includes naturally occurring infectious organisms and toxins. In 2012, the provisions were modified to include synthetic DNA and synthetic organisms.[108] The prescribed list now includes 'Nucleic acids that can produce infectious forms of any of the select agent viruses' and 'Recombinant and/or Synthetic nucleic acids that encode for the functional form(s) of any of the toxins listed'.[109] The list also includes select agents and toxins that have been genetically modified, and reconstructed particles of the 1918 influenza virus that are able to replicate.

The 2012 amendments show that the regulations have been modified to take into account the development of synthetic DNA technology. Under these provisions, possessing synthetic DNA comprising the whole genome or the toxin producing genes of one of the select agents would

---

[106] See 42 USC (Office of the Law Revision Counsel, 2016) § 262a and the regulations made under this: 42 CFR (n 58) Part 73.

[107] 42 USC (n 106) § 262a (i).

[108] Amended by *Possession, Use, and Transfer of Select Agents and Toxins; Biennial Review; Final Rule* 77 Federal Register 61110, (5 October 2012).

[109] 42 CFR (n 58) §§ 73.3, 73.4.

be unlawful. It would also be unlawful to possess a select agent which had been genetically modified using the techniques of synthetic biology.

Although unlawful, it would still be possible to obtain synthetic DNA with sequences derived from dangerous organisms without obtaining physical access to those organisms. It would also be possible to experiment with synthetic DNA from pathogens which are not on the Select Agents and Toxins List but are still dangerous, without encountering these controls. Thus there needs to be a method for detecting activities involving synthetic DNA which has a genetic sequence relating to dangerous organisms. Models for this detection are discussed below.

(c) Voluntary Department of Health and Human Services Screening Guidelines

In 2010, the US Department of Health and Human Services issued a Screening Framework Guidance for Providers of Synthetic Double-Stranded DNA (the Guidance).[110] This appears to have been developed from Church's early proposal for non-proliferation, the proposal by Church, Endy and collaborators for customer and sequence screening, and a NSABB recommendation to introduce monitoring of gene synthesis activities.[111]

The Government's Guidance was released after publication of a draft and a public consultation period. The goal of the Guidance is to: 'minimize the risk that unauthorized individuals or individuals with malicious intent will obtain access to toxins and agents of concern through the use of nucleic acid synthesis technologies, while simultaneously minimizing any negative impacts on the conduct of research and business operations'.[112]

The Guidance is intended to impose minimal costs on providers of synthetic DNA. It is also intended to be considered for implementation

---

[110] *Screening Framework Guidance for Providers of Synthetic Double Stranded DNA*, Federal Register **75**(197) 62820, 13 October 2010; Department of Health and Human Services, *Screening Framework Guidance for Providers of Synthetic Double Stranded DNA* <https://www.phe.gov/Preparedness/legal/guidance/syndna/Documents/syndna-guidance.pdf>, 3.

[111] Church, *A Synthetic Biohazard Non-proliferation Proposal* (n 42); National Science Advisory Board for Biosecurity, 'Addressing Biosecurity Concerns Related to the Synthesis of Select Agents' (n 52) 11.

[112] Department of Health and Human Services, *Screening Framework Guidance for Providers of Synthetic Double Stranded DNA* (n 110) 3.

internationally and has been designed to be able to be extended to cover international companies and US-based companies operating outside the US.

The Guidance sets out recommended procedures for three levels of scrutiny of orders for synthetic DNA: customer screening; sequence screening; and follow-up screening. Customer screening involves verifying information about the person ordering synthetic DNA, including their name, addresses and institutional or corporate affiliation.[113] An illustrative list of 'red flags' is provided. These are matters that should raise concern about the legitimacy of the order and include unclear identity of the person making the order, evasiveness about the information requested, no connection to biotechnology research, unusual payment methods and requests for confidentiality. If one or more of these red flags is present, the Guidance recommends that follow-up screening be conducted and that the provider should contact government agencies if they are unsure as to how to proceed.

The goal of sequence screening is to identify 'sequences of concern' within orders for synthetic DNA.[114] These are sequences derived from organisms on the Select Agents and Toxins List, and for foreign orders, the Commerce Control List under the Export Administration Regulations.[115]

The sequence screening method favoured in the Guidance is an approach called 'Best Match', though providers can choose to use other screening tools. The Best Match method involves comparing the requested genetic sequence to a database of known DNA sequences, such as Genbank. The goal is to check the requested sequence for genes that are unique to the organisms on the Select Agents and Toxins and Commerce Control Lists. A 'hit' occurs when a part of the requested sequence that is more than 200 bp long is more closely related to the genetic sequence of a listed pathogen than to any other sequence in the database. Scientists at the Joint Genome Institute in the US have developed software for sequence screening known as Black List Sequence Screening (BLiSS).[116] This software is part of a broader,

---

113 Ibid., 6.

114 Ibid., 4.

115 Ibid., 23.

116 L Simirenko et al, *Bliss: The Black List Sequence Screening Pipeline* Synthetic Biology Engineering Evolution Design Conference <http://synbioconference.org/2016/proceeding/paper/bliss-black-list-sequence-screening-pipeline>.

comprehensive review process developed by this Institute, which it suggests could be adopted by other institutions.[117]

If red flags are raised either in customer or sequence screening, follow-up screening occurs and involves gathering more information about the customer and the proposed use of the synthetic DNA, in order to assess the legitimacy of the order.[118] The Guidance recommends that the relevant biological safety officer, supervisor or director of research be contacted to verify the legitimacy of the order. If the customer is not affiliated with an institution, a literature review of their past research should be conducted. The contact information may be checked against standard directories, and the approval of the research through an institutional biosafety committee may be checked. If the provider is unsure as to whether they should fill the order, they are asked to contact government agencies such as the FBI, Department of Commerce, and agencies which implement the Select Agent and Toxins regulations.

The Guidance also recommends that gene synthesis companies keep records of customer orders, hits and follow-up screening for at least eight years.

7.4.3 Self-regulation by Gene Synthesis Companies

This section examines the key features of two separate codes of practice which have been developed for DNA synthesis oversight by groups of DNA synthesis companies.

(a) International Gene Synthesis Consortium Harmonized Screening Protocol: gene sequence and customer screening to promote biosecurity

A set of private sector guidelines for synthetic DNA sequence screening was agreed by the International Gene Synthesis Consortium (IGSC), comprised of five large gene synthesis companies in the US and Europe.[119] This Harmonized Screening Protocol: Gene Sequence and

---

[117] Nathan J Hillson, *Tools and Opportunities to Enhance Risk Analysis* National Academy of Sciences <http://nas-sites.org/biotech/files/2016/04/Tools_Risk_Analysis_Hillson.pdf>.

[118] Department of Health and Human Services, *Screening Framework Guidance for Providers of Synthetic Double Stranded DNA* (n 110) 7.

[119] International Gene Synthesis Consortium, *Harmonized Screening Protocol: Gene Sequence & Customer Screening to Promote Biosecurity* International Gene Synthesis Consortium <http://www.genesynthesisconsortium.org/wp-content/uploads/2012/02/IGSC-Harmonized-Screening-Protocol1.pdf>.

Customer Screening to Promote Biosecurity (the Protocol) requires sequence screening of an order for synthetic DNA against a 'Regulated Pathogen Database', which was to be constructed from the Select Agents and Toxins List, the Australia Group List of Biological Agents for Export Control and 'any other national list of regulated pathogens'.[120] According to the Venter Institute, this database has now been developed and is updated from time to time.[121] The sequence screening process is automated, but if it identifies potential pathogen or toxin sequences within the requested sequence, a human expert will review the order and decide whether to accept it, reject it or subject it to customer screening.

For customer screening, the Protocol requires that the customer provide identification including their name and address, institution, and contact details.[122] Customers will be screened against various official lists of persons, such as the Department of State's List of Statutorily Debarred Parties, as required by applicable national regulations. If DNA sequences from select agents are requested, additional customer screening will occur. These sequences will only be provided to government laboratories, universities, non-profit research institutions, and industrial laboratories 'demonstrably engaged in legitimate research'. A written description of the proposed use of the synthetic DNA is required. Companies will independently verify the identity of the researchers and the activity of their organisation and compare this with the proposed use of the synthetic DNA. The companies will only supply the synthetic DNA to an end-user which can demonstrate compliance with applicable regulations. They will only supply synthetic DNA to a redistributor if that company can identify the end-user and demonstrate their compliance with applicable regulations.

This consortium of companies states that it has relationships with law enforcement and intelligence authorities and will share information and report suspicious customers to these agencies.[123] It will collaborate with national governments to support effective oversight. Records of sequence screening will be kept for eight years. The IGSC will periodically revise its screening protocol.

---

[120] Ibid.
[121] Carter and Friedman (n 49) 10–11.
[122] International Gene Synthesis Consortium, (n 119).
[123] Ibid.

(b) International Association of Synthetic Biology Code of Conduct for Best Practices in Gene Synthesis

In 2009, the Industry Association for Synthetic Biology (IASB), a group of companies and organisations 'with a stake in synthetic biology', released The IASB Code of Conduct for Best Practices in Gene Synthesis (the Code of Conduct).[124] Under the Code of Conduct, the member gene synthesis companies would screen orders for synthetic DNA for reasonable sequence similarity to pathogens and review any hits against the Australia Group List of Biological Agents for Export Control, the Select Agents and Toxins List and relevant national organism lists. Procedures for checking hits and customer screening were also established.

More recently, the IASB website and the Code of Conduct seem to have disappeared from the Internet. It may be that this organisation has ceased to operate, and companies are instead using the IGSC protocol or their own procedures.

7.4.4 Comparisons between the Government and Industry Guidelines

Some commentators have argued that the industry codes of practice require greater involvement of human experts than the Government's Guidance and therefore that the government is promoting a lower standard of security assessment.[125] However, analysis shows that this is not accurate. The Government's Guidance requires the exercise of human discretion at various points. It states that the supplier should have 'expertise in-house to perform the sequence screenings, analyse the results and conduct the appropriate follow-up research to evaluate the significance of dubious sequence matches'.[126] The IGSC and government procedures both use automated sequence matching methods (this would be done with computer software as it is not feasible for a human expert to

---

[124] International Association for Synthetic Biology, *The IASB Code of Conduct for Best Practices in Gene Synthesis* International Association for Synthetic Biology <http://www.ia-sb.eu/tasks/sites/synthetic-biology/assets/File/pdf/iasb_code_of_conduct_final.pdf>, 1.

[125] Michael Eisenstein, 'Synthetic DNA Firms Embrace Hazardous Agents Guidance but Remain Wary of Automated "Best-Match"' (2010) **28** *Nature Biotechnology* 1225; Markus Fischer and Stephen M Maurer, 'Harmonizing Biosecurity Oversight for Gene Synthesis' (2010) **28**(1) *Nature Biotechnology* 20, 20–21; Heidi Ledford, 'Gene-synthesis Rules Favour Convenience' (2010) **467** *Nature* 898.

[126] Department of Health and Human Services, *Screening Framework Guidance for Providers of Synthetic Double Stranded DNA* (n 110), 21.

compare the requested sequence to all known genes in public databases). Under both procedures, analysis by a human expert occurs if there is a hit.

Therefore, comparison of the different screening guidelines does not indicate a reduced role for human experts under the Government's Guidance. It is not tenable to rely on the role of human experts to contend that the companies have agreed on a higher standard of regulation than the government has proposed, and should therefore be left to self-regulate.

Comparison shows that the Government Guidance is more detailed and thorough than the company guidelines. For example, the Government Guidance provides examples of red flags and potential screening scenarios. By contrast, the IGSC is more akin to a general plan for screening. There is greater room for interpretation by companies, which may regard commercial imperatives as preventing significant resources from being expended on sequence screening.

One aspect of the Government Guidance which can be argued to impose a lower standard is the narrower scope of pathogenic organisms and toxins against which orders for synthetic DNA will be screened. The Government's Guidance acknowledges that some DNA sequences which could pose security risks are not covered by the Select Agents and Toxins and Commerce Control Lists.[127] It states that the US Government has decided not to maintain a comprehensive list of agents against which sequences should be screened 'due to the complexity of determining pathogenicity because research in this area is ongoing and many such agents are not currently encompassed by regulations in the United States'.[128]

The companies using the IGSC Regulated Pathogen Database appear to be screening orders for synthetic DNA against a broader set of sequences. The IGSC Protocol requires, at a minimum, screening against the Select Agents and Toxins List and a list contained in a European Union Council Regulation. While it is not clear from the protocol itself, the Venter Institute states that the Regulated Pathogen Database includes a wider range of genetic sequences; 'whole genomes of pathogenic species (including those on the Select Agent List, the Commerce Control

---

127 Ibid., 9.

128 Ibid.

List, the Australia Group List and the European Union's Council Regulation' … as well as various other pathogen species and genes, and genes that are suspected to have a role in pathogenicity.[129]

In addition, the Venter Institute states that the IGSC companies review a long list of sequences with similarity to the ordered sequence, while the Government Guidance recommends that only the 'best match' is checked.

This means that a broader set of dangerous sequences can potentially be identified under the private sector procedures. Some scientists have expressed concern about known human pathogens which are not on the Select Agents and Toxins List, as these would not be identified using the Government's Guidance.[130] While the US Government has shied away from producing a more up-to-date, comprehensive database of pathogen sequences, it should collaborate with industry to do this in order to ensure that a consistent, up-to-date mechanism is used across the board. A 'centrally curated database' has also been suggested by the IGSC.[131] This requires a central organisation tasked with this responsibility. This should be coordinated and overseen by government.

This comparison of the sequence screening procedures suggests that a more effective regulatory intervention would result if the scope of the different sets of guidelines were combined. This would involve combining the broader pathogen screening scope of the company guidelines and the more detailed guidance on red flags and screening steps given in the Government Guidance. There would then be government oversight and management of the process of developing and maintaining the scientific databases necessary for screening.

7.5 SYNTHETIC BIOLOGY REQUIRES A DISTINCTIVE, ENFORCEABLE APPROACH

7.5.1 The Presidential Commission's Recommendations

The Commission observed a consensus among scientists and policy makers that in the synthetic biology context, security is a 'serious issue warranting ongoing and pro-active re-examination' as the science progresses.[132] As discussed, the Commission proposed recommendations for

---

[129] Carter and Friedman (n 49) 10–11.

[130] Eisenstein (n 125) 1226.

[131] Wagner, 'Evidence to The Presidential Commission for the Study of Bioethical Issues, Synthetic Biology' (n 70).

[132] The Presidential Commission for the Study of Bioethical Issues (n 1) 72.

addressing the risks of synthetic biology based on the principle of responsible stewardship. This encompasses the concept of prudent vigilance: the establishment of processes for assessing the risks and benefits of synthetic biology projects before and after they are undertaken, with the capacity to 'continue to assess safety and security as technologies develop' and to limit their use when necessary.[133]

The Commission also recommended initial and ongoing review of safety and security risks of synthetic biology as the technology develops.[134] It recommended that if significant 'unmanaged risks' are identified in the initial review, consideration be given to making certain existing regulatory measures mandatory, independent of the source of funding.

As in the Commission's approach to do-it-yourself biology, the principle of 'intellectual freedom' was a key element of its approach to biosecurity. The Commission emphasised that 'In academic communities, intellectual freedom is essential. The ability to explore ideas openly and freely – even controversial or unpopular ideas – is fundamental to the mission of education and research'.[135] It stated that 'A robust public policy regarding the responsible conduct of science must promote the creative spirit of scientists and unambiguously protect their intellectual freedom'.[136]

Thus the Commission showed a preference for minimal intrusion on intellectual freedom of scientists. It argued that intellectual freedom and responsibility requires that 'restrictions on research, whether by self-regulation among scientists or by government intervention, should limit the free pursuit of knowledge only when the perceived risk is too great to proceed without limit'.[137] However, it acknowledged that both biosafety and biosecurity risk assessment are difficult for synthetic biology due to its novelty and the potential for new organisms to be developed.

The principle of 'regulatory parsimony' was also emphasised, which requires 'only as much oversight as is truly necessary to ensure justice, fairness, security, and safety while pursuing the public good'.[138] The Commission saw this principle as important in relation to emerging technologies, because these are still developing and the 'blunt instruments' of legislation and regulatory control could inhibit beneficial

---

[133] Ibid., 4.
[134] Ibid., 147.
[135] Ibid., 142.
[136] Ibid., 141.
[137] Ibid., 12.
[138] Ibid., 141.

research.[139] It was concerned that regulation might operate so as to *reduce* security by preventing research on safeguards.

Thus the Commission's recommendations indicate a preference for self-regulation within the synthetic biology field. For example, it recommended 'fostering responsibility and accountability': 'The government should support a continued culture of individual and corporate responsibility and self-regulation by the research community, including institutional monitoring, enhanced watchfulness, and application of the National Institutes of Health Guidelines for Recombinant DNA Research'.[140]

Further, the Commission proposed that 'Creating a culture of responsibility in the synthetic biology community could do more to promote responsible stewardship in synthetic biology than any other single strategy.'[141] It argued that self-regulation creates a moral sense of ownership of the regulatory strategy within the scientific field, as part of a culture of responsibility.

This approach puts significant responsibility on the synthetic biology community. For example, enhanced watchfulness requires: 'the scientific community to recognize the varied risks associated with synthetic biology and develop internal processes to identify and respond to potential threats rapidly and effectively. Enhanced watchfulness reflects a relationship among scientists, citizens, and policy makers built on trust and mutual respect'.[142]

In a similar vein, the Commission recommended that 'the scientific community should take steps to carefully manage both scientific and social risks associated with synthetic biology as this field grows'.[143]

Overall, the Commission concluded that no special restrictions were warranted at that time.[144]

7.5.2 Voluntary Guidelines and Self-regulation are Insufficient

The Commission may have left its recommendations at this level because it had only a short time to prepare its report and no particular models for addressing the complex security problems of synthetic biology were

---

139 Ibid., 28.
140 Ibid., 13.
141 Ibid., 11.
142 Ibid., 145.
143 Ibid., 149.
144 Amy Gutmann, 'The Ethics of Synthetic Biology: Guiding Principles for Emerging Technologies' (2011) **41**(4) *Hastings Center Report* 17, 21.

presented to it. However, this approach of relying on the principles of responsible stewardship and intellectual freedom and responsibility to propose self-regulation and minimal intervention by government is a 'wait and see' approach which is not sufficiently cautious.

As I have argued, there are features of synthetic biology which indicate that a distinctive approach is required. These are the new construction approach to biology, the scale of modifications being made to cells, increasing gene synthesis capability, technology to build new genetic features and even whole genomes, and emphasis on standardisation and associated de-skilling. These differences from previous DNA technologies indicate that a voluntary guidelines approach, as has been taken in relation to recombinant DNA, is not sufficient. They also indicate that voluntary sequence screening guidelines or industry self-regulation are not appropriate as the major regulatory strategy. Instead, synthetic biology requires an enhanced, enforceable regulatory strategy for biosecurity.

While there is no simple answer to the biosecurity problem, and each strategy will have limitations, government action should be taken now to address the risks. As a starting point, procedures for screening orders for synthetic DNA should be made enforceable, rather than left to industry. The Commission's recommendations tend to emphasise the fostering of a culture of self-regulation and responsibility within synthetic biology to the extent that the role of government enforcement is given minimal consideration. As Church has pointed out, the gene synthesis industry has now developed principles and procedures, and these could be enforced by government. He commented that:

> This is something that would be nice to be backed up with regulation ... In terms of surveillance, I think one thing that was hard for the government to act until industry had shown what it was planning on doing. But at this point, I think they could tune into that and help it along and make it law and work internationally.[145]

He pointed out that at that time, there was an opportunity to regulate because if the technology for gene synthesis was licensed and synthesis activities monitored, it would be difficult for a would-be misuser of the technology to get around this system. This was because of the stage of development of DNA synthesis technology, which could not be done without specialised equipment.

[145] Church, 'Evidence to The Presidential Commission for the Study of Bioethical Issues, Synthetic Biology' (n 97).

Intellectual property expert Michael Geist has argued that in the context of Internet regulation, initially governments went along with the view that the online environment was not suited to government regulation, and allowed the private sector to take the lead in policy and self-regulation.[146] Later, it was observed that the 'emphasis is shifting ... from self-regulation to government regulation'.[147] A similar course should now be taken in synthetic biology, for reasons of regulatory effectiveness, legitimacy and connection to be set out below.

Even gene synthesis companies themselves have pointed to a role for government enforcement:

> Although we stand behind our self imposed regulation, there is no doubt that the government could act to improve its efficacy ... We have done our best to craft a screening list, but we believe that our governments should be able to provide the most up-to-date and accurate list of restricted sequences. Equally important to a comprehensive screening list is a plan for enforcement.[148]

(a) Effectiveness

Effective regulation is regulation which is fit for purpose and is 'making a difference'.[149] In proposing that reliance on industry self-regulation for gene synthesis screening will be effective, several arguments have been made. For example, political scientist Amy Smithson has argued that, first, industry is well-placed to develop self-regulation because it has the required knowledge of technology and business trends, while government and academia are ill-suited to the task.[150] Secondly, she has suggested that industry self-regulation can move more quickly and stay ahead of the potential misusers of the technology more effectively than government regulation. In this view oversight is most effectively accomplished

---

146 Michael Geist, 'Cyberlaw 2.0' (2003) **44** *Boston College Law Review* 323, 328–32.

147 Ibid., 324.

148 Jeremy Minshull and Ralf Wagner, 'Preventing the Misuse of Gene Synthesis' (2009) **27**(9) *Nature Biotechnology* 800, 800.

149 See Brownsword and Somsen (n 5) 11; Brownsword and Goodwin (n 5) 277.

150 Amy Smithson, 'Renovating Governance Strategies for Synthetic Biology and Other "Dual-Use" Technologies' (Paper presented at the 21st Century Borders/Synthetic Biology: Focus on Responsibility and Governance, 2013) <http://www.scienceforglobalpolicy.org/LinkClick.aspx?fileticket=VqAc-vBzvMk%3d&tabid=133>, 25–8.

through voluntary guidelines and information sharing arrangements developed by industry. It has also been suggested by economist and legal academic Stephen Maurer and chemical and biological weapons expert Jonathon B Tucker that due to market forces and public relations, gene synthesis companies will have to demonstrate that they are addressing security problems and cannot afford to breach sequence screening guidelines.[151]

Nonetheless, the uncertainty surrounding a biosecurity event using synthetic biology and the potential for catastrophic impact mean that a voluntary approach is not appropriate. Models such as the influential 'responsive regulation' theory of legal academic and economist Ian Ayres and sociologist of law John Braithwaite have been proposed as a way to regulate the behaviour of firms.[152] This theory proposes a 'pyramid' approach to enforcement strategies. The baseline strategy is industry self-regulation, escalating, for example, to 'enforced self-regulation', 'command regulation with discretionary punishment' and then 'command regulation with nondiscretionary punishment' as necessary.[153] Under this model, regulators should seek to cooperate with and persuade regulatees, in order to encourage effective self-regulation and compliance. The threat of more intrusive government action and escalating sanctions if regulated firms 'defect' is said to promote compliance. Similarly, Mandel proposes a model under which regulation only comes into effect if firms do not develop their own more rigorous arrangements.[154]

However, regulatory scholars Robert Baldwin and Julia Black argue that application of a 'responsive' model of escalating government intervention may not be appropriate in the context of potentially catastrophic risk.[155] Similarly, regulatory scholar Neil Gunningham has argued that use of the theory of 'smart regulation' – which involves initially 'experimenting' with less interventionist regulatory measures – is less

---

151 Maurer and Fischer (n 125) 22; Jonathon B Tucker, 'Double-edged DNA: Preventing the Misuse of Gene Synthesis' (2010) Spring **26**(3) *Issues in Science and Technology* 23. See also comments of conference attendees during debate on paper: Smithson (n 150) 30.

152 Ian Ayres and John Braithwaite, *Responsive Regulation: Transcending the Deregulation Debate* (Oxford University Press, 1992).

153 Ibid., 38–41.

154 Gregory N Mandel, 'Regulating Emerging Technologies' (2009) **1** *Law, Innovation & Technology* 75, 85–7.

155 Robert Baldwin and Julia Black, 'Really Responsive Regulation' (2008) **71**(1) *The Modern Law Review* 59, 62–3.

appropriate where there is potentially irreversible catastrophic harm.[156] Applying this to the synthetic biology biosecurity context, there is currently a degree of 'unmanaged risk' and it is not appropriate to wait and see how long self-regulation in the gene synthesis industry is successful. Because of the potentially severe consequences if this industry regulation fails, enforceable regulation should be the starting point. As health law specialist Lawrence O Gostin and international law academic David P Fidler have suggested, in the biosecurity context the issue is not *whether* there should be regulation of the biological sciences, but rather *how* this should be done.[157] In the biosecurity context, 'codes of conduct will not, without formal government action, be sufficient'.[158]

The security success of a screening regime for synthetic DNA depends on *all* DNA synthesis companies fully complying with the procedure.[159] If this does not occur, a person or organisation could work around the system by using a company which does not check the requested sequences against pathogen and toxin sequences. This suggests that enforcement is necessary. While there are public relations reasons which could push commercial gene synthesis providers to comply, there is also a cost to companies associated with employing human experts to check orders for synthetic DNA. Such experts would also require training. While this cost may be relatively low, there is concern that the cost of compliance may lead to a loss of stringent implementation of the voluntary system in a way that is difficult to detect. Further, over time the cost of the synthesis itself has decreased, so the proportional cost of compliance has increased.[160] In the context of potentially catastrophic risk, leaving compliance to market forces would not be sufficiently cautious. Because the gene synthesis industry is international in nature, international coordination in regulation of this field also needs to be considered to maximise the likelihood of all firms complying.

---

156 Neil Gunningham, 'Regulating Biotechnology: Lessons from Environmental Policy' in Hans Somsen (ed.), *The Regulatory Challenge of Biotechnology: Human Genetics, Food and Patents* (2007) 3, 7.

157 Fidler and Gostin (n 9) 12.

158 Ibid., 87 noting that 'the reverse also appears to be true'.

159 Stephen M Maurer et al, 'Working Paper – Making Commercial Biology Safer: What the Gene Synthesis Industry Has Learned About Screening Customers and Orders' (Goldman School of Public Policy and Boalt Law School, University of California, 2009) <http://gspp.berkeley.edu/assets/uploads/page/Maurer_IASB_Screening.pdf>, 10.

160 Carter and Friedman (n 49) 15.

Scholars such as Maurer have argued that the industry procedures require greater involvement of human experts than the Government's Guidance and therefore that the government is promoting a lower standard of security assessment.[161] However, this is not demonstrated in comparison of the procedures, and is not a valid reason to argue that the industry procedures will be more effective in addressing security risks.

The existence of an industry protocol as well as the Government's Guidance means that there are multiple rulebooks in play.[162] It has been suggested that compliance with screening procedures is now standard practice within the industry,[163] but it is not clear *which* procedures are being used. In evidence to the Commission, Wagner stated that the IGSC would incorporate the government's requirements into its protocol.[164] But the protocols are still different in various respects. Wagner also suggested that 'We definitely need a uniform screening practice and uniform screening agenda.'[165] If a government-enforced system were introduced, consistency across the gene synthesis industry could be promoted.

(i) Industry buy-in It has been suggested from various viewpoints that engagement and consultation with regulatees regarding regulation is likely to promote effective regulation and compliance. Ayres and Braithwaite have argued that companies may have more commitment to rules they have generated themselves.[166] Under the 'Really Responsive Regulation' framework, regulators are advised to be attentive to the culture and motivations of regulated firms, because these 'attitudinal settings' are relevant to compliance with the regulation, detection of non-compliance and the relationship between regulatees and regulators.[167]

Similarly, the Commission proposed that self-regulation promotes a sense of 'ownership' within a culture of responsibility.[168] While I have argued for an enforceable approach, this is not to suggest that there is *no* role for ongoing discussions and development and dissemination of best

---

161 See Eisenstein (n 125); Ledford (n 125).

162 Using the language of Ayres and Braithwaite, (n 152) 113–14.

163 Robert Friedman, 'Governance of Synthetic Biology' (Paper presented at the 21st Century Borders/Synthetic Biology: Focus on Responsibility and Governance, 2013) <http://www.scienceforglobalpolicy.org/LinkClick.aspx?fileticket=VqAc-vBzvMk%3d&tabid=133>, 8.

164 Wagner, 'Evidence to The Presidential Commission for the Study of Bioethical Issues, Synthetic Biology' (n 70).

165 Ibid.

166 Ayres and Braithwaite (n 152) 113.

167 Baldwin and Black (n 155) 69–70, 77–94.

168 The Presidential Commission for the Study of Bioethical Issues (n 1) 28.

practices within the synthetic biology field. The synthetic biology field has a culture of early consideration of issues such as safety and security and proactive development and sharing of strategies for addressing these risks within the field. Scientists are keen to work together on these goals, as observed in debates at the Institute on Scientists for Global Policy conference on synthetic biology regulation in 2012.[169] The gene synthesis industry has also been proactive in agreeing on strategies to monitor gene synthesis. This culture should be encouraged and government agencies should seek to engage with scientists and industry and seek consensus on regulatory measures where possible.

However, pursuit of the goals of maintaining positive relationships with regulated companies – and generating consensus regarding the content of regulation – should not mean that voluntary or self-regulatory measures are the default position. The gene synthesis industry's demonstration that it is attuned to biosecurity risks provides a positive basis for discussion, as does the synthetic biology community's engagement with these issues. The gene synthesis industry has also had several years to become used to sequence screening procedures. If there is effective engagement with these industry stakeholders, it may be possible to develop an enforceable approach while minimising potential damage to the attitudes of companies to the regulation.

(ii) Technical limitations of sequence screening The tools available for checking requested sequences of synthetic DNA against sequences from pathogens and toxins have some technical limitations, as the function of a sequence of DNA and its harmful potential can depend on the biological context in which the DNA is present.[170] Accordingly, sequence screening can only form one aspect of a security risk strategy.[171] No single measure is going to provide a comprehensive answer to the biosecurity problems associated with synthetic biology.

Therefore, policy makers and scientists should continue looking for ways to address the biosecurity risks posed. In addition to government regulation of the gene synthesis industry, other strategies are needed.

---

[169] Institute on Science for Global Policy, '21st Century Borders/Synthetic Biology: Focus on Responsibility and Governance' (2013) <http://www.scienceforglobalpolicy.org/LinkClick.aspx?fileticket=VqAc-vBzvMk%3d&tabid=133>.

[170] Relman, 'Evidence to The Presidential Commission for the Study of Bioethical Issues, Synthetic Biology' (n 39).

[171] Ibid.

These strategies should include education and engagement[172] with academic scientists, synthetic biology companies and do-it-yourself biologists about potential dual-use issues, and use of institutional review mechanisms for oversight of research.

(b) Legitimacy

Self-regulatory approaches lack legitimacy, one of the key criteria for strategies for regulating emerging technologies.[173] Procedural legitimacy requires regulators to be transparent, accountable and inclusive, and an essential part of this is public participation.[174] The Commission itself highlighted the importance of 'robust public participation' in democratic deliberation, stating that 'Public discussion and debate promote the legitimacy of whatever outcomes are reached'.[175] However, it gave the role of regulatory development mainly to scientists. Such self-regulation would lack transparency and accountability.

To promote legitimacy, there should be opportunities for those who are interested to express their views and decision-making should be responsive to those views.[176] While both industry and academic scientists in synthetic biology have proactively sought to address biosecurity risks, it is important to consider that those outside the group of scientists and policy makers familiar with synthetic biology also have concerns and views about regulation. Just as scientific expertise should not be the only perspective considered in developing regulation,[177] nor should the perspective of industry or academia be the exclusive determinant of a regulatory strategy. Self-regulation within the synthetic biology industry lacks a mechanism to take public concerns into account. Only industry views are considered, and there is very little information publicly available about the actual practices of IGSC companies. The 2015 report by the Venter Institute has performed a valuable role in shedding light on the development of the screening database and procedures. For reasons of

---

172 Ibid.

173 Roger Brownsword, 'So What Does the World Need Now? Reflections on Regulating Technologies' in Roger Brownsword and Karen Yeung (eds), *Regulating Technologies: Legal Futures, Regulatory Frames, and Technological Fixes* (2008) 23, 31–2.

174 Brownsword and Goodwin (n 5) 48; Brownsword and Somsen (n 5) 17–19.

175 The Presidential Commission for the Study of Bioethical Issues (n 1) 29, 155.

176 Roger Brownsword, *Rights, Regulation and the Technological Revolution* (Oxford University Press, 2008), 128.

177 Brownsword and Somsen (n 5) 18–19.

public legitimacy, it is important that perspectives beyond industry are sought in development of regulation. Broad stakeholder engagement is crucial in developing effective regulation for emerging technology generally.[178]

The ETC Group objected to the efforts of synthetic biologists to self-regulate. They did not view the scientists' efforts as legitimate because the scientists were seen to have vested interests. Attaining public legitimacy – that is legitimacy outside the group of scientists, companies and social scientists working in synthetic biology – will be difficult for self-regulatory approaches given the risks that are posed and the widespread concern in the wake of the synthetic cell announcement.

Fidler and Gostin have proposed that biosecurity policy, including supervision of biological research, should be grounded in a rule of law framework.[179] The rule of law is a 'philosophy of governance' which emphasises 'the supremacy of legal rules'.[180] They argue that the need for the rule of law framework is generated by new policy challenges and 'new governance contexts' of biosecurity regulation.[181] Biosecurity regulation will require political power to be exercised in new contexts and a rule of law framework is important for 'embedding the exercise of political power in a framework of rules'.[182] These rules determine 'how such power is wielded and for what purposes'.[183]

Further, the need for a rule of law framework arises from the difficult tensions and competing considerations outlined above in Part 7.2.[184] For example, there is potential for biosecurity regulation to infringe upon individual rights and liberties and to limit scientific or intellectual freedom.[185] This can be seen as an issue of limitation of individual freedom because scientific freedom and open discussion and dissemination of research are concepts which arise from the individual rights of freedom of speech and association.[186] In designing regulation for dual-use technologies, there is a tension between the principles of scientific freedom and the regulatory goal of security risk minimisation.[187] There

---

178 Mandel (n 154) 90–91.
179 Fidler and Gostin (n 9) 8, 13–15.
180 Ibid., 191–2.
181 Ibid., 188–9.
182 Ibid., 188–213.
183 Ibid., 191.
184 Ibid., 73–80, 188–203.
185 Ibid., 8–15, 188–95, 199.
186 Ibid., 198–9.
187 Ibid., 12–13.

are concerns that regulation of research for security purposes could impede the development of beneficial synthetic biology applications.[188] A rule of law framework supplies a way to begin addressing these difficult tensions.[189]

This rule of law framework should be adopted for biosecurity regulation of synthetic biology because it exhibits all the regulatory challenges of biosecurity regulation of life sciences research and the difficult tensions and competing considerations as identified above in Part 7.3. While Fidler and Gostin do not directly state that the rule of law framework requires government oversight of biological research, this is implicit throughout their argument regarding the challenges of biosecurity policy. For example, they point out the need for government to introduce legislation to give it authority to regulate science, among other biosecurity objectives.[190] They also observe a consensus that 'regulating science to ensure new developments, techniques and technologies do not lead to biological weapons is prudent'.[191] Accordingly, applying the rule of law framework to synthetic biology biosecurity regulation would mean that there is government oversight of synthetic biology research. This should include government oversight of gene synthesis activities.

Government intervention based on this rule of law framework requires that the limitations placed on scientific research are designed to prevent a severe and probable threat, scientifically effective, and the least restrictive measures which could achieve the biosecurity objective.[192] While the likelihood of synthetic biology being used to create a novel pathogen may be relatively low, it is certainly possible that gene synthesis services could be used to gain access to the DNA of dangerous organisms, which would reduce biosecurity. The screening of orders for synthetic DNA can provide a scientifically rigorous approach to minimising this risk, as part of a suite of biosecurity measures. It would only limit the freedom to engage in biological research and innovation where a particular order for synthetic DNA poses a security threat. Thus an enforceable oversight mechanism for monitoring gene synthesis activities can be supported by the rule of law framework.

The rule of law framework also requires transparency and accountability in biosecurity policy, transparency is seen as a core feature of

---

188 See, e.g., The Presidential Commission for the Study of Bioethical Issues (n 1) 28.

189 Fidler and Gostin (n 9) 79.

190 Ibid., 195.

191 Ibid., 259.

192 Ibid., 79.

effective surveillance of biological research.[193] Transparency and accountability are not present in industry self-regulation or voluntary standards for gene synthesis. Thus addressing biosecurity risks of synthetic biology within a rule of law framework would require that the oversight of biological research be done by government through a transparent, accountable mechanism.

Codes of conduct developed within scientific fields can contribute an additional form of accountability in which scientists keep watch on biological research.[194] However, *external* accountability is also required.[195] This requires government enforcement of biosecurity regulation to promote accountability to those outside science and industry. If external accountability were present, this would promote a public perception of legitimacy of the regime. Public trust in relation to an emerging technology and confidence in the regulation is a key factor in the success of the technology.[196] Thus it is in the interests of industry, academic scientists and government to promote this trust and confidence.

7.5.3 Adaptation is Important for Regulatory Connection and Effectiveness

As with the Commission's recommendations regarding environmental risk and biosafety, a key positive aspect of its recommendations in relation to biosecurity is the acknowledgement that the risks will need to be periodically reviewed and regulatory interventions will need to be adjusted over time.

Synthetic biology requires particular consideration of the adaptability of regulation due to the pace of progress in this new field. The speed of progress was impressed upon the Commission in the evidence it received from scientists. For example, Endy noted that the team at the Venter Institute had demonstrated a 100-fold improvement in their ability to construct DNA over the last six years.[197] This fast pace makes the prediction of developments in synthetic biology and biosecurity risk assessment more difficult, increasing the need for continuing review and adaptation. As Posner has pointed out: 'Law's faltering struggle to cope

---

193 Ibid., 79, 193.

194 Ibid., 213.

195 Zhang et al (n 90) 4.

196 Mandel (n 154) 90.

197 Endy, 'Evidence to The Presidential Commission for the Study of Bioethical Issues, Synthetic Biology' (n 43).

with the onrush of science … is an old story, but a true one, and becoming more worrisome by the day'.[198]

Proactive and continuing review of the performance of the regulatory strategy and seeking of information about the state of the science are necessary to promote both regulatory effectiveness and regulatory connection. In the synthetic biology context, any regulatory strategy adopted for biosecurity is time limited because it will fall behind the stage of technological development. Any oversight of DNA synthesis can only be effective while DNA synthesis technology is at a particular stage of development. For example, it is becoming possible for laboratories to purchase their own DNA synthesiser rather than outsource to gene synthesis companies. In 2015, SGI-DNA launched a benchtop synthesiser known as BioXP 3200. This currently retails for about $50,000 and can produce DNA fragments of up to 1.8 Kb. This means that there is already a gap in the scope of the government guidance, as the Venter Institute has pointed out.[199] This will need to be addressed, perhaps by a licensing regime associated with the equipment and enforceable mechanisms for oversight of sequences that are produced with licensed equipment.

For these reasons, enforceable regulation in the security area should be developed and implemented now, and information about its effectiveness and how it is impacting upon the conduct of legitimate synthetic biology research should be proactively sought, along with information about the progress of science in the field. This knowledge should be applied in a process of institutional learning in order to update the regulation over time.

7.6 CONCLUSION

This chapter has examined the complicated biosecurity issues raised by the emerging science of synthetic biology, and the regulatory challenges these present. The early proactive engagement of synthetic biologists with security issues is commendable and should continue, though the efforts to self-regulate by developing a declaration were not successful. The existing legislative provisions for biosecurity leave some of the risks posed by synthetic biology unaddressed. The voluntary Government Guidance and self-regulatory private sector processes for DNA synthesis screening are broadly similar, but some differences remain.

[198] Posner (n 8) 8.
[199] Carter and Friedman (n 49) 16.

The unique features of synthetic biology and the complexity of the security risks create distinctive aspects of the social and scientific context in which regulation of synthetic biology occurs. A voluntary guidelines approach is not sufficiently cautious. A government-enforced, coordinated and adaptive system is needed for reasons of regulatory effectiveness, legitimacy and connection. This proposed enforceable regulation of DNA synthesis will fit into the trend of US policy arguably becoming more precautionary in the area of biosecurity.[200]

These changes to the existing regulatory environment will better enable regulation of synthetic biology to manage the biosecurity risks posed.

---

200 Jonathan B Wiener, 'The Rhetoric of Precaution' in Jonathan B Wiener et al (eds), *The Reality of Precaution: Comparing Risk Regulation in the United States and Europe* (Routledge, 2011) 3, 20.

PART III

Regulating for the benefits

8. How will patents affect synthetic biology?[1]

8.1 INTRODUCTION

In synthetic biology, Venter's Synthetic Genomics Inc. and other commercial entities are pursuing patent rights. Following the synthetic cell announcement, Venter claimed 'this is clearly the first life form totally developed out of a computer and by humans, so it is much closer to a human invention'.[2] This pursuit of patent protection is in contrast to the open, sharing approach of the BioBricks community.[3]

Commentators have queried how synthetic biology will interact with the patent system.[4] The anti-technology ETC Group has strongly opposed some synthetic biology patent applications, arguing that they could create monopolies on fundamental technology in synthetic biology.[5]

---

[1] The findings in this chapter are also presented in A McLennan, 'Patent Law and the Emerging Science of Synthetic Biology: An Examination of Principle and Practice' (2017) **36**(2) *Biotechnology Law Report* 59; A McLennan, 'Patent Law and the Emerging Science of Synthetic Biology: An Examination of Principle and Practice, Part 2' (2017) **36**(3) *Biotechnology Law Report* 105.

[2] Prepared Statement of J Craig Venter, 'Evidence to United States House of Representatives Committee on Energy and Commerce, Effects of Developments in Synthetic Genomics and Implications for Energy and Health' (27 May 2010) Washington DC.

[3] Henk van den Belt, 'Synthetic Biology, Patenting, Health and Global Justice' (2012) **7**(3) *Systems and Synthetic Biology* 1, 1.

[4] For example, Rutz raised the question 'is synthetic biology really so different from classical biotechnology that further provisions governing patentability are justified?' in Berthold Rutz, 'Synthetic Biology and Patents: A European Perspective' (2009) **10**(Special Issue) *EMBO reports* S14, S14.

[5] ETC Group, *Extreme Monopoly: Venter's Team Makes Vast Patent Grab on Synthetic Genomes* (8 December 2007) <http://www.etcgroup.org/en/node/663>; ETC Group, *Backgrounder: J. Craig Venter Institute's Patent Application on World's First Human-Made Species* (2007) ETC Group <http://www.etcgroup.org/sites/www.etcgroup.org/files/publication/pdf_file2/etc_synthbckgrnder_final2.pdf>.

This chapter considers patenting in synthetic biology with a view to answering the question of whether novel legal issues arise as compared to previous biotechnologies, and whether law reform or new policies are required to manage synthetic biology patents. One of the key roles of regulation in relation to emerging technologies is to promote desirable innovation.[6] The goal of the regulatory intervention represented by the patent system is to incentivise innovation. If the patent system instead operates so as to hinder research and innovation, as has been argued in some areas of the life sciences, this would be a problematic unintended consequence.[7] This chapter assesses whether the patent system is likely to operate in such an unintended, inhibitory fashion in relation to synthetic biology research.

One of the major objections raised to patents in the life sciences is that these do not fall within the scope of patentable subject matter. Hence this chapter examines the application of case law on patentable subject matter in *Bilski et al v Kappos* (*Bilski*); *Mayo Collaborative Services Inc. v Prometheus Inc.* (*Mayo*); *Alice Corporation v CLS Bank International* (*Alice*); and *Association for Molecular Pathology v United States Patent and Trademark Office (USPTO)* (*Myriad*) to synthetic biology.[8] The likely patentability of various tools and techniques is considered in order to illuminate the potential impact of patents on synthetic biology research. I argue that the key processes underpinning synthetic biology will meet the tests for patentable subject matter under the recent case law. The patentability of synthetic DNA constructs and standard biological parts is more complex, but these tools are likely to meet the tests set out in these cases. There is also potential for patents on genetic sequences to impact upon synthetic biology research, and the patentability of these is

---

[6] See Roger Brownsword and Han Somsen, 'Law, Innovation and Technology: Before We Fast Forward – A Forum for Debate' (2009) **1** *Law, Innovation and Technology* 1, 2.

[7] See Roger Brownsword and Morag Goodwin, *Law and the Technologies of the Twenty-First Century: Text and Materials* (Cambridge University Press, 2012), 290–95.

[8] *Bilski v Kappos*, 130 US (S Ct, 2010) 3218 (*Bilski*); *Mayo Collaborative Services Inc. v Prometheus Laboratories Inc.*, 132 US (S Ct, 2012) 1289 (*Mayo*); *Association for Molecular Pathology v United States Patent and Trademark Office*, 702 F Supp 2d (SDNY, 2010) 181 (*Myriad 1*); *Association for Molecular Pathology v United States Patent and Trademark Office*, 653 F 3d (CA Fed (NY), 2011) 1329 (*Myriad 2*); *Association for Molecular Pathology v United States Patent and Trademark Office*, 689 F 3d (CA Fed (NY), 2012) 1303 (*Myriad 3*); *Association for Molecular Pathology, et al v Myriad Genetics Inc., et al*, 133 US (S Ct, 2013) 2107 (*Myriad*).

discussed. I argue that it is important for the development of synthetic biology that standard biological parts should be accessible. This access to basic tools will promote research and development in the field.

The chapter then presents a case study of the examination practices of the USPTO in relation to patent families filed by scientists at the Venter Institute. This section analyses whether the patent system is operating to deal adequately with applications in synthetic biology. I argue that the case study shows that the patent system is working to narrow down the scope of broad claims and avoid granting patents for technology too early in its development.

8.2 POLICY DEBATES OVER PATENTING IN THE LIFE SCIENCES

Patent law in the US has been controversial in its application to life sciences. Various commentators have expressed concern about the potential impact of patenting in biotechnology and biomedical research and this debate is ongoing. The patenting of tools and applications has caused a furore in relation to gene patents, pharmaceutical drugs, genetically modified crops and embryonic stem cell research.[9] The debates have centred on issues of access to research tools and research areas, access to and costs of health care, and ownership of living things. There have been debates about whether the technological neutrality of the patent system can be maintained in light of emerging biotechnologies, and whether *sui generis* rules are required.

There are two major concerns about patenting in the life sciences which run through this debate. Some of these controversies have centred on the interpretation and application of the criteria for a valid patent, such as the debate over whether genes should constitute patentable subject matter. Other controversies have centred on how patent holders use their rights, for example the granting of exclusive licenses. Some patent holders have engaged in licensing or enforcement behaviour that has earned them the nickname 'patent trolls'. A patent troll is an inventor

---

[9] See, e.g., Matthew Rimmer, *Intellectual Property and Biotechnology: Biological Inventions* (Edward Elgar, 2008); Matthew Rimmer and Alison McLennan (eds), *Intellectual Property and Emerging Technologies: The New Biology* (Edward Elgar, 2012).

(or patent holder) who is not using or commercialising their invention.[10] In particular, it refers to a patent holder who is not intending to use the invention, but has acquired the patent in order to hold those wishing to use it to ransom by charging high fees.[11] In addition, there has been concern that changes in innovation policy in the US have altered academic scientific culture so that there are now greater commercial incentives associated with university research, meaning that scientists have become less likely to share their research with each other.[12]

8.2.1 Broad Patents on Foundational Technologies

In literature regarding the impact of patents on biomedical research, it has been argued that patents can become problematic if they are broad and granted over foundational or enabling technologies and tools, which would be needed in a broad range of projects for future innovation and development in a field.[13] Such patents could create a situation in which one entity could control or direct scientific progress in that field, possibly limiting progress.[14] There is also evidence from a number of different fields of technology showing the hindering impact of very broad, foundational patents.[15]

In some cases, it is argued that such a broad patent over a foundational tool or technology should not have been granted because it does not fall within the scope of patentable subject matter.[16] Under section 101 of the US Patent Act: 'Whoever invents or discovers any new and useful

---

[10] Richard A Epstein and F Scott Kieff, 'Questioning the Frequency and Wisdom of Compulsory Licensing for Pharmaceutical Patents' (2011) **78** *University of Chicago Law Review* 71, 89.

[11] See Joachim Henkel and Stephen M Maurer, 'Parts, Property and Sharing' (2009) **27**(12) *Nature Biotechnology* 1095, 1096.

[12] See Timothy Caulfield et al, 'Evidence and Anecdotes: An Analysis of Human Gene Patenting Controversies' (2006) **24**(9) *Nature Biotechnology* 1091, 1091; Arti K Rai, 'Regulating Scientific Research: Intellectual Property Rights and the Norms of Science' (1999) **94**(1) *Northwestern University Law Review* 77.

[13] Arti K Rai and Rebecca S Eisenberg, 'Bayh-Dole Reform And The Progress Of Biomedicine' (2003) **66**(Winter/Spring) *Law and Contemporary Problems* 289, 296; Arti Rai and James Boyle, 'Synthetic Biology: Caught Between Property Rights, the Public Domain and the Commons' (2007) **5**(3) *PLoS Biology* 0389, 0390.

[14] Rai and Eisenberg (n 13) 301.

[15] Rai and Boyle (n 13) 0390.

[16] See, e.g., *Myriad* (n 8).

process, machine, manufacture, or composition of matter, or any new and useful improvement thereof, may obtain a patent therefor, subject to the conditions and requirements of this title.'[17] While section 101 is interpreted broadly, products of nature, laws of nature, physical phenomena and abstract ideas are not patentable.[18] The reasons for limiting the scope of patentable subject matter were discussed in the US Supreme Court's recent decision in *Mayo Collaborative Services Inc. v Prometheus Inc.*, Breyer J held that: 'Phenomena of nature … mental processes, and abstract intellectual concepts are not patentable, as they are the *basic tools of scientific and technological work*.'[19] He pointed out that monopolisation of such basic tools could impede rather than promote innovation.[20] Further, the 'basic underlying concern' that 'patent law not inhibit further discovery by improperly tying up the future use of laws of nature' had been repeatedly stated by the Supreme Court.[21] This danger was seen as particularly acute where the claim was merely an 'instruction to "apply the natural law," or otherwise foreclosed more future invention than the underlying discovery could reasonably justify'.[22]

It has been argued for more than a decade that patents on human genetic sequences are foundational patents which are likely to have a damaging impact on genetics research, though there is debate about their actual impact. In the *Myriad* litigation, it was argued that sequences of human DNA fall within the judicial exclusion of products of nature and therefore are not patentable subject matter.[23] The Association for Molecular Pathology (AMP) and other plaintiffs had brought a challenge against various patents held by Myriad Genetics Inc. (Myriad) relating to the breast and ovarian cancer genes BRCA1 and BRCA2. The presence of certain mutations in these genes increases a woman's average chance of developing breast and ovarian cancers.[24]

Myriad held patents relating to these genes and methods of genetic testing. It had issued cease and desist letters to other providers of genetic tests and became the sole provider in the US. The plaintiffs claimed that

---

17 35 USC (Office of the Law Revision Counsel, 2016) § 101.

18 *Myriad 2* (n 8) 1348, quoting *Diamond v Chakrabarty*, 447 US (S Ct, 1980) 303.

19 *Mayo* (n 8) 1293, 1301, quoting *Gottschalk v Benson*, 409 US (S Ct, 1972) 63, 67.

20 *Mayo* (n 8) 1301.

21 Ibid.

22 Ibid.

23 See *Myriad 1* (n 8), 181.

24 *Myriad 2* (n 8) 1339.

Myriad's patents on isolated sequences of DNA, stretches of 'complementary DNA' (cDNA) and methods of comparing and analysing genetic sequences were invalid.[25]

The plaintiffs claimed that these patents impeded research and data sharing in the scientific field.[26] Further, they claimed that the patents compromised patient care, and hindered improvements to testing methods for these cancers. Patients could not seek a second test result to confirm that provided by Myriad. The stakes in this litigation were thus very high for patients, researchers and the biotechnology industry.

Arguments about the problematic nature of broad patents have also been made on the basis that such patents are sometimes granted where the scope of the claims is overly broad, and not adequately supported by the disclosure of information in the patent specification. For example, in the *Myriad* litigation in the Federal Circuit, Justice Bryson discussed patent breadth in his partial dissent. He considered the scope of the claims and the actual contribution of the inventor relative to the claims and their impact.[27] He noted that 'substantial adverse effects on research and treatment' were likely to occur if the claims were upheld.[28] The claims for short stretches of DNA and cDNA were framed so broadly that they covered sequences found in BRCA genes as well as in many other genes. Bryson J held that Myriad had not contributed to the understanding of other genes. Such broad claims could impede innovation in genetics.

For these reasons, rigorous examination of patent applications by the USPTO to make sure they meet these tests is important to avoid impeding innovation. The USPTO's examination of patents in synthetic biology is considered in the case study in this chapter.

8.2.2 Patent Thickets and the Anticommons

In terms of patent impact, Heller and Eisenberg warned of a potential 'tragedy of the anticommons' in which '[a] proliferation of intellectual property rights upstream may be stifling life-saving innovations further downstream in the course of research and product development'.[29] In this

---

25 *Myriad 1* (n 8) 185.

26 Ibid., 209–11.

27 *Myriad 2* (n 8) 1373.

28 Ibid., 1374.

29 Michael A Heller and Rebecca S Eisenberg, 'Can Patents Deter Innovation? The Anticommons in Biomedical Research' (1998) **280**(5364) *Science* 698, 699.

theory, an anticommons may develop where there is fragmentation of proprietary rights in a field, so that there are too many patents of narrow scope.[30] This can make it costly to negotiate all the licenses needed to create freedom to operate in a particular area, and there is potential for a particular patent holder to 'hold up' those who wish to operate in a field where access to that patent is needed.[31] A 'crowded' patent environment like this is often described as a 'patent thicket'.[32] The anticommons argument has been raised as a concern in relation to patents on DNA sequences, as there are many patented sequences within the human genome.[33]

However, there is also scholarship criticizing the theory of anticommons development in biomedical research and suggesting that it is not borne out by empirical research. For example, studies in various jurisdictions have shown that while there are many patents on genetic sequences, anticommons problems are not often observed.[34] Some studies have shown that genetics researchers use strategies such as inventing around a patent or ignoring patents and infringing them.[35]

Nonetheless, the controversial and high-stakes litigation over the breast cancer gene patents and the concern of scientists to use open access and contractual sharing mechanisms in synthetic biology demonstrate that there is ongoing unease about the impact of biotechnology patents and access to patented biotechnology.

---

30 Ibid.; see also Rai and Boyle (n 13).

31 Ibid., 699; Rai and Boyle (n 13) 0390.

32 See e.g., Rai and Boyle, ibid.

33 Caulfield et al (n 12) 1091; Naomi Hawkins, 'The Impact of Human Gene Patents on Genetic Testing in the United Kingdom' (2011) **13**(4) *Genetics in Medicine* 320, 320.

34 Caulfield et al (n 12) 1092; Hawkins ibid; John P Walsh et al, 'View from the Bench: Patents and Material Transfers' (2005) **309**(5743) *Science* 2002. However, it has been suggested that there is a need for more research, and policy attention to this area: see, E Richard Gold et al, 'Gene Patents – More Evidence Needed, but Policymakers Must Act' (2007) **25**(4) *Nature Biotechnology* 388. For an Australian empirical study see Dianne Nicol and Jane Nielsen, 'Patents and Medical Biotechnology: An Empirical Analysis of Issues Facing the Australian Industry' (Centre for Law and Genetics Occasional Paper No. 6, Centre for Law and Genetics, 2013) <http://www.ipria.org/publications/reports/BiotechReportFinal.pdf>.

35 Caulfield et al (n 12) 1093; Hawkins (n 33) 323; Walsh et al (n 34) 2002.

8.3 SYNTHETIC BIOLOGY PATENT ENVIRONMENT AND POTENTIAL ISSUES

8.3.1 Patent Environment

In the US, patent applications for synthetic biology have fallen into two broad categories: (1) biological tools, methods and products; and (2) computer programs.[36] The computer program category includes software for design of biological devices and genetic circuits,[37] use in DNA assembly,[38] design of synthetic nucleic acids[39] and analysis of biochemical activity within cells.[40]

This chapter focuses on the biological category. In that category, some patent applications have focused on the construction of a synthetic cell. Scientists at the Venter Institute have filed applications for patents on technology including a minimal bacterial genome, DNA assembly, methods for cloning and manipulating genomes, methods of installing a genome into a cell and methods for producing a synthetic cell.[41] There are also patents for methods of DNA synthesis.[42]

Other patent applications in the US have involved the creation of useful biological products from cells, such as Jay D Keasling and

---

[36] See also Alison McLennan and Matthew Rimmer, *Inventing Life: Patent Law and Synthetic Biology* (2012) The Conversation <http://theconversation.com/inventing-life-patent-law-and-synthetic-biology-5178>.

[37] For example, Chunbo Lou, et al (2012) 'Synthetic Biology Tools', United States Patent Application No 13/489205; Andrew Nicholas, John Brojer Phillips and Michael David Pedersen (2009) 'Creating Genetic Devices', US Patent Application No 12/552829.

[38] For example, Nathan J Hillson (2012) 'Scar-Less Multi-Part DNA Assembly Design Automation', US Patent Application No 13/364285.

[39] For example, Mark Welch and Claes Gustafsson (2012) 'Design of Synthetic Nucleic Acids for Expression of Encoded Proteins', US Patent No 8,126,653.

[40] For example, Ricardo E Paxson and Joseph F Hicklin (2010) 'Method and Apparatus for Integrated Modeling, Simulation and Analysis of Chemical and Biological Systems Having a Sequence of Reactions, Each Simulated at a Reaction Time Determined Based on Reaction Kinetics', US Patent No 7,769,576.

[41] For example, Daniel G Gibson, et al (2008) 'Assembly of Large Nucleic Acids', US Patent Application No 12/247126.

[42] For example, Lei Young (2010) 'Synthesis of Error-minimized Nucleic Acid Molecules', United States Patent No 7,704,690; Peter J Belshaw, et al (2011) 'Method for the Synthesis of DNA Sequences', US Patent No 8,008,005.

colleagues' production of artemisinic acid – an anti-malarial drug precursor – in a genetically modified cell.[43] The spin-out company Amyris now has a granted patent for Production of Isoprenoids through biosynthetic pathways using engineered bacteria, including artemisinic acid.[44] This same platform technology is being used in biofuels development.

There are now many patents for biofuels. For example, Synthetic Genomics has patents on methylbutanol and dimethyloctane as advanced biofuels.[45] Synthetic Genomics also has a patent for 'Production of Branched-Chain Alcohols by Photosynthetic Microorganisms'.[46] Amyris has patents on jet fuel and methods for making it, as well as molecules and fuel compositions of various kinds and methods for making them.[47]

There has already been some patent infringement litigation in synthetic biology in the areas of DNA synthesis and design technologies. In 2007, Codon Devices, Duke University and MIT sued Blue Heron Biotechnology, claiming it had infringed five patents related to DNA synthesis. The dispute was later settled. In 2007, Febit Biotech sued Codon Devices, claiming infringement of a patent on a method of producing synthetic nucleic acids. In 2012, DNA synthesis company DNA 2.0 sued Genome Compiler Corporation for infringement of its genetic sequence design software.

8.3.2 Potential Patent Problems

It has been suggested that poor quality 'bad' patents could impede innovation and that 'Overly restrictive licensing and smotheringly broad patent interpretations could make a shambles of synthetic biology'.[48] Intellectual property experts James Boyle and Arti K Rai have raised

---

[43] See Amyris, *Breakthrough Science* (2013) Amyris <http://198.55.195.109/Innovation/155/BreakthroughScience>.

[44] Neil Stephen Renninger, et al (2010) 'Production of Isoprenoids', US Patent No 7,659,097.

[45] Kevin V Martin, et al (2012) 'Dimethyloctane as an Advanced Biofuel', US Patent No 8,309,323; Stephen Picataggio, et al (2012) 'Methylbutanol as an Advanced Biofuel', US Patent No 8,114,641.

[46] Paul Gordon Roessler, et al (2012) 'Production of Branched-Chain Alcohols by Photosynthetic Microorganisms', US Patent No 8,124,400.

[47] See, e.g, Neil Stephen Renninger, et al (2010) 'Fuel Compositions Comprising Farnesane and Farnesane Derivatives and Method of Making and Using Same', US Patent No 7,846,222.

[48] The Editors, 'How to Kill Synthetic Biology' (2006) **294**(6) *Scientific American* 4.

concern about the potential for both broad foundational patents and patent thickets to impede innovation in synthetic biology.[49] In her evidence to the Commission, Rai argued that there should not be an attempt to eliminate patenting in synthetic biology, as patents provide important incentives.[50] However, she highlighted two important policy areas to consider: quality of patents and licensing behaviour of patent holders. A 'high quality' patent is one that properly meets the statutory requirements for patentability.[51]

Because synthetic biology is actually comprised of several approaches to biological engineering, several different patent problems could theoretically arise. Broad patents are a potential concern in synthetic biology.[52] At this stage of synthetic biology's development there is potential for broad patents over foundational technologies and processes of synthetic biology to cause problems for innovation in the field. Broad patents on foundational technologies could result in fewer end-products being developed in synthetic biology. This could occur if, for example, a broad patent was granted over the minimal genome developed at the Venter Institute. This patent application had the potential to cover a foundational technology.[53]

Potentially, a patent thicket could arise if many patents are granted on the basic building blocks used in synthetic biology, standard biological parts. Potential synthetic biology products have been likened to complex

---

[49] Rai and Boyle (n 13); see also Sapna Kumar and Arti Rai, 'Synthetic Biology: The Intellectual Property Puzzle' (2007) **85** *Texas Law Review* 1745; James Boyle, *The Public Domain: Enclosing the Commons of the Mind* (Yale University Press, 2008) 174–7.

[50] Arti K Rai, 'Evidence to The Presidential Commission for the Study of Bioethical Issues, Synthetic Biology' (13 September 2010) Philadelphia Penn.

[51] See Arti Rai et al, 'Patent Reform: Unleashing Innovation, Promoting Economic Growth and Producing High-Paying Jobs – A White Paper from the US Department of Commerce' (US Department of Commerce, 13 April 2010) <https://www.esa.doc.gov/sites/default/files/reports/documents/patentreform_0.pdf>, 2–6.

[52] Rai, 'Evidence to The Presidential Commission for the Study of Bioethical Issues, Synthetic Biology' (n 50).

[53] Alain Pottage, 'Protocell Patents: Property Between Modularity and Emergence' in Mark A Bedau and Emily C Parke (eds), *The Ethics of Protocells: Moral and Social Implications of Creating Life in the Laboratory* (MIT Press, 2009) 165, 173–5.

technologies which contain many components, such as those in electronics and information technology.[54] An analogy can be drawn with electrical circuits, as synthetic biologists are attempting to build 'genetic circuits' comprised of units of DNA.[55]

Also, a human-designed synthetic cell would be a complex entity which might make use of a number of components that could be subject to patent protection themselves.[56] This complex nature means that it is unlikely one company would hold all the intellectual property rights necessary for a particular research project in synthetic biology.[57] Negotiating many licenses could become costly and there is high potential for unintentional patent infringement.[58] This complexity has led some commentators to stress that the introduction and fostering of openness and sharing in synthetic biology is critical.[59]

Patents on genetic sequences also have the potential to impact upon research in synthetic biology. Many standard biological parts have been placed in the public domain and are being shared by synthetic biologists via the Registry of Standard Biological Parts. However, Rai gave evidence to the Commission's inquiry that there were (at that time) many patents on genetic sequence parts in the Registry that could potentially touch on many existing patents.[60] She explained that patent thickets could therefore become problematic in synthetic biology, similar to problems which occurred in the information technology industry. Accordingly, more recent case law regarding patentability of DNA sequences is an important issue for synthetic biology research and is analysed below.

---

54 Henkel and Maurer 'Parts, Property and Sharing' (n 11) 1095.

55 Christina Smolke, 'Evidence to The United States National Academies, The Organisation for Economic Co-operation and Development, and The Royal Society, Opportunities and Challenges in the Emerging Field of Synthetic Biology' (9 July 2009) Washington DC, 142.

56 Jennifer Kuzma and Todd Tanji, 'Unpackaging Synthetic Biology: Identification of Oversight Policy Problems and Options' (2010) **4**(1) *Regulation & Governance* 92, 99–100.

57 Henkel and Maurer 'Parts, Property and Sharing' (n 11) 1095.

58 Ibid., 1095–6.

59 Ibid., 1098; Stephen M Maurer, 'Before it's too Late: Why Synthetic Biologists Need an Open-Parts Collaboration – and How to Build One' (2009) **10**(8) *EMBO Reports* 806.

60 Rai, 'Evidence to The Presidential Commission for the Study of Bioethical Issues, Synthetic Biology' (n 50).

8.4 SYNTHETIC BIOLOGY AND PATENTABLE SUBJECT MATTER

Synthetic biology involves multiple types and fields of technology. Therefore, applying patent case law to synthetic biology is more complex than in some preceding areas of technology, such as genetic testing or embryonic stem cell research. This section breaks these down into synthetic biology processes, synthetic DNA and standard biological parts. The patentability of DNA sequences is also important in the synthetic biology context.

8.4.1 Processes

The patentability of processes has been discussed in the decisions of *Bilski*; *Mayo*; *Alice*; and *Myriad*. In *Bilski*, the Supreme Court had to decide whether the claims were for a patentable process or just an abstract idea. The disputed patent was for a method of hedging against risk when buying and selling commodities in the energy market. The claims included a method and a reduction of this to a formula. There were also claims to the use of this formula in the energy market.

In the decision below, the Court of Appeal had held that the test to determine 'whether a process claim is tailored narrowly enough to encompass only a particular application of a fundamental principle rather than pre-empt the principle itself' is whether '(1) it is tied to a particular machine or apparatus, or (2) it transforms a particular article into a different state or thing'.[61]

On appeal, the Supreme Court held that this 'machine or transformation test' was a 'useful and important clue, an investigative tool', but 'is not the sole test for deciding whether an invention is a patent-eligible process'.[62] Further, under *Gottschalk v Benson* 'transformation and reduction of an article "to a different state or thing"' is a clue where the claimed method does not involve a machine.[63] This decision was subsequently applied in the decisions in *Mayo* and *Myriad*.

In *Mayo*, the disputed patents (licensed to Prometheus) were for methods of using the concentration of certain metabolites of a drug in the blood of patients to determine whether the dosage should be adjusted.[64]

---

61 *In re Bilski*, 545 F 3d (Fed Cir, 2008) 943, 954.

62 *Bilski* (n 8) 3227.

63 *Gottschalk v Benson* (n 19) 70, quoted in *Bilski* (n 8) at 3227.

64 *Mayo* (n 8) 1294–5.

The metabolism of the drug in the body was known to vary between patients. The level of metabolites was an indicator of whether the drug dosage was too low, when the drug could be ineffective, or too high, causing harmful effects. The patents claimed a method of optimising drug dosage using these correlations. Justice Breyer delivered the Supreme Court's opinion and held that precedents require that 'a process that focuses upon the use of a natural law [must] also contain other elements or a combination of elements, sometimes referred to as an "inventive concept," sufficient to ensure that the patent in practice amounts to significantly more than a patent upon the natural law itself'.[65]

An *application* of a law of nature may be patentable, and so the question is 'do the patent claims add *enough* to their statements of the correlations [that is, the laws of nature] to allow the processes they describe to qualify as patent-eligible processes that *apply* natural laws?'[66] Merely adding some 'well understood, routine, conventional activity' to the process would not generally be sufficient to meet the test.[67] Further, just limiting an idea to one field of use was not sufficient to make it patentable.

In relation to the disputed patents in *Mayo*, Breyer J held that the additional step of 'administering' the drug merely indicated the audience of the invention (doctors administering this class of drug) and the step of 'determining' the metabolite levels comprised 'well understood, routine, conventional activity' which did not add to the process significantly.[68] Together, the claims amounted to telling doctors to 'apply the law', and such patents had been held to be invalid in the past.[69] Thus the claims did not sufficiently add to the law of nature to make the claimed process patent-eligible.[70]

Therefore under *Mayo*, process claims in synthetic biology which utilise processes of nature such as genetic recombination within a cell would have to add 'enough' to those natural processes that they could be described as *applying* the processes.

In 2014, the Supreme Court affirmed and applied the decision in *Mayo* in *Alice*. The patent claims in that case concerned a 'computer implemented scheme for mitigating "settlement risk" (that is, the risk that only one party to a financial transaction will pay what it owes) by using a

---

65 Ibid., 1294.
66 Ibid., 1297, 1298–9.
67 Ibid., 1298.
68 Ibid.
69 Ibid., 1299–301.
70 Ibid., 1299–302.

third-party intermediary'.[71] The court viewed the decision in *Mayo* as laying out a two-step test for 'distinguishing patents that claim laws of nature, natural phenomena and abstract ideas from those that claim patent-eligible applications of those concepts'.[72] First, the court considers whether the claims encompass one of these patent-ineligible concepts. If they do, then the court considers what has been added to the patent-ineligible concept, and whether this is sufficient to transform it into an eligible application.

In *Alice*, the court found that the claims were directed to an abstract idea. Similar to the 'risk hedging' claimed in *Bilski*, the use of 'intermediated settlement' was a longstanding economic practice. This could not be transformed into a patent-eligible application by adding the use of a 'generic computer' to implement the abstract idea. The claims for a method, a computer system and a computer-readable medium all failed for these reasons.

In *Myriad*, the plaintiffs claimed that Myriad's patents on methods of comparing DNA sequences were invalid as they claimed the abstract idea of comparing gene sequences and pre-empted a natural phenomenon, the correlation between certain genetic mutations and pre-disposition to cancer.[73] Lourie J, with whom Bryson J and Moore J concurred on this issue, found that the claims were invalid as they claimed an abstract mental process. They could not be made valid by limiting the claim to breast cancer genes or certain mutations.

The plaintiffs also claimed that Myriad's patents for a method of establishing the effectiveness of potential cancer drugs were not eligible for patent protection.[74] Lourie J held that the method of screening potential cancer drugs was patent-eligible under the 'machine or transformation' test. The steps of 'growing' cells and 'determining' growth rate were transformative. Further, the claim was not so abstract as to claim a scientific principle; it was tied to a type of cell, a type of therapeutic compound and a specific method of determining growth rate. Thus where methods involve steps which use or create novel, 'transformed' subject matter, this indicates that the method is patent-eligible.

In 2012, the Supreme Court ordered the Court of Appeal for the Federal Circuit to reconsider its *Myriad* decision in light of the Supreme Court's decision in *Mayo*. The Court of Appeal came to the same result

---

71 *Alice* (n 8) 2351–2.

72 Ibid., 2355.

73 *Myriad 2* (n 8) 1355.

74 Ibid., 1334.

as above, for essentially the same reasons.[75] Applying *Mayo*, Lourie J held that the claims for a method of screening drugs included steps of growing 'transformed', man-made cells with altered BRCA genes in the presence and absence of the drugs. These transformed cells had 'enhanced function and utility'.[76] The process was patent-eligible: 'performing operations, even known types of steps, on, or to create, novel, i.e., transformed subject matter is the stuff of which most process or method invention consists'.[77] Further, the claims were restricted to certain cell types transformed with certain genes and the impact on growth of a specific drug type was to be tested.

Methods such as generating a synthetic genome by assembling smaller pieces of DNA or standard components, or producing a synthetic cell by installing such a synthetic genome, are likely to be patentable subject matter under these tests. While they *utilise* natural processes such as genetic recombination within a cell, those processes of nature are only part of the methods claimed. These methods go beyond the processes of nature to transform existing matter into something which does not exist in nature. There are various additional steps such as treating DNA and cells with different enzymes and designing DNA sequences that can be joined together. These methods also go beyond adding routine, conventional, well understood activity to a process of nature because they transform the existing matter into a new product with different utility. These methods are thus distinguishable from the law of nature claimed in *Mayo*, where nothing was added other than the instruction to apply the law. They are more analogous to the transformed cells which were found to be patentable subject matter in *Myriad.* Thus the processes would meet the tests in *Mayo* and *Myriad* (Federal Circuit).[78]

Synthetic biology processes are also likely to meet the machine or transformation test, because they would have transformative steps. For example, a method such as installing a genome into a cell involves steps such as preparing an isolated genome or synthesising a genome, preparing a recipient cell and installing the genome.[79] These would

---

[75] See *Myriad 3* (n 8).

[76] See ibid., 1336.

[77] Ibid.

[78] In *Myriad*, the decision on Myriad's process patents was not appealed to the Supreme Court, thus the Supreme Court only decided an appeal on the composition of matter patents.

[79] See, e.g., John I Glass, et al (2006) 'Installation of Genomes or Partial Genomes into Cells or Cell-Like Systems', US Patent Application No 11/644,713; Daniel G Gibson et al, 'Creation of a Bacterial Cell Controlled by a

transform 'a particular article into a different state or thing' as required in *Bilski* and *Gottschalk* – the resulting cell with a new genome inside. Similarly, while DNA assembly processes utilise the natural action of cells and enzymes, they are transformative because a new nucleic acid molecule is generated.

It is important to note that if there were claims for synthetic biology processes which did involve natural phenomena, the cases show that these could not become patent-eligible merely by adding the use of a computer to perform or implement the process. Computer software for designing DNA cassettes for assembly or other processes would need to add something more to the natural phenomena than 'applying it' on a computer.

Patent-eligible processes in synthetic biology would also have to meet the other criteria for patent validity. Some of the basic processes for which patents have been sought by the Venter Institute discussed below have been rejected under other criteria for patent validity such as obviousness, lack of novelty, lack of enablement and failure to provide a sufficient written description.

8.4.2 Genetic Sequences

Patentability of DNA is an important issue in the context of BioBricks and other standard biological parts, because it affects the question of whether such parts could infringe existing genetic sequence patents.[80]

In *Myriad*, the plaintiffs objected to both the patents on isolated sequences of DNA and stretches of cDNA. cDNA is made in the laboratory for particular experimental purposes. Within the genomes of natural organisms, there are stretches of the genetic code that do not contain coding information for protein production. cDNA 'contains the same protein-coding information found in a segment of natural DNA but omits portions within the DNA segment that do not code for proteins'.[81]

In 2010, Sweet J of the District Court, Southern District of New York, decided that both the isolated DNA sequences and cDNA sequences were

---

Chemically Synthesized Genome' (2010) **329**(5987) *Science* 52; and Lartigue C et al, 'Creating Bacterial Strains from Genomes That Have Been Cloned and Engineered in Yeast' (2009) **325** *Science* 1693.

[80] Andrew W Torrance and Linda J Kahl, 'Synthetic Biology Standards and Intellectual Property' (The National Academies, 2012) <http://synberc.org/sites/default/files/Torrance%20&%20Kahl_Syn%20Bio%20Standards%20and%20IP_NAS%20Report%202012.pdf>, 19.

[81] *Myriad* (n 8) 2111.

not patentable subject matter.[82] The Court of Appeal for the Federal Circuit reversed these findings.[83] Lourie J for the majority noted that for compositions of matter, the relevant test is drawn from *Diamond v Chakrabarty* (*Chakrabarty*): 'whether human intervention has given "markedly different" or "distinctive" characteristics' from what is found in nature.[84] All three judges agreed that the cDNA molecules were patentable, as they have a different structure and genetic sequence from DNA found in nature.

The Supreme Court gave its decision in 2013. The applicant had asked the Supreme Court to decide the question: 'Are human genes patentable?'[85] Myriad maintained that the isolated DNA had a different structure from the natural genomic DNA. The US Government as *amicus curiae* argued that isolated DNA should not be patentable. However, it argued, cDNA should be considered patent-eligible, because it is 'an artificial creation in the laboratory'.[86] Also, patents on cDNA would not stop the use of the original gene by others for medical or scientific research.

Thomas J for the Supreme Court held that 'a naturally occurring DNA segment is a product of nature and not patent-eligible merely because it has been isolated, but that cDNA is patent-eligible because it is not naturally occurring'.[87] Thomas J concurred with the discussion in *Mayo* of the policy behind exceptions to patentability under section 101. He reiterated that courts have long held that section 101 contains important implicit exceptions: '"Laws of nature, natural phenomena, and abstract ideas are not patentable." Rather, "they are the basic tools of scientific and technological work" that lie beyond the domain of patent protection'.[88] Without this exception, patents could 'tie up' the use of those basic tools and inhibit innovation. This would be 'at odds with the very point of patents, which exist to promote creation'.[89] Again referring to *Mayo*, he held that 'patent protection is a delicate balance between

---

82 *Myriad 1* (n 8) 220.

83 *Myriad 2* (n 8) 1334.

84 Ibid., 1351, referring to *Chakrabarty* (n 19), *American Fruit Growers v Brogdex Co*, 283 US (S Ct, 1931) 1, and *Hartranft v Wiegmann*, 121 US (S Ct, 1887) 609.

85 Association for Molecular Pathology et al, (2012), 'Brief for Petitioners' in *Association for Molecular Pathology v United States Patent and Trademark Office*, appeal docketed No. 12-398 (US S Ct, heard 15 April 2013).

86 *Myriad* (n 8), transcript of oral argument, 25–6.

87 *Myriad* (n 8) 2111.

88 Ibid., 2116, referring to *Mayo* (n 8) (references omitted).

89 *Myriad* (n 8) 2116, referring to *Chakrabarty* (n 18).

creating “incentives that lead to creation, invention and discovery” and “imped[ing] the flow of information that might permit, indeed spur invention”’.[90]

Thomas J held that Myriad’s ‘principal contribution was uncovering the precise location and sequence’ of the breast cancer genes. Distinguishing this case from *Chakrabarty*, Thomas J held that: ‘Myriad did not create anything ... it found an important and useful gene, but separating that gene from its surrounding genetic material is not an act of invention ... Groundbreaking, innovative, or even brilliant discovery does not by itself satisfy the § 101 inquiry.’[91]

Although the location of the genes was previously unknown and required extensive research to uncover, ‘extensive effort alone is insufficient to satisfy the demands of § 101’.[92]

Thomas J rejected the argument that the severing of chemical bonds to isolate the DNA created a non-naturally occurring molecule. The patents relied on the genetic information encoded in the genes, not the chemical structure of the molecule. Unlike the Court of Appeals, the Supreme Court was not deferential to the existing USPTO practice. Thomas J held that Congress had not specifically endorsed the USPTO’s practice in legislation.

By contrast, it was held that ‘cDNA does not present the same obstacles to patentability as naturally occurring, isolated DNA segments’ because it is a molecule that is not naturally occurring ‘the lab technician unquestionably creates something new when cDNA is made. cDNA retains the naturally occurring exons of DNA, but it is distinct from the DNA from which it was derived. As a result, cDNA is not a “product of nature” and is patent-eligible under § 101.’[93]

Thus the Supreme Court in *Myriad* has altered the approach to patentability of isolated DNA molecules with naturally occurring sequences, and held these to be beyond the scope of patentable subject matter. Prior to the Supreme Court’s decision in *Myriad*, use of standard biological parts might have infringed patents where the sequence of a part encompassed a sequence that is within a patented stretch of DNA.[94] Thus under the Court of Appeal’s decisions in *Myriad* and the existing USPTO practice, users of standard biological parts could have been

---

[90] Ibid., 2116.
[91] Ibid., 2117.
[92] Ibid., 2118.
[93] Ibid., 2119.
[94] Torrance and Kahl (n 80) 19.

infringing patents. This potential access problem had been noted in the synthetic biology context.[95]

However, the new approach to isolated DNA under *Myriad* means that although standard biological parts may cover sequences of patented DNA, those patents can no longer be enforced. The infringement potential has now been negated under *Myriad*. Therefore, a thicket of DNA patents is unlikely to hinder synthetic biology research.

8.4.3 Synthetic DNA

The patentability of synthetic DNA is important because in future DNA synthesis is likely to increasingly become the method used for producing human-designed DNA constructs such as standard biological parts. The costs of DNA synthesis are decreasing and the efficiency is improving.[96]

In analysing the likely patentability of synthetic DNA and standard biological parts, it is necessary to consider how much they differ from naturally occurring DNA. DNA is a unique type of molecule which has a physical structure and form, but also carries encoded information in the genetic sequence of the DNA. There is an important distinction between the physical instantiation (the tangible form) of a stretch of DNA and its informational origin. The physical instantiation refers to how the actual physical DNA has been created. For example, it could have been created by using the polymerase chain reaction to copy an existing piece of natural DNA. More recently it could have been synthesised in a laboratory using the four chemical base building blocks of DNA.

The piece of DNA will also have an informational origin. This refers to the origin of the information encoded by the DNA, its genetic sequence. The sequence could have its origin in a natural organism, for example a sequence which represents a gene from the bacteria *E. coli*. Scientists are now able to design genetic sequences using computers, and then build the designed sequence. Such DNA has its informational origin in a computer, and its physical instantiation through gene synthesis. A particular piece of DNA built in a laboratory could have its informational origin in a natural organism or a computer.

Courts are beginning to contemplate DNA molecules with their physical instantiation being synthesis in a laboratory. In the Supreme Court opinion in *Myriad*, Thomas J held that in addition to extracting natural

---

95 Rai, 'Evidence to The Presidential Commission for the Study of Bioethical Issues, Synthetic Biology' (n 50).

96 Rob Carlson, *Biology is Technology: The Promise, Peril, and New Business of Engineering Life* (Harvard University Press, 2010), 66–7.

DNA from cells, 'It is also possible to create DNA synthetically through processes similarly well-known in the field of genetics', and gave the example of the method of cDNA production.[97] He described cDNA as 'synthetic DNA created in the laboratory'.[98] The methods of complete *de novo* synthesis of DNA from its chemical units used in synthetic biology were not considered in the decision. However, the Supreme Court decision focused on the genetic sequence information contained in DNA rather than its chemical structure. It also focused on the creation of a non-naturally occurring molecule. Thus it is likely that a piece of synthetic DNA with the same genetic sequence as a stretch of naturally occurring DNA would not be patent-eligible under this decision.

Nonetheless, if the sequence of the synthetic DNA created in the laboratory is different from any naturally occurring sequence, that molecule *may* be found patentable under *Myriad*. There would be two issues in relation to synthetic DNA, whether it has '"markedly different characteristics" from that found in nature' and whether it has 'the potential for significant utility'.[99] A claim which met these tests would be directed to patentable subject matter.[100]

It remains to be seen how a court would draw the line as to what degree of change to a naturally occurring informational sequence is enough to make a piece of synthetic DNA 'markedly different' under the test in *Chakrabarty*, applied in *Myriad*. The Supreme Court's comments in *Myriad* do not indicate how different a sequence must be from a naturally occurring sequence to be considered not a product of nature. Indeed the Supreme Court stated:

> Scientific alteration of the genetic code presents a different inquiry, and we express no opinion about the application of § 101 to such endeavors. We merely hold that genes and the information they encode are not patent-eligible under § 101 simply because they have been isolated from the surrounding genetic material.[101]

Some synthetic DNA constructs are very similar to naturally occurring genetic sequences. For example, the genome of the Venter Institute's 'synthetic cell' was almost identical to the naturally occurring genome sequence of the bacterium *M. mycoides*. The genome was assembled from stretches of synthetic DNA, but the sequence only differed from the

---

97 *Myriad* (n 8) 2112.
98 Ibid.
99 *Myriad 3* (n 8) 1340.
100 Ibid.
101 *Myriad* (n 8) 2120.

natural version in several places.[102] Precisely that genetic sequence cannot be found in nature; indeed the changes were made to distinguish the natural and synthetic versions of the cell.[103]

While *Mayo* involved patented processes, in *Myriad* Moore J and Bryson J (in dissent) held that the reasoning also applies to compositions of matter.[104] In *Mayo*, Breyer J held that the court's precedents required that a claimed process include sufficient additional elements – or 'inventive concept' – that the patent would not just be a patent over the law of nature. Applying *Mayo* to the disputed patents in *Myriad*, Bryson J (in dissent) considered whether there was 'an inventive concept that involves more than merely incidental changes to the naturally occurring product'.[105] Bryson J held that the isolated DNA at issue in *Myriad* did not meet these tests. Isolating the DNA did not add any inventive character. The 'critical aspect' was the genetic sequence, which remained the same.

Applying these decisions to synthetic DNA, under *Mayo* there would need to be sufficient additional elements beyond a naturally occurring sequence (a product of nature), or an inventive concept, such that the claims were not merely for the product of nature. If the synthetic DNA merely reproduced a naturally occurring genetic sequence, it is unlikely that this would be found to have 'inventive character'. If a non-naturally occurring synthetic DNA was claimed, this may meet the test if there was something inventive or additional in the alteration of the sequence that distinguished it from the naturally occurring sequence. For example, if the synthetic DNA had a particular new utility, this may establish a sufficiently inventive concept or contribution.

In relation to utility, in *Myriad* Moore J held that an invention should have an '"enlargement of the range of … utility" as compared to nature' or 'potential for significant utility'.[106] The cDNA met this test because while the molecules were 'inspired by nature', they had additional utility, for example in protein expression.[107] Though the coding regions that carry the genetic information were the same as in the natural DNA, the removal of the non-coding sections and the additional utility in protein

---

[102] Gibson et al (n 79).

[103] Ibid., 52.

[104] See, e.g., *Myriad 3* (n 8) 1340. However, Moore J held that the claims in *Chakrabarty* and *Funk* were more analogous to the claims at issue in *Myriad.*

[105] Ibid., 1355.

[106] *Myriad* 2 (n 8) 1359–60, referring to *Funk Bros Seed Co v Kalo Inoculant Co*, 333 US (S Ct, 1948) 127 and *Diamond* (n 17).

[107] *Myriad 2* (n 8) 1364.

production were enough to give cDNA a 'distinctive character and use, with markedly different chemical characteristics'.[108] Some of the claimed isolated DNA molecules also had an enlarged range of utility, for example the shorter sequences could be used as primers or probes. This meant they were patentable subject matter.

Moore J distinguished the claims for short sequences of DNA in *Myriad* from the un-patentable claims in *Mayo*, which merely set out a law of nature.[109] In *Myriad*, the short sequences could be used for genetic testing. Thus the claims were not directed to a relationship between the sequences of DNA and cancer, but to 'a tool that can be used to determine if that relationship exists'.[110] This tool had 'different and beneficial utility', rendering it patentable.[111]

Where the sequence of synthetic DNA does not match a naturally occurring sequence, it is likely to meet these tests requiring utility. Synthetic DNA would be designed with specific purposes in mind that cannot be accomplished with naturally occurring DNA.

A piece of synthetic DNA which reproduces naturally occurring DNA could also potentially foreclose innovation as it would effectively be a patent over a naturally occurring isolated genetic sequence. In *Mayo*, Breyer J considered the question 'how much future innovation is fore-closed relative to the contribution of the inventor?'[112] This was of concern to Bryson J (in dissent) in considering the patents for isolated naturally occurring human genetic sequences in *Myriad.* He was swayed by the potential for broad patents to foreclose more future innovation than is justified by the underlying invention.[113] He concluded that Myriad was not entitled to such broad protection. Thus it is unlikely that a synthetic stretch of DNA which is merely analogous to a naturally occurring genetic sequence would be found to be patentable subject matter, because it could have the same potential to foreclose innovation.

Overall, it is likely that synthetic, computer-designed DNA will meet the product of nature exclusion tests more easily than earlier bio-technologies, as the genetic sequence will be written by humans.[114] The

---

108 *Myriad 3* (n 8) (Moore J).

109 Ibid., 1341.

110 *Myriad 3* (n 8) 1342.

111 Ibid.

112 *Mayo* (n 8) 1303.

113 *Myriad 3* (n 8) 1356.

114 Andrew W Torrance, 'Synthesizing Law for Synthetic Biology' (2010) **11**(2) *Minnesota Journal of Law, Science and Technology* 629, 638–42.

resulting products will not occur in nature, but will be clearly produced by human intervention.[115]

The USPTO may be well placed to issue guidelines regarding synthetic DNA. Currently, the Manual of Patent Examining Procedure does not expressly refer to synthetic DNA or synthetic nucleic acids. In the absence of a court decision on synthetic DNA, the USPTO may need to consider the issue and put in place guidelines as to the patentability of synthetic DNA. While the USPTO does not have authority to make substantive rules, it does issue guidelines for examination and revise these from time to time, so it can take some role in policy development.[116]

8.4.4 Standard Biological Parts

Standard biological parts are key building blocks in the parts agenda approach to synthetic biology. Currently, many parts are constructed from sequences which have their informational origin in natural organisms. For example, the Registry of Standard Biological Parts contains promoters from a number of microorganisms. Under the Supreme Court decision in *Myriad*, standard biological parts are unlikely to be patentable where they just reproduce a naturally occurring sequence. However, parts are often engineered to have minor changes from the natural sequence. For example, to use the BioBricks Assembly Protocol, small changes would have to be made to the naturally occurring genetic sequence.[117] Under *Myriad*, this may be enough to make them patent-eligible. Standard biological parts have utility in construction of biological devices, so that they are likely to meet the requirement for enlarged utility over naturally occurring DNA. Thus they are analogous to the cDNA which was found to be patentable subject matter in *Myriad*.

In addition, composite parts or biological devices are being produced which combine several basic parts together to produce a biological device which can conduct a function within a cell. While the genetic sequences of the basic parts might be derived from naturally occurring sequences, the combination of these sequences could result in a stretch of DNA which is not identical to any naturally occurring genetic sequence. This

---

115 Ibid.

116 See Arti K Rai, 'Engaging Facts and Policy: A Multi-Institutional Approach to Patent System Reform' (2003) **103**(5) *Columbia Law Review* 1035, 1131–3.

117 Barry Canton et al, 'Refinement and Standardization of Synthetic Biological Parts and Devices' (2008) **26**(7) *Nature Biotechnology* 787, 788.

takes synthetic biology even further from naturally occurring phenomena, and it is unlikely that such a device would be regarded as a product of nature.

Due to the strong culture of openness and sharing in the synthetic biology community, it is unlikely that many within that community would seek patents on parts or devices. However, the field of synthetic biology encompasses bio-entrepreneurs and commercially funded ventures, as well as research scientists working within an academic setting and a culture of sharing. Therefore it is possible that parts will be patented. It is certainly possible that a patent troll could disrupt research in synthetic biology, including within the standard biological parts community, if it withheld access to a patented element needed for this research or sought to use reach-through licenses to control the commercialisation of that research.

Thus it is important that open sharing arrangements have been created under which scientists can volunteer to contribute these parts to a 'commons' and share them with other researchers.[118] This has been going on for a number of years and thousands of parts are already in the public domain.

Access to standard biological parts is also important due to synthetic biology's emphasis on standardisation. Synthetic biology's parts agenda was founded by leading synthetic biologist Drew Endy and collaborators to address the fact that biological engineering was 'still an expensive, unreliable and *ad hoc* research process'.[119] Endy and colleagues blamed biology's lack of standards for biological function and other measures for costs and inefficiencies.[120] Thus, they saw a need for 'the development of technologies and promulgation of standards that support the definition, description and characterization of the basic biological parts, as well as standard conditions that support the use of parts in combination'.[121]

In the parts agenda, standardisation is achieved through the development, definition and distribution of standard biological parts themselves, and through the development of technical standards for use in parts research. The technical standards being developed relate to matters such

[118] See also Boyle (n 49) 176–8; Kumar and Rai (n 49) 1763–7; Rai and Boyle (n 13) 0391–2.

[119] Drew Endy, 'Foundations for Engineering Biology' (2005) **438**(7067) *Nature* 449, 449.

[120] Ibid., 450.

[121] Ibid.

as defining and describing parts, assembling them in the laboratory and measuring the activity of molecules encoded by parts.[122]

Standardisation efforts can be seen to distinguish synthetic biology from other life sciences research.[123] Specific patent problems could arise from this standardisation.[124] For example, if standard parts or devices are patented, there is the potential for issues to arise if these parts or devices are then used in standard tools or techniques in synthetic biology. Problems can occur when a standard becomes widely used in a field, but incorporates patented elements.[125] The patent holder could withhold access to the patent needed to use the standard, which is problematic if the standard is already widely used.[126] This is particularly problematic if it is difficult to subsequently change the standard.[127] Rai has proposed that it will be helpful to look at the mechanisms used in the information technology industry and translate this to the synthetic biology context.[128]

Although a great deal of work has been focused on standard-setting in synthetic biology, no particular technical standard has become mandatory in the field as there is no single body with authority to impose such standards.[129] For example, there are several different protocols which can be used for the assembly of biological parts.[130] As there are no mandatory standards and multiple techniques for different tasks in synthetic biology, the problems associated with a patented element used in a standard does not seem likely to arise at present. However, patent issues such as the possibility of infringing patents and mechanisms for addressing such patents should be considered as part of standard-setting efforts in synthetic biology.[131]

---

[122] Adam Arkin, 'Setting the Standard in Synthetic Biology' (2008) **26**(7) *Nature Biotechnology* 771; Canton et al (n 117).

[123] Rai, 'Evidence to The Presidential Commission for the Study of Bioethical Issues, Synthetic Biology' (n 50).

[124] Ibid; Arti K Rai, 'Law and Technology Unstandard Standardization: The Case of Biology' (2010) **53**(1) *Communications of the ACM* 37, 38–9.

[125] Kumar and Rai (n 49) 1758; Rai, 'Evidence to The Presidential Commission for the Study of Bioethical Issues, Synthetic Biology' (n 50); see also Rutz (n 4) S16.

[126] Kumar and Rai (n 49) 1758.

[127] Ibid.

[128] Rai, 'Evidence to The Presidential Commission for the Study of Bioethical Issues, Synthetic Biology' (n 50); Rai, 'Unstandard Standardization: The Case of Biology' (n 124) 39.

[129] Torrance and Kahl (n 80) 11.

[130] Ibid., 6–7.

[131] Rai, 'Unstandard Standardization: The Case of Biology' (n 124) 39.

In addition, there are economic reasons why certain parts may become particularly valuable. It has been argued that use of a biological part will increase its value, because this generates knowledge about how the part actually worked, reducing the effort and cost involved in future use of that part.[132] Therefore, scientists would be more inclined to use parts that have already been successfully used by others.[133] Problems could then arise if a particular part was not accessible due to expensive license fees.[134] For these reasons, economists Joachim Henkel and Stephen Maurer (also a legal academic) have advocated for openness of parts.[135]

For all these reasons, it is important for the development of synthetic biology that standard biological parts should be accessible. This access to basic tools will promote research and development in the field, potentially leading to beneficial innovation.

8.5 PATENT EXAMINATION CASE STUDY: VENTER PATENTS

Scientists at the Venter Institute have filed several families of patent applications for tools and techniques in synthetic biology. These include a minimal bacterial genome, methods for creating synthetic genomes and cells, and methods of DNA assembly. This section examines the scope of the claimed inventions and how they have been treated by the USPTO. Analysis of patent examining practices is important because it reveals whether the validity criteria are applied such that high-quality patents are being granted. High-quality patents – which meet the statutory requirements for patentability – can help to drive innovation, while poor-quality patents can hinder it.[136] Ensuring high-quality patents is important for promoting innovation in synthetic biology.[137] Thus both courts and the USPTO should stringently apply the patent-eligible subject matter tests and validity criteria.

---

132 Joachim Henkel and Stephen M Maurer, 'The Economics of Synthetic Biology' (2007) **3**(117) *Molecular Systems Biology* 1, 1–2; Henkel and Maurer, 'Parts, Property and Sharing' (n 11) 1096.

133 Henkel and Maurer, 'The Economics of Synthetic Biology', ibid.; Henkel and Maurer, 'Parts, Property and Sharing' (n 11) 1096.

134 Henkel and Maurer, 'Parts, Property and Sharing' (n 11) 1096.

135 Ibid.; Maurer (n 59).

136 Rai et al (n 51) 2–6.

137 Rai, 'Evidence to The Presidential Commission for the Study of Bioethical Issues, Synthetic Biology' (n 50).

This particular case study was chosen because there has been controversy about the patents filed by the Venter Institute in synthetic biology. For example, ETC Group stressed that the Venter Institute's patent application claiming synthetic genomes was broadly framed to seek monopoly over all synthetic genomes and could 'entangle the entire field of synthetic biology'.[138]

The original claims were indeed very broad, and could have covered many aspects of synthetic biology. Some of the claims for the Venter Institute patents had characteristics which were of concern to the Supreme Court in *Mayo*. In that case, the claims involved 'highly general language', and threatened to impede the development of more refined methods.[139] The claims were also broad enough to cover processes not yet discovered.[140]

Concerns about patent impact were raised again following the synthetic cell announcement in 2010. Boyle wrote: 'We do not want a monopolist over the code of life.'[141] He reasoned that patents will promote investment and commercialisation in synthetic biology applications such as biofuels development, but if they are granted 'at too fundamental a layer', this could damage science and innovation.[142] Intellectual property issues were also prominent in the consultations of the Commission which followed the synthetic cell announcement.

For its part, the position of Venter's company Synthetic Genomics Inc. is that:

> intellectual property is important in the synthetic genomics/biology space, as it is one of the best means to ensure that this important area of basic science research will be translated into key commercial products and services for the benefit of society. SGI intends to provide licenses to its synthetic genomics patents.[143]

---

[138] ETC Group, *Extreme Monopoly: Venter's Team Makes Vast Patent Grab on Synthetic Genomes* (n 4).

[139] *Mayo* (n 8) 1302.

[140] Ibid.

[141] James Boyle, 'Monopolists of the Genetic Code?' (28 May, 2010) *The Public Domain* <http://www.thepublicdomain.org/2010/05/28/monopolists-of-the-genetic-code/>.

[142] Ibid.

[143] J Craig Venter Institute, *First Self Replicating Synthetic Bacterial Cell: Frequently Asked Questions* <http://www.jcvi.org/cms/research/projects/first-self-replicating-synthetic-bacterial-cell/faq/>.

This case study was also chosen because the approach of Venter to patent issues has been marked with controversy over more than a decade. For example, Venter was associated with the NIH's patent applications over expressed sequence tags, which created controversy.[144] Venter subsequently left the NIH to pursue human genome sequencing efforts in a privately funded arena, creating the possibility for human genome sequence data to be owned by a private company. There was much concern in the human genetics community regarding terms of access to the data, and distrust in relation to whether the data would be made freely available.[145]

Thus, it is important to consider whether the patent applications could result in the use of the intellectual property system in a way that impedes beneficial innovation in synthetic biology. Analysis of patent examining practices for patents filed by the Venter Institute – a key player in synthetic biology – is an important aspect of understanding the potential impact of synthetic biology patents on beneficial innovation in the field.

For this case study, I have chosen families of patent applications which appear to cover basic tools and processes in the construction approach to synthetic biology, in order to assess how the examination of these may impact upon research in the field.

8.5.1 Patent Validity Criteria

Before analysing the USPTO's examination of the patent applications in this case study, it is first necessary to understand the criteria on which these examinations are made. To meet the criteria for a valid US patent, an invention must be a new and useful 'process, machine, manufacture,

---

[144] See Robert Cook-Deegan, *The Gene Wars: Science, Politics and the Human Genome* (Norton, 1994), 316–33; J Craig Venter, *A Life Decoded, My Genome: My Life* (Viking, 2007) 135–47, 175. See also Matthew Rimmer, 'The Sorcerer II Expedition: Intellectual Property and Biodiscovery' (2009) **6** *Macquarie Journal of Intellectual Property* 147, 151. I acknowledge that Venter has stated that he did not support patents on expressed sequence tags or commercial monopoly over all of the human genome data: Venter, ibid., 130–33, 135–7, 150, 158.

[145] Venter has stated that he wished the human genome data to be publicly available, and patents to be limited to those genes which clearly had therapeutic or diagnostic value: Venter, ibid., 180, 260. However, in relation to the concerns see for example Bruce Alberts and Aaron Klug, 'The Human Genome Itself Must be Freely Available to all Humankind' (2000) **404**(6776) *Nature* 325; Declan Butler, 'US/UK Statement on Genome Data Prompts Debate on "Free Access"' (2000) **404**(6776) *Nature* 324.

or composition of matter' where the invention is novel, not obvious, and contains a written description sufficient to enable a person skilled in the relevant field to make and use the invention ('enablement').[146] In order to be novel, the invention must not already be found in any existing patent, patent application or publication, or in public use, on sale, or otherwise available to the public.[147] The test for obviousness is whether 'the differences between the claimed invention and the prior art are such that the claimed invention as a whole would have been obvious before the effective filing date of the claimed invention to a person having ordinary skill in the art'.[148] For example, 'The combination of familiar elements according to known methods is likely to be obvious when it does no more than yield predictable results'.[149] Determining whether the claims are obvious is an objective analysis which involves determining the scope and content of the prior art, the differences between the prior art and the claims at issue, and the level of ordinary skill in the art.[150]

In order to be 'useful', the invention must have a real-world utility in its present form.[151] The utility of a claimed invention must be 'substantial' and 'specific', showing a 'significant and presently available benefit to the public'.[152]

The 'written description' and 'enablement' requirements are distinct.[153] The first requires that the applicant demonstrate that they were in possession of the invention at the filing date.[154] The application must also contain claims particularly pointing out and distinctly claiming the subject matter which the applicant regards as their invention.[155] The second requirement is that the applicant provide sufficient information that a person with ordinary skill in the field could make and use the invention without engaging in undue experimentation first.[156]

The results of this case study are summarised in three tables below. The examination of patents which were not granted is summarised in

---

146 35 USC (n 17), §§ 101, 102, 103, 112.

147 Ibid., § 102(a), (b), (d).

148 Ibid §103; *KSR Intern Co v Teleflex Inc*, 550 US (S Ct, 2007) 398.

149 *KSR v Teleflex* ibid., 416.

150 *Graham v John Deere Co of Kansas City*, 383 US (S Ct, 1966) 1, 17–18.

151 *In re Dane K Fisher v Lagudi*, 421 F 3d (Fed Cir, 2005) 1365, 1371.

152 Ibid.

153 *Ariad Pharmaceuticals Inc. v Eli Lilly and Company*, 598 F 3d (CA Fed (Mass), 2010) 1336, 1340.

154 *Vas-Cath Inc. v Mahurkar*, 935 F 2d (Fed Cir, 1991) 1555, 1563; *Ariad v Eli Lilly* ibid., 1351.

155 35 USC (n 17), § 112(b).

156 Ibid., § 112(a); *In re Wands*, 858 F 2d (CA Fed, 1988) 731, 736–7.

Table 8.1. The examination of patents which were granted is summarised in Table 8.2. Table 8.3 shows examples of claims which were narrowed down through the examination process.

8.5.2 Minimal Bacterial Genome

An application titled 'Minimal Bacterial Genome' was filed in 2006.[157] This application originally had claims covering the minimal set of about 400 genes required for *Mycoplasma genitalium* to grow and reproduce in a rich culture medium; that is, the set of genes which the researchers had identified as essential for its survival in a laboratory environment. The application also claimed a minimal genome with the addition of genes involved in producing ethanol or hydrogen, a non-naturally occurring organism containing this set of genes and a method of using this organism to produce ethanol or hydrogen.

In response to the USPTO's initial examination of the application, the claims were cancelled and new claims filed. These were for a specific set of genes encoding at least 351 listed proteins, and this set of genes with various additions such as genes from a different organism useful for producing a drug, vaccine or energy source. The application further claimed a non-naturally occurring organism with the specified gene set and a cell culture of this organism. The scope of these claims is significant because it is very broad and could potentially have impacted upon synthetic biology research in a variety of areas.

Most of the claims were rejected as failing to meet the written description requirement. The claims were also found to be obvious and there was a failure to point out and distinctly claim. The claim relating to production of drugs and energy sources was rejected because it was unclear what was encompassed by this claim. The applicant subsequently amended the claims, but the remaining claims were rejected for being directed to non-patentable subject matter under the Supreme Court decision in *Myriad*. The claims for a set of bacterial genes were rejected as the DNA was naturally occurring. The applicant could not overcome this objection by argument. The objection on the grounds of failure to point out and distinctly claim was withdrawn, but the claims were rejected on novelty, obviousness and enablement grounds. The claims were too broad, they were only considered to be enabling in relation to

---

[157] John I Glass, et al (2006) 'Minimal Bacterial Genome', US Patent Application No 11/546364.

Mycoplasma genitalium, not for other bacteria. The application was later abandoned by the applicants.

8.5.3 Genome Assembly Methods

(a) Synthetic genomes (now 'method of nucleic acid cassettes assembly')

The patent application originally titled Synthetic Genomes was filed in 2006.[158] It originally claimed a method of constructing a range of large DNA cassettes of 4–7 kilobases in length containing adjacent portions of a genome overlapping by about 50 nucleotides and assembling these to produce a synthetic genome. Originally the application was very broad, claiming 'a synthetic genome' and 'a synthetic cell comprising a synthetic genome'. The method was claimed where the synthetic genome was either bacterial or non-bacterial in origin. The method was also claimed where the whole genome had been constructed from chemically synthesised DNA. The application further claimed a synthetic genome produced by the method above which also contained genes that produce a 'product of interest', in particular an energy source.

The claims were narrowed so that only methods for producing synthetic genomes were claimed rather than genomes themselves, and only bacterial genomes were claimed. The USPTO rejected the claims on the grounds of obviousness and anticipation. The applicant was able to overcome the novelty objection by amending the claims. The claims were also rejected due to insufficient written description and on enablement grounds. The method of making a synthetic cell and method of energy production were removed from the claims. The rejections were then maintained on the same grounds.

The application has been amended and rejected several more times. Some obviousness objections were overcome by argument, but others were maintained. Some of the written description objections were overcome by amending the claims, but the new claims generated new written description objections. The applicant has made the key claim more detailed in order to try to answer the rejections. Most recently, a non-final rejection has been issued in 2016, based on written description and obviousness objections.

---

[158] J Craig Venter, et al (2006) 'Method of Nucleic Acid Cassette Assembly (originally Synthetic Genomes)', US Patent Application No 11/635355.

(b) Gibson Assembly

The patent application titled 'Methods For In Vitro Joining and Combinatorial Assembly of Nucleic Acid Molecules' was filed in 2009.[159] The original application claimed a method for joining a set of DNA molecules, where molecules which will be adjacent to each other have overlapping sequences, into a single DNA molecule in a one-step reaction.[160] This method has become known in the field as Gibson Assembly™. The patent applications relating to the method are significant because the process is an important and widely used technology in synthetic biology.[161]

In 2011, the applicants filed a divisional application for a kit for conducting Gibson Assembly in response to a restriction requirement.[162] A novelty rejection was subsequently withdrawn, but the claims were rejected on obviousness grounds through several rounds of examination. Eventually, the patent application for the kit was abandoned.

The claims for the assembly method were rejected for failing to point out and distinctly claim, lack of novelty and obviousness. The novelty rejection was withdrawn but the failure to point out and distinctly claim and obviousness were maintained in further examination. The applicant was subsequently able to overcome these objections by argument, and the application for a patent was allowed in 2015.

(c) Assembly of large nucleic acids

A further patent application for methods of assembling DNA was filed in 2008.[163] It claimed a 'method for the synthesis of a desired nucleic acid' by combining DNA cassettes of up to 5 MB in length having overlapping sequences *in vitro*, and then assembling them *in vivo* (inside a host cell).[164] This was claimed where the assembled DNA represents a

---

[159] Daniel G Gibson, et al (2009) 'Methods For In Vitro Joining and Combinatorial Assembly of Nucleic Acid Molecules [Method]', US Patent Application No 12/371543.

[160] Ibid., original claims.

[161] An empirical survey by Kahl and Endy found that the majority of respondents used the Gibson Assembly method. It is not exclusively licensed to these researchers: Linda J Kahl and Drew Endy, 'A Survey of Enabling Technologies in Synthetic Biology' (2013) **7**(13) *Journal of Biological Engineering* 1, 5–12.

[162] Daniel G Gibson, et al (2011) 'Methods For In Vitro Joining and Combinatorial Assembly of Nucleic Acid Molecules [Kit]', US Patent Application No 13/035,699.

[163] Gibson et al 'Assembly of Large Nucleic Acids' (n 41).

[164] Ibid., original claims.

naturally occurring genome, or a 'desired metabolic pathway'. Other methods and the resulting molecules were claimed also.

The USPTO required the applicant to restrict the application to one method or product and the applicant chose to claim the main method of DNA assembly. The USPTO rejected the application for lack of novelty and obviousness. Various methods existed in the prior art, and it was routine to modify the parameters in the way claimed. Amended claims were rejected for failure to point out and distinctly claim the subject matter of the invention. Essential steps in the process were omitted from the specification. The novelty and obviousness objections were withdrawn after further amendments and removal of some of the claims. However, objections on the grounds of failure to point out and distinctly claim the subject matter of the invention have been maintained and a further rejection was issued in 2016.

(d) In vitro recombination methods

A patent application filed in 2006 claimed 'Method for In Vitro Recombination'; a method for assembling DNA molecules *in vitro* (inside host cells).[165] The claimed method involved using DNA molecules with shared sequences at their ends which would join to each other in the presence of various enzymes. The method was claimed for joining two to 'at least eight' DNA molecules along with a kit for conducting this method. The applicant was required to elect the methods or a kit for conducting the methods and chose to pursue the method claims. The claims were rejected for indefiniteness, failure to point out and distinctly claim, novelty and obviousness. By presenting arguments against these rejections and adding detail to the claims specifying further parameters of the method, the applicant was able to overcome these objections. The patent was granted in 2010.

A further application titled 'In Vitro Recombination Method' was also filed in 2006.[166] It claimed an *in vitro* method for joining two DNA molecules with shared sequences at their ends, using purified proteins. A method for joining a plurality of DNA molecules in a defined order was also claimed. The applicant was required to elect methods or products, and chose the methods. The claims were rejected for failure to point out and distinctly claim, novelty and obviousness. By specifying the particular length of the region of shared sequence identity, use of purified

[165] Daniel Glenn Gibson and Hamilton O Smith (2011) 'Method For In Vitro Recombination', US Patent Application No 11/502624.

[166] Lei Young, et al (2010) 'In Vitro Recombination Method', US Patent No 7,723,077.

enzymes and other details, the objections were overcome. The patent was granted in 2010.

8.5.4 Installation of Genomes into Cells

A patent application titled 'Installation of Genomes or Partial Genomes into Cells or Cell-like Systems' was filed in 2006.[167] This application originally claimed 'a method for making a synthetic cell, the method comprising: obtaining a genome that is not within a cell; and introducing the genome into a cell or cell-like system'. This method was claimed where the genome was made of naturally occurring, man-made or hybrid DNA. The application also claimed a method of producing a product of interest, using synthetic cells produced by this method. Further it claimed the synthetic cell produced by the method. Following a restriction requirement the application was amended to claim the method only, using a man-made genome.

The USPTO rejected the claims under the enablement requirement as the applicant did not provide working examples for the methods and because the results of the methods were unpredictable. There was failure to point out and distinctly claim. A novelty rejection was overcome by amending the claims, but the other objections were maintained.

Further amended claims were still considered too broad and were rejected for lack of enablement. The claims were not considered to be enabling for installation of genomes into any *Mycoplasma* bacterium, only the specific species for which installation had been demonstrated in the specification. The applicant overcame this objection by narrowing the claims to only cover those particular species.

The 'failure to point out and distinctly claim' objection was overcome by adding more detail and specificity about the steps that formed part of the method claimed. The patent was allowed in 2016.

A second application for installation of genomes was filed in 2010.[168] It claimed a method involving preparing a donor genome from one species, preparing a recipient cell from another species of the same genus, and installing the donor genome in the recipient cells so that a phenotypic transformation occurs (the recipient cell becomes the same species as the donor cell). This was rejected on enablement grounds.

---

[167] John I Glass et al (2006) 'Installation of Genomes or Partial Genomes into Cells or Cell-Like Systems', US Patent Application No 11/644713.

[168] John I Glass, et al (2010) 'Methods of Genome Installation in a Recipient Host Cell', United States Patent Application No 12/598414.

The applicant was not able to overcome the lack of enablement by argument alone. As with the first installation patent, the applicant narrowed the claims to only cover the method when used with particular species, as well as including more specific detail in the process set out. The patent was allowed in 2016.

8.5.5 Synthetic Cell ('Methods for Cloning and Manipulating Genomes')

In 2010 two patent applications were filed claiming Methods for Cloning and Manipulating Genomes.[169] The first claimed 'a method for cloning a donor genome' by obtaining a donor genome, and introducing it into a heterologous host cell.[170] The method was claimed using various donor genome types including bacterial, fungal, algal, cyanobacterial, viral or synthetic. The method was claimed using eukaryotic or prokaryotic host cells. Further claims described introducing the donor genome into a recipient cell of various types and removing the genome of the recipient cell. This method was claimed where the recipient cell takes on the characteristics of the donor genome. Cells produced in this way were also claimed.

In response to a restriction requirement the applicant chose to restrict its claims to the main method for cloning a genome. The claims for various cells produced by the methods were cancelled. The claims were rejected as being indefinite, anticipated and obvious. It was deemed obvious to modify methods already within the prior art in the way claimed, with a reasonable prospect of success. Later versions of the claims were also rejected on enablement grounds.

The novelty and enablement objections were overcome by amending the claims. The novelty objection was overcome by adding a further step to the method, namely transferring the donor genome from a host cell into a recipient yeast cell to produce a self-replicating synthetic cell. To overcome the enablement objection, more detail and specificity were added to the claims. By providing a declaration from Gibson including arguments regarding the knowledge a person in the art could be assumed to have, the 'failure to point out and distinctly claim' objection was

---

[169] Gwynedd A Benders, et al (2010) 'Methods for Cloning and Manipulating Genomes [Continuation in Part]', US Patent Application No 12/783489; Gwynedd A Benders, et al (2010) 'Methods for Cloning and Manipulating Genomes [Original]', US Patent Application No 12/718911.

[170] Benders, et al (2010) 'Methods for Cloning and Manipulating Genomes [Original]' (n 169).

overcome. This declaration also provided detailed arguments as to the differences between the application and the prior art, and the inventiveness shown in overcoming various technical problems to successfully implement the claimed method. This assisted in overcoming the obviousness objections. The application was allowed in 2015.

A continuation-in-part of this application claimed a 'method for creating a synthetic cell' by assembling a synthetic donor genome, introducing it into a host cell, isolating it and then introducing it into a recipient cell, as well as a method for making a cell where the phenotype is encoded by a synthetic genome.[171] The claims were restricted to methods only, using a synthetic donor genome. The USPTO rejected the claims as incomplete for omitting essential steps, and indefinite. These claims were found to be anticipated by Gibson's previous work and obvious in light of this work and previous patents. It would have been obvious to combine these methods with a reasonable prospect of success.

The novelty objection was overcome by specifying that the method was for assembling a synthetic genome, introducing it into a host cell, recovering it from that host cell and transferring it to a recipient cell to produce a self-replicating cell. The applicant successfully argued that the previous Gibson work did not demonstrate this method.

Enablement objections arose due to the large class of donor genomes which were encompassed by the claimed method. These were overcome by limiting the scope of the claimed method to certain cell types and removing options for how the donor genome might be assembled. New claims were added claiming the use of the method with specific Mycoplasma species.

Obviousness and indefiniteness objections continued to be raised through numerous rounds of amendment of the claims and re-examination. Many of the claims were withdrawn through this process. Eventually, by modifying the claims and making arguments regarding the knowledge a person in the art could be assumed to have, all of the indefiniteness objections were overcome.

As in the related synthetic cell application above, the Gibson declaration assisted to overcome the obviousness objection. It provided detailed arguments as to the differences between the application and the prior art, and the inventiveness shown in overcoming various technical problems to successfully implement the claimed method.

---

171 Benders, et al (2010) 'Methods for Cloning and Manipulating Genomes [Continuation in Part]' (n 169) original claims.

Table 8.1 Rejected patent applications by scientists of the Venter Institute for basic technologies

Patent Application Name	Number	Filing Date	Federal Funds	Status	Status Date	Reason for Rejections
Method of Nucleic Acid Cassette Assembly (originally Synthetic Genomes)	11/ 635,355	6 December 2006	Yes	Non-final rejection	15 June 2016	Enablement Obviousness Failure to point out and distinctly claim Written description Written description – new matter Novelty
Minimal Bacterial Genome	11/ 546,364	10 December 2006	Yes	Non-final rejection, Abandoned	27 July 2015	Obviousness Written description Failure to point out and distinctly claim – overcome Novelty Non-patentable subject matter – product of nature
Assembly of Large Nucleic Acids	12/ 247,126	7 October 2002	No	Final rejection	21 June 2016	Failure to point out and distinctly claim Obviousness Novelty
Methods for In Vitro Joining and Combinational Assembly of Nucleic Acid Molecules (Gibson Assembly, Kit)	13/ 035,699	25 February 2011	No	Final rejection, Abandoned	20 March 2015	Novelty – overcome Obviousness

Table 8.2 Granted patents to scientists of the Venter Institute for basic technologies

Patent Application	Number	Filing Date	Federal Funds	Status	Status Date	Earlier Reasons for Rejection
In Vitro Recombination Method	7,723,077	11 August 2006	Yes	Issued	25 May 2010	Failure to point out and distinctively claim Novelty Obviousness
Method for In Vitro Recombination	7,776,532	11 August 2006	Yes	Issued	17 August 2010	Failure to point out and distinctively claim Novelty Obviousness
Methods for In Vitro Joining and Combinational Assembly of Nucleic Acid Molecules (Gibson Assembly, Method)	8,968,999	13 February 2009	No	Issued	3 March 2016	Obviousness Written description – new matter Failure to point out and distinctly claim Novelty – overcome
Methods for Cloning and Manipulating Genomes (the synthetic cell)	9,273,310	3 May 2010	No	Issued	1 March 2016	Failure to point out and distinctly claim – some overcome Improper dependent claims – some overcome Novelty Obviousness

Patent Application	Number	Filing Date	Federal Funds	Status	Status Date	Earlier Reasons for Rejection
Methods for Cloning and Manipulating Genomes (the synthetic cell) continuation in part	9,267,132	19 May 2010	No	Issued	23 Feb 2016	Failure to point out and distinctly claim Enablement Novelty - Obviousness
Installation of Genomes or Partial Genomes into Cells or Cell-like Systems	9,434,974	22 December 2006	Yes	Allowed, terminal disclaimer	6 September 2016	Enablement Failure to point out and distinctly claim Novelty
Methods of Genome Installation into a Recipient Host Cell	9,481,880	22 April 2010	Yes	Allowed, terminal disclaimer	12 October 2016	Enablement

Table 8.3 Examples of narrowed scope of patent claims

Patent Application	Broad Claims	Narrowed/Refined Claims
Minimal Bacterial Genome (abandoned)	A minimal gene set defined in various ways with reference to lists of genes (e.g., defined as including or lacking certain numbers of listed genes) The minimal gene set plus genes for producing hydrogen or ethanol The minimal gene set when contained in a free-living organism A free-living organism with that gene set A method of producing hydrogen or ethanol using that organism A method of determining the function of genes by manipulating those genes in an organism containing the minimal gene set A method of producing a DNA library	A minimal set of genes encoding at least certain numbers of proteins with listed functions The minimal gene set where the genes do not encode for a 'dispensable protein' (a list of such proteins is included in the claims) That gene set where the encoded proteins are from *M. genitalium* A minimal gene set plus genes for energy (hydrogen or ethanol), drug, enzyme or vaccine production

Table 8.3 (continued)

Patent Application	Broad Claims	Narrowed/Refined Claims
Method of Nucleic Acid Cassettes Assembly (Synthetic Genomes) (rejected)	'A method for constructing a synthetic genome comprising: assembling nucleic acid cassettes that comprise portions of the synthetic genome, wherein at least one of the nucleic acid cassettes is constructed from nucleic acid components that have been chemically synthesized, or from copies of chemically synthesized components' That method, where the synthetic genome is • a eukaryotic (higher organism) genome • a bacterial genome • a minimal genome • a non-naturally-occurring genome • substantially identical to a naturally-occurring genome • made from chemically synthesized components, or • a synthetic cellular genome The method, automated The method where the genome is rationally designed The method where the synthetic genome also includes genes for producing a product of interest, including an energy source Methods of producing products of interest, such as energy *A synthetic genome* [emphasis added] *A synthetic cell comprising a synthetic genome* [emphasis added] Various methods for constructing a synthetic cell A synthetic cell made by one of the claimed methods A synthetic cell comprising a microorganism of one species which has had the genome removed, and a synthetic genome that exhibits at least one property that is different A method comprising: designing and constructing a synthetic genome, introducing it into a biological system and expressing it	'A method for assembling a nucleic acid construct comprising: (a) Generating a plurality of double-stranded nucleic acid cassettes from chemically synthesized oligonucleotides or copies thereof, wherein the cassettes comprise adjacent regions of a nucleic acid construct to be assembled and wherein each of the cassettes overlap with one or more other cassettes by at least 50 nucleotides and each is between about 4 kb and 7 kb in length, not including overlaps; and (b) utilizing an enzyme having 3' or 5' exonuclease activity, a DNA polymerase, and a ligase in a chew back and repair step to simultaneously assemble the plurality of nucleic acid cassettes *in vitro* whereby an assembled nucleic acid construct is produced'. The method, where the assembled nucleic acid is non-naturally occurring The method, where the cassettes are made from chemically synthesized DNA The method, automated The method, including a specific process for generating the cassettes The method, further involving introducing the assembled nucleic acid construct into a vesicle

Patent Application	Broad Claims	Narrowed/Refined Claims
Installation of Genomes or Partial Genomes into Cells or Cell-Like Systems (granted)	'A method for producing a synthetic cell, comprising: obtaining a genome that is not within a cell; and introducing the genome into a cell or cell-like system' • with either a naturally occurring genome or a man-made genome, or where the genome is taken from a cell of one species and inserted into the genome of another species • where the genome is inserted into various recipient systems A method of producing a product of interest using a synthetic cell produced by the method above A synthetic cell produced by the method above	'A method for making a bacterial synthetic cell comprising: preparing a *Mycoplasma capricolum* recipient cell for installation of a bacterial genome and isolating a genome from a *Mycoplasma mycoides* cell by suspending the isolated genome in an agarose suspension; and introducing this into the *Mycoplasma capricolum* recipient cell by contacting the recipient cell with the genome, wherein the genome of the *Mycoplasma capricolum* recipient cell is removed or inactivated to produce the recipient cell prior to, or after, introduction of the isolated genome by means of a physical, genetic or selective method or a combination thereof; wherein the introduced genome is self-replicating; and the introduced genome does not recombine with a genome of the recipient cell, and the recipient cell takes on the phenotype programmed by the introduced genome thereby producing a synthetic cell'.

Table 8.3 (continued)

Patent Application	Broad Claims	Narrowed/Refined Claims
Synthetic Cell ('Methods for Cloning and Manipulating Genomes') (granted)	'A method for cloning a donor genome comprising: obtaining a donor gene from a donor cell or synthesising a donor genome as one or more fragments'; and introducing the donor genome and a host vector into a host cell of a different species, to produce a host cell containing the donor genome which incorporates the host vector. This method, where the donor genome is a bacterial, fungal yeast, cyanobacterial, viral or synthetic genome (among other genome types listed) The method, where the host cell is a eukaryotic or prokaryotic cell The method, comprising various further steps such as modifying the donor genome within the host cell, recovering it from the host cell, and introducing it into a recipient cell The method, including introducing the donor genome into a recipient cell, where that recipient cell is bacterial, yeast, fungal, insect, plant or algal Cells produced by the various methods A process for making a cell which has the characteristics of the donor cell, by introducing the donor genome into a host cell, recovering it, and introducing it into a recipient cell A cell which comprises a donor genome and exhibits characteristics encoded by the donor genome and not otherwise exhibited by the cell	'A method for cloning a donor genome comprising: obtaining a donor genome from a bacterial or cyanobacterial or a microalgal donor cell', and introducing it into a yeast host cell, and propagating the donor genome in the yeast host cell, and preparing the donor genome for transplantation into a bacterial, cyanobacterial, or microalgal recipient cell by methylating it, and preparing the recipient cell by inactivating certain enzymes, and recovering the donor genome from the host cell and introducing it into the recipient cell, where the recipient cell allows the donor genes to be expressed, and where the donor genome can sustain viability and self-replication of the recipient cell The method, where introducing the donor genome effectively transforms the characteristics of the recipient cell into those of the donor cell, incorporating any modifications

8.5.6 Discussion of Patent Examination

This case study shows that the USPTO mainly used the written description, enablement, non-obviousness and novelty requirements to require the applicants to narrow down the scope and more specifically refine their claimed inventions (see examples in Table 8.3), and ultimately to reject the claims in some cases. In addition, there was one instance where the USPTO rejected an application on the basis that it claimed naturally occurring DNA, which is non-patentable subject matter following Myriad.

Imposing a high written description requirement makes it difficult to successfully claim a whole class or 'genus' of things, and generates more confined claims.[172] For example, most claims of the Minimal Bacterial Genome application were rejected as failing to meet the written description requirement. Even though the applicant had narrowed its application to claim a gene set encoding particular proteins, limited examples were provided in the specification, which provided only general teachings and lacked 'guidance as to the individual components of the product'.[173] The specification only covered *Mycoplasma genitalium*, but the claims covered a much larger scope of nucleic acids from any bacterial organism. The USPTO described the genus of claimed nucleic acids as 'vast' and 'variant'.[174] It held that the applicants had failed to provide representative examples of the class of things covered by the claim, or specify the structural features of things in that class, in order to demonstrate it was in possession of that class.

Written description was also used to reject claims in the Synthetic Genomes application. The USPTO held that the applicants were not in possession of the invention at the time of filing. They did not publish how to assemble the synthetic genome of *M. genitalium* until two years later, and did not describe how to assemble genomes of other bacteria, yeast or other organisms. They did not describe any minimal genomes or non-naturally occurring genomes, or their assembly, nor sequences for producing an energy source or synthetic cells obtained by inserting a synthetic genome into a vesicle. Thus the application did not support the claimed subject matter. Some claims were deleted, while others were refined, such as removing the reference to 'constructing a synthetic cell' so that the application claim read 'introducing the synthetic genome or portion into a vesicle'.[175] Nonetheless, the rejection was maintained.

Many of the claims in the above applications were rejected on enablement grounds. Many were filed too early; they were effectively research plans rather than enabled inventions. For example, the Synthetic Genomes specification was held to be enabling for assembly of bacterial genomes, but not minimal, cellular or non-naturally occurring genomes,

---

[172] Brenda M Simon and Christopher T Scott, 'Unsettled Expectations: How Recent Patent Decisions Affect Biotech' (2011) **29**(3) *Nature Biotechnology* 229, 230.

[173] Glass, et al (2006) 'Minimal Bacterial Genome' (n 157) rejection of 2 July 2010, 7.

[174] Ibid.

[175] Venter, et al (2006) 'Method of Nucleic Acid Cassette Assembly (originally Synthetic Genomes)' (n 158) amended claims 28 July 2010.

production of an energy source or a synthetic cell. There was a lack of working examples; the specification was merely an invitation to try to make the claimed invention. The USPTO held that an 'extremely large' amount of experimentation would be required before a person skilled in the art could apply this technology.[176] It held that new technologies and years of inventive effort would be required, making the level of experimentation undue.

The USPTO emphasised the breadth of the claims in its enablement rejection of the first installation of genomes patent. The applicants had provided an example of transferring a genome from one species of *Mycoplasma* into another. However, the claims were too broad; they covered making synthetic cells in this way with *any* donor or *any* recipient cells. The status of the art showed that transferring DNA from one organism to another had unpredictable effects and it was unpredictable whether a genuine change in the host cell's characteristics could be caused in this way. While the applicant had showed how to use the method with one pair of donor and recipient species of *Mycoplasma* bacteria, undue experimentation would be required to test compatibility between other pairs of cell types because there are many unpredictable aspects and technical problems. It was even unpredictable using the method between closely related species of the *Mycoplasma* genus. This objection was only overcome by narrowing the claims to the particular species with which the method was demonstrated in the specification. More detail was also added to the method claimed.

Similarly, the second genome installation application was rejected for lack of enablement as the claims were so broad as to encompass any species of donor and recipient cell, but the application was not regarded as enabling for any species other than one specified pair of donor and recipient species. Transfer of genomic DNA was unpredictable and this, combined with the breadth of the claims and the lack of specific guidance or working examples, meant that undue experimentation would be required. The applicant overcame the objection by only claiming the method when used with particular species of bacteria and adding more specific detail to the claims.

The early claims for the synthetic cell patents were also rejected on enablement grounds. For the first patent, more detail was added to the claims, such as specifying that the donor genome would be from a bacterial, cyanobacterial or microalgal donor cell. The method was

[176] Ibid, rejection of 1 October 2010, 17.

fleshed out in various ways, such as specifying modifications to the donor genome and recipient cell that facilitated the transfer of the genome.

Similarly, the early claims for the synthetic cell method continuation-in-part were very broad. They referred to methods involving a 'synthetic donor genome' and a 'recipient cell' without specifying any particular type of genome or cell. They were so broad as to cover all types of organisms, including single-celled organisms, plants and animals. By the time the claims were granted, they had been narrowed to specifically refer to 'a synthetic bacterial, cyanobacterial or microalgal donor genome' and 'a bacterial, cyanobacterial or microalgal recipient cell'. Claims for broader categories of donor genome and recipient cell were rejected. Detail was also added to the method regarding preparing the donor genome and recipient cells.

The novelty requirement resulted in a refinement of claims in the Synthetic Genomes application, where more detail was added to the main method claim to avoid the anticipation rejection. This also occurred in the first installation of genomes application and the synthetic cell applications. Once these details were added, the novelty objections were withdrawn.

The obviousness requirement was also stringently applied to reject many of the applications. For example, the Synthetic Genomes method was rejected as obvious because it would have been obvious to a person with ordinary skill in the art to combine several existing references and they would have had a reasonable expectation of success.[177] Subsequently, the USPTO held that varying the sizes of the nucleic acid cassettes and their overlap regions to obtain the optimal parameters through 'routine optimization' was not inventive.[178] The USPTO referred to *In re Aller*, where it was held that 'where the general conditions of a claim are disclosed in the prior art, it is not inventive to discover the optimum or workable ranges by routine experimentation'.[179] It is unlikely that this finding could be challenged by relying on the decision of the Court of Appeals in *In re Kubin*.[180] In *In re Kubin*, the court held that the reasoning in *In re O'Farrell* on the issue of 'obvious to try' had been re-instated by *KSR*.[181] In *In re O'Farrell* it was held to be not obvious to: 'vary all parameters or try each of numerous possible choices until one

177 Ibid, rejection of 15 September 2009.
178 Ibid, rejection of 28 January 2010, 24–5.
179 *In re Aller*, 220 F 2d 454, 456.
180 *In re Marek Z Kubin*, 561 F 3d (Fed Cir, 2009) 1351.
181 Ibid., 1359.

possibly arrived at a successful result, where the prior art gave either no indication of which parameters were critical or no direction as to which of many possible choices is likely to be successful.'[182]

However, under *KSR* obviousness is suggested where a person skilled in the art pursues 'known options' from 'a finite number of identified, predictable solutions'.[183] This is closer to the situation in the Synthetic Genomes patent application because the prior art indicated that the optimum method would involve particular ranges of length for nucleic acid cassettes and overlap regions.[184]

This discussion and the tables above show that the patent examining criteria of written description, enablement, novelty and obviousness have been stringently applied to applications originating from the Venter Institute. Some applications have been rejected and for granted patents the claims have been refined and narrowed through repeated examination. Overly broad patents on foundational technologies have not been granted in the sample discussed in this chapter. Therefore, I would not expect to see the same kind of high-stakes, drawn-out infringement litigation for patents on basic tools arising from these applications as has occurred in relation to tools such as gene patents. However, various rejections are still under dispute, and litigation regarding the validity of the applications could ensue. In future, litigation could also arise relating to infringement of Synthetic Genomics' patents for biofuels as end-products are produced and commercialised.

Though a small sample, this study suggests that in the case of these patent applications, the patent system is working well to preserve patent quality because a rigorous approach has been taken at the USPTO in relation to patent applications for these basic technologies in synthetic biology. It is possible that the approach of the USPTO to life sciences patents has been tightened in response to controversies about stem cell and gene patents beginning in the 1990s and still continuing today. It is also possible that particularly rigorous attention was paid to these applications due to the controversy associated with Venter and the high public profile of Venter and his Institute. A larger examination of patent examining practices in synthetic biology would be needed to confirm these results.

---

182 *In re O'Farrell*, 853 F 2d (CA Fed, 1988) 894, 903.

183 *KSR v Teleflex* (n 149) 421.

184 As held in J Craig Venter, et al (2006) 'Method of Nucleic Acid Cassette Assembly (originally Synthetic Genomes)', US Patent Application No 11/635355, rejection of 28 January 2010.

8.5.7 Presidential Commission's Discussion of Patent Issues

In its 2010 report on synthetic biology, the Commission did not make any specific recommendations regarding intellectual property impact or policies for promoting access to research in synthetic biology. While it identified potential intellectual property issues in synthetic biology, the Commission stopped short of making specific recommendations, stating that it 'offers no specific opinion on the effectiveness of current intellectual property practices and policies in synthetic biology'.[185] The report is disappointing in this respect, because the Commission had been asked to 'develop recommendations about any actions the Federal Government should take to ensure that America reaps the benefit of this developing field of science while identifying appropriate ethical boundaries and minimizing identified risks'.[186] This clearly brought intellectual property issues within its remit. As the Commission itself noted, it had a unique opportunity to consider the issues raised by synthetic biology early in the development of the field.

Instead, the Commission gave a broad summary of potential intellectual property issues and the views about these that were raised in its consultations. It recommended that 'the government should keep careful watch on this field' and suggested investigation into 'whether current research licensing and sharing practices are sufficient to ensure that basic research results involving synthetic biology are available to promote innovation, and if not, whether additional policies and best practices are needed'.[187] It could have gone further, for example by investigating the sorts of patents which are being applied for and granted in synthetic biology, considering their potential impact and looking at potential access mechanisms.

The Commission suggested that the concerns in relation to patenting of synthetic biology are similar to those surrounding gene patents and patents on genetic tests.[188] The potential negative impacts of broad

---

185 The Presidential Commission for the Study of Bioethical Issues, 'New Directions: The Ethics of Synthetic Biology and Emerging Technologies' (The Presidential Commission for the Study of Bioethical Issues, December 2010) <http://bioethics.gov/sites/default/files/PCSBI-Synthetic-Biology-Report-12.16.10_0.pdf>, 121.

186 Barack Obama, 'Letter from President Barack Obama to Dr Amy Gutmann' (20 May 2010).

187 The Presidential Commission for the Study of Bioethical Issues (n 185) 118–22.

188 Ibid., 120.

foundational patents and patent thickets are similar. However, some issues of patent law and patent impact are actually quite different in the synthetic biology context. Because synthetic biology involves building or engineering something, rather than identifying information that is already in existence in nature, it will not be problematic in quite the same way. If the building of a genome or cell using synthetic biology produces a useful outcome, this may be patentable because it is an outcome that is not found in nature. For example, Keasling and colleagues have patented their metabolic engineering technology for producing isoprenoids.[189] This is quite different from identifying and patenting a gene involved in a particular human cancer, as in *Myriad.* Because that information was already in existence in nature, akin to a law or product of nature, granting a monopoly on the information could potentially impact negatively upon a whole field of research. There is no opportunity for other researchers (or patients) to negotiate around the patent by coming up with alternative tools or techniques or accessing these alternatives through other providers.

By contrast, various tools and techniques have been developed for some synthetic biology technologies, such as DNA synthesis and assembly. Thus if the patent validity criteria are applied to only grant high-quality synthetic biology patents, damaging impacts upon whole fields of research should not be observed. Synthetic biology therefore has a different relationship with the existing patent system than earlier biotechnology such as genes, genetic tests and human stem cells.

However, the Commission rightly pointed out that particular intellectual property issues surrounding synthetic biology arise from the parts agenda.[190] It noted synthetic biology's emphasis on standardisation, and 'the use of standardized modular parts' so that 'access to those standard components could be especially critical to the development of the field'.[191]

8.6 CONCLUSION

This chapter has considered the patenting of synthetic biology in light of one of the key roles of regulation of emerging technologies, that of

189 Renninger, et al 'Production of Isoprenoids' (n 44).

190 The Presidential Commission for the Study of Bioethical Issues (n 185) 120.

191 Ibid., 120–21.

facilitating 'desirable technological innovation'.[192] Therefore, this chapter has investigated whether patenting may have unintended, inhibitory effects on synthetic biology research as has been argued in other areas of the life sciences.

I have argued that the two problems theoretically most likely to arise at this stage of synthetic biology research are those of patent thickets and broad patents on foundational technologies. The discussion of recent case law regarding patentable subject matter has shown that the basic processes of constructive synthetic biology, synthetic DNA and standard biological parts are likely to form patentable subject matter. The patent examination case study has shown that for patent applications for basic tools and techniques in synthetic biology emanating from research at the Venter Institute, the statutory criteria of written description, enablement, non-obviousness and novelty have been stringently applied. It does not appear that overly broad, foundational patents were granted in the sample studied, though the impact of the patents will need to be observed over time.

This investigation has found that as yet it is not clear that the patents system is having an unintended, inhibitory effect on innovation in synthetic biology research. There is, however, potential for patent-troll behaviour and patent thickets on standard parts. Exploitation of patents which cover standard building blocks in synthetic biology could be problematic if it occurs. The next chapter will consider the mechanisms established by leading scientists in the BioBricks community to avoid these problems.

---

192 See Brownsword and Somsen (n 6) 2.

9. Building with BioBricks: a commons for sharing synthetic biology research[1]

9.1 INTRODUCTION

In 2013, Drew Endy attended a ceremony at the White House in which he was honoured as an 'open science champion'. Endy was recognised for his efforts to make synthetic biology openly accessible through the BioBricks Foundation.[2] The scientists honoured in this ceremony were lauded for their work 'to generate, promote, and use open scientific data as fuel for new products, successful businesses, and game-changing scientific insights'.[3]

Endy's open, sharing approach to synthetic biology is in contrast to the patenting approach discussed in the previous chapter. This chapter examines the approach to intellectual property, sharing and commons development of Endy and collaborators in the 'parts agenda'. Scientists working in this field have created a 'commons' for their research.[4] This is a response by those synthetic biologists to the perceived potential

---

[1] This chapter updates and expands upon Alison McLennan, 'Building with Biobricks: Constructing a Commons for Synthetic Biology Research' in Matthew Rimmer and Alison McLennan (eds), *Intellectual Property and Emerging Technologies: The New Biology* (Edward Elgar Publishing, 2012) 176. A short online article was also published summarising some of the arguments in that book chapter: Alison McLennan and Matthew Rimmer, *Inventing Life: Patent Law and Synthetic Biology* (2012) The Conversation <http://theconversation.com/inventing-life-patent-law-and-synthetic-biology-5178>.

[2] Synthetic Biology Engineering Research Centre, *Endy Honored by White House as an Open Science Champion* Synthetic Biology Engineering Research Centre <http://synberc.org/highlight/endy-open-science-champion>.

[3] Ibid.

[4] James Boyle, *The Public Domain: Enclosing the Commons of the Mind* (Yale University Press, 2008), 178; Sapna Kumar and Arti Rai, 'Synthetic Biology: The Intellectual Property Puzzle' (2007) **85** *Texas Law Review* 1745, 1763–7; Arti Rai and James Boyle, 'Synthetic Biology: Caught Between

problems of the patent system. This chapter explores the motivations for creating this commons and whether it is likely to be successful, with a view to building on the understanding of interactions between synthetic biology and the intellectual property system obtained in Chapter 8.

The commons is supported by the BioBricks Foundation, the Registry of Standard Biological Parts (the Registry), and the International Open Facility Advancing Biotechnology (the BIOFAB). The BioBricks Foundation and the Registry each have a strategy for sharing of standard biological parts, which I call the 'BioBricks model' and the 'Registry model' respectively. Together, these strategies are referred to as the 'standard biological parts commons'. As explained in Chapter 2, I use the term 'parts agenda' to refer to the branch of synthetic biology which focuses on using standard biological parts.[5] This is an aspect of the use of engineering principles such as standardisation and modularity in synthetic biology.

First, this chapter will consider the establishment and nature of the standard biological parts commons and how it fits into discussions of the concepts of the 'commons', the 'public domain' and 'common pool resources'. The standard biological parts commons is then compared to the models of open source software, open biotechnology and democratising innovation. The chapter then evaluates the access arrangements which underpin the standard biological parts commons. The chapter argues that the model of the 2011 BioBrick Public Agreement (BPA), combined with the leadership and community building activities of the BioBricks Foundation, and the education and enculturation activities of the iGEM Foundation, are together likely to successfully create a commons for synthetic biology researchers. This commons will operate so as to promote innovation in synthetic biology research.

As discussed in the previous chapter, at this stage of synthetic biology's development the actual impact of intellectual property in terms of restricting or promoting innovation in synthetic biology is not yet clear. However, it is important for the development of synthetic biology that standard biological parts should be accessible, to promote research and development in the field. Therefore, this chapter argues that public sector agencies should support the sharing mechanisms of the standard biological parts commons. Potential mechanisms for doing this are

---

Property Rights, the Public Domain and the Commons' (2007) **5**(3) *PLoS Biology* 0389, 0391–2.

5 This term has been used by Joachim Henkel and Stephen M Maurer. See, e.g., Joachim Henkel and Stephen M Maurer, 'Parts, Property and Sharing' (2009) **27**(12) *Nature Biotechnology* 1095.

considered in the final section, along with patent pools and the defence of experimental use.

9.2 BUILDING A STANDARD BIOLOGICAL PARTS COMMONS

Synthetic biology's parts agenda involves combining engineering principles with molecular and cell biology in order to reduce the complexity of biological engineering and increase its predictability and reliability.[6] Standardisation is a key engineering principle underlying the parts agenda. This standardisation involves defining and using well-characterised standard biological parts and developing technical standards for using these parts in research. It also involves the development of 'legal standards' for sharing parts.[7]

A 'BioBrick™ standard biological part' is a particular type of biological part, one that 'meets the technical and legal standards set forth by the BioBricks Foundation'.[8] It is a 'nucleic acid encoded-molecular biological function (e.g., turn on/off gene expression), along with the associated information defining and describing the part'.[9]

This parts agenda is supported by the BioBricks Foundation and the Registry, as well as the BIOFAB. The BioBricks Foundation is a public benefit organisation established by the founders of the parts agenda in 2006.[10] The Registry was formerly an institution located at MIT, where it also ran the iGEM competition. The iGEM Foundation has now been formed as a separate independent non-profit organisation, which runs both the competition and the Registry.[11] The BIOFAB is a 'biological design-build' facility which will produce standard biological parts and make these freely available for use in synthetic biology research.[12] It aims to 'produce thousands of free standardized DNA parts to shorten the

---

[6] Drew Endy, 'Foundations for Engineering Biology' (2005) **438**(7067) *Nature* 449.

[7] See ibid., 450.

[8] The BioBricks Foundation, *FAQ* <http://www.biobricks.org/FAQ.php>.

[9] Registry of Standard Biological Parts, *Help: About the Registry* <http://partsregistry.org/Help:About_the_Registry>; The BioBricks Foundation (n 8).

[10] The BioBricks Foundation, *The BioBricks Foundation* (2013) <http://www.biobricks.org/index.php>.

[11] iGEM Foundation, *The iGEM Foundation* iGEM <http://igem.org/About>.

[12] BIOFAB, *About the BIOFAB* BIOFAB <http://biofab.org/about>.

development time and lower the cost of synthetic biology for academic or biotech laboratories'.[13]

As well as their technical scientific objectives, the BioBricks Foundation and Registry also aim to create an openly accessible resource for sharing of biological parts and associated information. In fact, it can be argued that they aim to create a 'commons' for synthetic biology researchers.[14] The BioBricks Foundation and Registry have created online space for the sharing of experimental data, knowledge and ideas, and a system for the sharing of genetic sequence information relating to parts and physical biological materials. They have also worked to build a community of synthetic biologists with a shared philosophy and common goals. Specific commons-related objectives of the BioBricks Foundation have been to 'improve the process of scientific discovery by lowering the technical barriers to sharing and dissemination of knowledge in biological research, building a community of researchers that values, practices, and innovates the open sharing of information'.[15]

The BioBricks Foundation's website provides its vision for a commons that benefits not only scientist but all people. It believes in:

> a future where there is a free-to-use language for programming life that benefits everyone. A future where people around the world collaborate and communicate to create local biological solutions to meet global needs. When people have the tools and infrastructure to work with one another, we can meet global needs for food, medicine, shelter, clear water and air. To achieve this better future, BioBricks Foundation supports the open and ethical development of biotechnology that benefits all people and the planet.[16]

To support this vision, the BioBricks Foundation has established a contractual tool for sharing of BioBricks, the BPA. This contains intellectual property and attribution conditions for both the contribution of parts and the use of those contributed parts. Users and contributors are required to register in order to use this legal tool.

By contrast, the Registry is an open repository of standard biological parts which is not underpinned by any legal arrangements. This Registry is an online catalogue of parts and information for participants in the

---

[13] The Biofab, *Biofab Launches World's First Biological Design-Build Facility* (2010) Biofab <http://biofab.org/datasheet>.

[14] See Boyle (n 4) 176–8; Kumar and Rai (n 4) 1763–7; Rai and Boyle (n 4) 0391–2.

[15] The BioBricks Foundation, *Education Program* (2012) <http://biobricks.org/programs/education-program/>.

[16] The BioBricks Foundation, *The BioBricks Foundation* (n 10).

iGEM competition and academic labs.[17] Physical samples of the DNA encoding parts can also be provided. It is a 'continuously growing collection of genetic parts that can be mixed and matched to build synthetic biology devices and systems'.[18]

This online resource now contains over 10 000 parts.[19] Users can search for parts in the Registry's online catalogue under various characteristics, such as biological function. The catalogue provides information about each part including the individual who designed it, and whether it has been used by other users. Designers of parts can record notes or instructions, and users can record their experience using parts so that this can be shared by other researchers.

Users of this Registry do not have to agree to a contract, but there is an expectation that they will submit any new parts they create with Registry materials back to the Registry and share their experiences with other users.[20] Thus iGEM describes this philosophy as 'get and give (and share)'.[21]

Researchers must register in order to participate in the Registry. When a researcher wishes to access a part, they can select it from one of the 'kits' of parts that are distributed regularly, request it from the Registry, assemble it from small strips of natural DNA in their lab, or have the DNA commercially synthesised.[22]

9.2.1 The 'Commons', the 'Public Domain' and Collective Action to Manage 'Common Pool Resources'

The definitions of 'the public domain' and 'the commons' have been the subject of a scholarly discussion spanning the last 30 years. Boyle has referred to two 'enclosure movements', the first being enclosure of farming land as private property.[23] This enclosure avoided the 'tragedy of the commons', in which a resource that is 'open to all', such as a pasture

---

17 iGEM Foundation, *Registry of Standard Biological Parts* Registry of Standard Biological Parts <http://parts.igem.org/Main_Page>.

18 Ibid.

19 The BioBricks Foundation, *Frequently Asked Questions (BioBricks Public Agreement)* (2016) The BioBricks Foundation <http://biobricks.org/bpa/faq/>.

20 iGEM Foundation, *Our Philosophy* Registry of Standard Biological Parts <http://parts.igem.org/Help:Philosophy>.

21 Ibid.

22 See e.g., Registry of Standard Biological Parts, *Part:BBa_I742152: Get Part* <http://partsregistry.org/partsdb/get_part.cgi?part=BBa_I742152>.

23 James Boyle, 'The Second Enclosure Movement and the Construction of the Public Domain' (2003) **66**(1–2) *Law and Contemporary Problems* 33, 35.

for grazing cattle, ends up being overused and under-cared-for.[24] This shows the use of the term 'commons' to refer to tangible resources such as land. It is also used in relation to intellectual property. Boyle regards the recent expansion of intellectual property rights to 'enclose' things that were previously regarded as 'common property or uncommodifiable' as a second enclosure movement.[25] Examples include the sequence of the human genome.[26]

The public domain has been seen as 'the "outside" of the intellectual property system, the material that is free for all to use and to build upon'.[27] By contrast, intellectual property and media law scholar David Lange has argued that the public domain should be seen as something apart from intellectual property, an 'opposing concept' which requires its own recognition.[28]

Other scholars, who reject the idea of the public domain being simply material in which there are no intellectual property rights, are copyright expert Jessica Litman and information law expert Yochai Benkler.[29] Litman describes the public domain as 'a commons that includes those aspects of copyrighted works which copyright does not protect'.[30] Benkler defines the public domain as 'the range of uses of information that any person is privileged to make absent individualized facts that make a particular use by a particular person unprivileged'.[31]

In contrast to the public domain, the 'enclosed domain is the range of uses of information as to which someone has an exclusive right, and that no other person may make absent individualized facts that indicate permission from the holder of the right, or otherwise privilege the

---

[24] The tragedy of the commons was described by biologist Garrett Hardin in Garrett Hardin, 'The Tragedy of the Commons' (1968) **162**(3859) *Science* 1243, 1244.

[25] James Boyle, 'Foreword: The Opposite of Property?' (2003) **66**(1–2) *Law and Contemporary Problems* 1, 3; Boyle, 'The Second Enclosure Movement' (n 23) 37.

[26] Boyle, 'The Second Enclosure Movement' (n 23) 37.

[27] Boyle, 'Foreword' (n 25) 1.

[28] David Lange, 'Reimagining the Public Domain' (2003) **66**(1–2) *Law and Contemporary Problems* 463, 465–6.

[29] Boyle, 'The Second Enclosure Movement' (n 23) 61.

[30] Jessica Litman, 'The Public Domain' (1990) **39** *Emory Law Journal* 965, 968.

[31] Yochai Benkler, 'Free as the Air to Common Use: First Amendment Constraints on Enclosure of the Public Domain' (1999) 74 *New York University Law Review* 354, 362.

specific use'.[32] This conception of the public domain includes those things that are protected by intellectual property, but in which permission to access is implicit.[33] This view regards the public domain as 'broadly usable information resources', which encompasses a wide range of internet resources.[34]

Boyle describes a 'commons' as 'material that is not subject to individual control; rather it is controlled, if at all, by some larger group'.[35] Law and technology expert Lawrence Lessig describes a commons as:

> a resource that is free. Not necessarily zero cost, but if there is a cost, it is a neutrally imposed or equally imposed cost … No permission is necessary, no authorization may be required … The point is not that no control is present, but rather that the kind of control is different from the control we grant to property.[36]

In commons literature, property rights provide individual, legal control, the opposite of which is the 'well-run commons, a realm of collective, and sometimes informal, controls that avoids the tragedy of the commons without need for single party ownership'.[37]

I use 'the public domain' to refer to material or information that has been revealed to the public, for example via the internet. That is, it is not private or secret. It can be seen and accessed by anyone. This may include material in which there are patent rights, as well as unpatented material. Extending Litman's conception of the public domain to apply to patents as well as copyright, the public domain includes (as well as unpatented material) those aspects of patented inventions which patents do not protect because the patent holder has made the information publicly available. This is broader than the BioBricks Foundation's own description of the public domain, in which: 'creative works and inventions are considered to

---

32 Ibid., 362.

33 Pamela Samuelson, 'Enriching Discourse on Public Domains' (2006) **66**(4) *Duke Law Journal* 783, 798. Samuelson refers here to copyright, but other forms of intellectual property protection, such as patents, could apply to material that is none the less in the public domain.

34 Ibid., 798.

35 Boyle, 'Foreword' (n 25) 30.

36 Lawrence Lessig, 'The Architecture of Innovation' (2002) **51** *Duke Law Journal* 1783, 1788.

37 Boyle, 'Foreword' (n 25) 8.

be in the "public domain" if they are not covered by any current intellectual property rights (such as patent or copyright)'.[38]

The BioBricks Foundation has used the language of both 'open' and 'free'. 'Freedom' has become the preferred rhetoric, with BPA referring to 'free use', 'free-to-use', and inviting scientists to 'Get your freedom'.[39] 'Free' can have a range of meanings.[40] The closest to that offered by the BioBricks Foundation is the freedom Boyle describes as 'free trade in expression and innovation, as opposed to monopoly'.[41]

The literature suggests that 'public domain' is a broad, general concept, whereas 'commons' is more specific. For example, a particular commons is created by Creative Commons licenses for copyrighted works. To participate in that commons, certain rules have to be agreed to. I have chosen the language of the 'commons' to describe the sharing approaches in the parts agenda because a distinct term is needed from the 'public domain', where there is no control over how information may be used. I also use the term 'commons' because it connotes the existence of a community of participants. I use it to describe a shared space for synthetic biologists to access research and tools developed by others, and to conduct and share their research collaboratively.

Elinor Ostrom's work on collective action to manage 'common-pool resources' is relevant in understanding and evaluating this commons. Ostrom defines a common-pool resource as 'a natural or man-made resource system that is sufficiently large as to make it costly (but not impossible) to exclude potential beneficiaries from obtaining benefits from its use'.[42] Other scholars have assumed that external government regulation or private property rights are needed to manage such resources successfully.[43] Ostrom pointed out that these policy perspectives have limitations, and that some groups have been able to self-regulate to manage common-pool resources through collective action.[44] She explained that if those using a common-pool resource act independently of each other, they will usually get a lower return than if they have a

---

[38] The BioBricks Foundation, *FAQ (BioBricks Public Agreement)* (2016) (n 19).

[39] See The BioBricks Foundation, *The BioBrick™ Public Agreement* (2013) The BioBricks Foundation <http://biobricks.org/bpa/>.

[40] Boyle, 'The Second Enclosure Movement' (n 23) 61–3.

[41] Ibid., 58.

[42] Elinor Ostrom, *Governing the Commons: The Evolution of Institutions for Collective Action* (Cambridge University Press, 1990), 30.

[43] Ibid., 8–15.

[44] Ibid., 13–25.

coordinated strategy for sharing the resource, and they could even risk destroying the resource.[45] If they take collective action, they will gain a higher joint return.[46] In order to self-regulate successfully, the users of the resource must overcome the problem of how to 'organize and govern themselves to obtain continuing joint benefits when all face temptations to free ride, shirk, or otherwise act opportunistically'.[47] Their specific problems include how to supply themselves with a set of rules for managing the resource, how to gain commitment to these rules and how to mutually monitor compliance.[48]

Ostrom wanted to understand why some groups of users have been able to overcome these challenges successfully and others have not.[49] She studied a variety of systems of collective management of natural resources including fisheries, meadows and water sources. From this empirical work, Ostrom developed a set of design principles for successful management of common pool resources.[50]

The discussion of a 'commons' in this chapter makes use of these design principles. However, there is an important variation from Ostrom's theory of collective action to manage common-pool resources. A key feature of common-pool resources is that resource units '*are not subject to joint use or appropriation*. The fish harvested by one boat are not there for someone else'.[51] Accordingly, those sharing the resource need to work out how to preserve the resource system over time.[52] Similarly, Boyle's concept of a 'well-run commons' is one which enables the tragedy of the commons to be avoided.[53]

In the synthetic biology context, the nature of the resource is different. Information such as the DNA sequence of a standard biological part is a non-rivalrous resource. Unlike a grazing pasture or a fishery, use by one person cannot deplete the resource units that are available to others. Therefore, the tragedy of the commons does not arise. Instead, the theoretical tragedy which the BioBricks commons seeks to avoid is that of the 'anticommons' in biological research. According to the anticommons theory, patented inventions can end up being *under*-used because of

---

45 Ibid., 38–9.
46 Ibid., 39.
47 Ibid., 29.
48 Ibid., 42–5.
49 Ibid., 45.
50 Ibid., 90.
51 Ibid., 31 (emphasis in original).
52 See ibid., 36–50.
53 Boyle, 'Foreword' (n 25) 31.

difficulties accessing them, and research is thus slowed down.[54] The synthetic biologists want to promote the use of standard biological parts, but within a framework of rules and norms.

Nonetheless, the theoretical work of Boyle, Ostrom and others in understanding how a resource can be effectively shared is important in the synthetic biology context. Some of the key organisational problems faced by the resource users in Ostrom's work are also faced by the synthetic biologists. These include the problems of negotiating rules, gaining commitment and promoting compliance with these rules. Like the users of a common-pool resource, synthetic biologists stand to benefit from sharing their resources in a coordinated way. What they can accomplish as a collective is likely to be greater than what individual scientists could achieve in isolation, as will be seen below.

One of Ostrom's design principles is that individuals, who have the right to access the shared resource, as well as the boundaries of the resource itself, must be clearly defined.[55] In the standard biological parts commons the resource is clearly defined: those parts that have been contributed under the BPA or to the Registry. For parts that have been contributed under BPA, registration is needed to use the online parts. This requires entering an email address and password. As Lessig envisioned, there is no entity with authority to grant or deny access to any would-be user. However, use of these resources requires agreement to the rules in the BPA. In this sense, would-be users are excluded from accessing the shared resources if they are not prepared to agree to the rules. Access to the Registry is slightly different. Participants in iGEM teams have to provide a range of information about themselves, as do laboratories who wish to access the Registry. Registration to use the Registry has to be approved by the iGEM Foundation. In both cases, actors who have authority to access the resource are only those prepared to agree to the rules.

Boyle has argued that 'The key to the well-run commons is frequently that it has informal systems of collective control.'[56] In a similar vein, Ostrom's design principles state that the rules should at least be partly designed by the users themselves and that it is important that 'Most operators affected by the operational rules can participate in modifying

---

[54] See Michael A Heller and Rebecca S Eisenberg, 'Can Patents Deter Innovation? The Anticommons in Biomedical Research' (1998) **280**(5364) *Science* 698. However, this theory has been criticised and has generally not been borne out by empirical studies: see Chapter 8.

[55] Ostrom (n 42) 90.

[56] Boyle, 'Foreword' (n 25) 31.

the rules.'[57] With this in mind, I use the term 'commons' because it implies group control rather than individual control over the resource. There are controls on how the resources may be used. While the development of these controls is led by the BioBricks Foundation and the iGEM Foundation, other members of the commons can comment. In particular, the legal controls in the BPA are developed through an open process on which scientists and other interested parties are invited to comment. Members of the synthetic biology community have taken up this opportunity. Accordingly, those who will be affected by the rules can participate in their design and propose modifications. The technical standards for the actual scientific work are developed in a similarly open way.

For both the BPA and the Registry, the resources are 'free' in the sense Lessig describes; there is no cost associated with accessing the information and only a 'neutrally imposed' cost would apply to manufacture and shipping of physical biological parts.

Boyle has argued that the literature shows that successful commons have layers of 'norms' as well as 'freedom'.[58] Similarly, Ostrom's research shows that successful collectives to manage common-pool resources have norms that define 'proper' behaviour, helping the users to live interdependently while minimising conflict.[59] The significance of community norms reflected in the community of synthetic biologists, which is supported by the norms I described in Chapter 3. These are the norms of 'sharing', making a 'positive contribution' through the development of beneficial applications of synthetic biology, and 'engagement' with the issues raised by synthetic biology research and with policy making processes.

9.2.2 Operation of the Commons: Community Building, Leadership and Education

The BioBricks Foundation and the iGEM Foundation engage in a number of community building, leadership and education activities. The importance of this sharing and community building in synthetic biology is a theme in various published articles and testimonies to government committees by proponents of the parts agenda. For example, Endy has stated that 'vibrant, open research communities and strategic leadership

57 Ostrom (n 42) 90, 186.
58 Boyle, 'The Second Enclosure Movement' (n 23) 66.
59 Ostrom (n 42) 88.

are necessary to ensure that the development and application of biological technologies remains overwhelmingly constructive'.[60]

The BioBricks Foundation is a focal point for community discussion and development of standards (including legal mechanisms) for research involving BioBrick parts. Any member of the synthetic biology community can access and comment on draft standards through the BioBricks Foundation's website. The BioBricks Foundation encourages the use of these openly developed BioBricks standards. The BioBricks Foundation has also convened the large, international Synthetic Biology X.0 conferences, bringing synthetic biologists and other scientists together with lawyers, ethicists and government representatives. In addition to scientific achievements, ethical, regulatory and public engagement issues have been discussed at these interdisciplinary meetings.

The iGEM competition, a synthetic biology competition for undergraduate students and some high school students, is also a key community building institution for synthetic biology. iGEM has grown from a small competition featuring a handful of teams in 2004 to an international event participated in by 5 600 participants from 42 countries in 2016. iGEM teams are made up of undergraduates led by experienced researchers at their university. They are given a set of biological parts, and over the summer they attempt to design and construct a functioning biological device.[61] In addition to lab-based experiments, teams can also include a 'human practices' element in their project, allowing them to investigate issues of 'societal implications', risk, regulation and public awareness raised by synthetic biology.[62] Teams have looked into issues such as patentability and public understanding of synthetic biology.

At the end of iGEM, the participants come together for a jamboree. They present their work, and judging of the competition takes place. Leading up to iGEM, there is a programme of workshops and training, in which students and scientists can learn from leaders in the field. iGEM also allows students to learn about the broader issues surrounding synthetic biology, such as through outreach on biosecurity issues by

---

[60] Endy (n 6) 449.

[61] For example, in 2007 the University of California Berkeley's iGEM team attempted to construct a genetic device that would allow bacteria to make haemoglobin, generating 'Bactoblood': see University of California Berkeley 2007 iGEM Team, *Bactoblood* <http://parts.mit.edu/igem07/index.php/Berkeley_UC>.

[62] iGEM Foundation, 'iGEM: Practices Statistics and Stories 2008, 2009, 2010' iGEM Foundation, <http://igem.org/wiki/images/1/16/IGEM_Practices_Digest_%26_Extracts_2008-2010.pdf>.

members of the FBI. Teams have to submit safety forms which require them to consider the risks to laboratory members, the public, the environment and security associated with their experiments.

Through the excitement and enthusiasm generated by iGEM, future synthetic biologists are created. iGEM has thus become a mechanism for enculturation of a whole new generation of synthetic biologists in the open, sharing philosophy of the BioBricks Foundation and the Registry, and in understanding of public engagement, security, safety and intellectual property issues. Participants in iGEM have already gone on to become scientists in the field.[63] Endy has commented that 'we have now, in a bottom-up, grass-roots fashion, de facto installed a genetic-engineering curriculum for the future of our field in 120 schools worldwide'.[64]

Thus through these important community building institutions, the communal norms of the parts agenda community and the rules for participating in the commons have been developed and fostered.

9.2.3 Operation of the Commons: Access Arrangements

In 2009, draft legal agreements titled 'BioBrick Public Agreement' were released for comment, and an updated version was released in 2010.[65] This BPA was further updated and then 'went live' during the Synthetic Biology 5.0 Conference in 2011.[66] The BPA was developed by a team including legal academics and lawyers from a private firm specialising in intellectual property and technology.[67] The BPA contains a Contributor Agreement and a User Agreement.[68] These are web-based click-through agreements. These agreements focus on the contributed 'Materials',

---

[63] iGEM Foundation, *iGEM: Synthetic Biology Based on Standard Parts* iGEM <http://ung.igem.org/Main_Page>.

[64] Jon Mooallem, 'Do-It-Yourself Genetic Engineering', *The New York Times Online* (Online), 10 February 2010 <http://www.nytimes.com/2010/02/14/magazine/14Biology-t.html?_r=1&pagewanted=1>; ibid.

[65] See Drew Endy et al, *The BioBrick™ Public Agreement: DRAFT Version 1a* (January 2010) <http://dspace.mit.edu/bitstream/handle/1721.1/50999/BPA_draft_v1a.pdf?sequence=1>.

[66] The BioBricks Foundation, *The BioBrick™ Public Agreement* (2013) (n 39).

[67] See Drew Endy, *Open Biotechnology and the BioBrick Public Agreement* (2009) <http://openwetware.org/images/f/fd/Why_the_BPAv1.pdf>.

[68] The BioBricks Foundation, *The BioBrick™ User Agreement Version 1.0* (2011) The BioBricks Foundation <http://biobricks.org/bpa/users/agreement/>; The BioBricks Foundation, *The BioBrick™ Contributor Agreement Version 1*

defined as 'the particular standardized genetic material(s), their uses, and any associated sequence or functional information' specified in the Contributor Agreement. The system also provides a tool for searching for 'free-to-use' parts contributed under the BPA.

The rules that have been developed by the BioBricks community for sharing parts under the BPA reflect another of Ostrom's design principles, that rules need to 'relate to the specific attributes of the CPR [common-pool resource] and the community of appropriators using the CPR'.[69] The rules in the BPA are very specific to the nature of standard biological parts as both information and a physical resource, the types of users who would wish to access parts, the practical and legal issues raised by this access, and the concerns that users are likely to have about conditions of access, such as attribution and intellectual property issues. The details of these operational rules are set out below.

The contributor is defined as the 'person, company, institution, or other entity submitting the Materials', and who enters into the contributor agreement.[70] They can 'sign' the contributor agreement through the BioBricks Foundation's website under a link titled 'Give a Part: Contribute DNA for free use.' Making a contribution does not require that any physical DNA be sent to a particular collection point, as many parts can be synthesised from the sequence information.[71] Contributors are invited to consider sending physical samples of more complex parts to a repository such as the Registry.[72]

Contributors must provide information about all intellectual property relating to the contributed materials by disclosing any patents or patent applications, copyright or data rights.[73] This includes rights relating to genetic sequences. The contributor can indicate that to the best of their knowledge there are no relevant intellectual property rights. By entering into the contributor agreement, the contributor agrees to permit users to use the contributed materials under the terms of the user agreement. If the contributor does have intellectual property rights in the contributed

---

(2010) The BioBricks Foundation <http://biobricks.org/wp-content/themes/bbf/bpa-sample.php>.

69 Ostrom (n 42) 186.

70 The BioBricks Foundation, *The BioBrick™ Contributor Agreement Version 1* (2010) (n 68).

71 The BioBricks Foundation, *Frequently Asked Questions (BioBricks Public Agreement)* (2016) (n 19).

72 Ibid.

73 The BioBricks Foundation, *The BioBrick™ Contributor Agreement Version 1* (2010) (n 68).

materials, they must 'irrevocably agree' not to assert or threaten to assert these patents or property rights (or future rights from patent applications), 'not to invite to license', and 'not to enforce the rights in any manner against or otherwise adverse' to the BioBricks Foundation or the users.

The contributor also agrees not to charge fees for use of the parts apart from nominal costs of manufacture and shipping. They can request that users attribute the parts to the contributor when they distribute or commercialise material arising from their use of the parts. The contributor agreement does not prevent contributors from entering into a separate agreement with users or others on different terms, so long as any such separate agreement does not 'diminish or derogate from' the rights of the user in the materials provided under the contributor agreement.

The user is defined as 'the person who may request and use the Materials, or the company, institution, or other entity on whose behalf [the] User is authorized to act'.[74] They access the User Agreement via a link titled 'Get your Freedom: Sign agreement and get your parts'. Once they click 'agree', users 'have the right to use all parts contributed under the Agreement'. Users agree that if the contributor has requested attribution, the user will make reasonable efforts to acknowledge the contributor when they make available, commercialise or otherwise distribute the contributed materials. If they undertake any of these distribution activities, including distributing the materials in a modified form, they must also make reasonable efforts to acknowledge that the materials were used under the BPA by conspicuously including the BPA logo 'in all packaging or product inserts, publications, and grant related materials'.

From these agreements, it appears that a new contributor agreement would need to be executed each time a particular contributor donates new materials to the Registry. However, an individual user would only need to agree to the user agreement once for that agreement to apply to all their use of material contributed under any contributor agreements.[75]

For the Registry, there are no contractual conditions, but when users access the repository the expectation is that 'They'll *give* back to the Registry the new parts they've made, as well as data and experience on new and existing parts. Finally, users will *share* experience and collaborate in the Registry's open community through their wikis, the forums,

[74] The BioBricks Foundation, *The BioBrick™ User Agreement Version 1.0* (2011) (n 68).

[75] Ibid.

and other social tools'.[76] While the rules are less defined and detailed, they are still specific to the nature of the resource and the users, complying with Ostrom's design principle.

9.2.4 Why Create a Commons?

The idea of an open access scheme for biological parts was developed by the founders of the parts agenda, including the three scientists above, in about 2004. The perceived importance of open access is clear in Endy's seminal 2005 article Foundations for Engineering Biology.[77] There Endy explained the challenges facing synthetic biology, including intellectual property problems.

The Frequently Asked Questions (FAQs) for the BPA explain the reasons for taking an open access approach:

> it is difficult to share and reuse genetically encoded functions due to high transaction costs associated with patent-based licensing (i.e., time and money) … we believe that biotechnology must move towards a free-to-use 'dictionary' of biological functions that allow many people to benefit from all the potential creative and constructive uses of biology … There are few tools that help people to partner in biotechnology in achieving more than any one person or organization can alone. We therefore decided to develop some of the needed tools.[78]

In his book *The Public Domain*, Boyle revealed part of the motivation for establishing the open access system. He explained that synthetic biologists came to him to ask how the categories of intellectual property would apply to their research, and how they could keep the science open.[79] Boyle suggested to them that an option would be to 'simply put all their work into the public domain immediately'.[80] In his subsequent book, he noted that this was occurring through the Registry.[81] This meant that patents could not be sought for the publicly disclosed material.[82] He saw this as demonstrating a new development, the creation of a 'privately

---

[76] iGEM Foundation, *Our Philosophy* Registry of Standard Biological Parts <http://parts.igem.org/Help:Philosophy>.

[77] Endy, 'Foundations for Engineering Biology' (n 6).

[78] The BioBricks Foundation, *FAQ (BioBricks Public Agreement)* (2016) (n 19).

[79] Boyle, *The Public Domain: Enclosing the Commons of the Mind* (n 4) 174.

[80] Ibid., 177.

[81] Ibid.

[82] Ibid.

constructed commons where the public domain created by the state does not give you the freedom that you believe creativity needs in order to thrive'.[83]

The founders of the parts agenda, Endy, Tom Knight and Rettberg, all came from engineering backgrounds. They were 'inspired' by open source software,[84] and may be trying to replicate the success of this approach in the field of synthetic biology research in order to ameliorate their concerns about intellectual property. Indeed, open biotechnology in general can be seen to have grown from the open source software movement.[85]

Prior to the development of the BPA, concerns about intellectual property problems getting in the way of synthetic biology research could be observed in comments the founding scientists had made in various fora. For example, Endy had noted some practical problems in synthetic biology research, stating that legal standards were needed to define the way in which parts from many different sources can be shared.[86] In his view, negotiating Material Transfer Agreements in order to get access to biological materials is time consuming and if a researcher can access the DNA sequence of the part they are interested in, it may be more efficient to have it synthesised rather than negotiate a material transfer agreement. Endy concluded that 'new low- or zero-cost legal strategies that facilitate and protect the exchange of the genetic information defining parts, and the use of parts in combination, are needed'.[87]

Fellow founder of the parts agenda, Rettberg, has also expressed concerns about how patents could affect synthetic biology research. He has argued that patenting in biotechnology plays out differently from in the computer industry.[88] In the computer industry no particular individual patents seemed to be 'critical', and access to patented inventions could occur through cross-licensing.[89] However in the biotechnology industry, Rettberg was concerned about the ability of individual patents to impact

---

[83] Ibid., 178.

[84] Ibid., 176.

[85] Yann Joly, 'Open Biotechnology: Licenses Needed' (2010) **28**(5) *Nature Biotechnology* 417, 417.

[86] Endy, 'Foundations for Engineering Biology' (n 6) 450.

[87] Ibid.

[88] Randy Rettberg, 'Evidence to The United States National Academies, The Organisation for Economic Co-operation and Development, and The Royal Society, Opportunities and Challenges in the Emerging Field of Synthetic Biology' (10 July 2009) Washington DC, 55–6.

[89] Ibid., 56.

upon the intellectual property environment. He cautioned that in biotechnology:

> individual patents could have extreme financial value. And that, I believe has produced a tendency in the academic community to use patents as lottery tickets ... And therefore a patent is put in place which acts as an intellectual block to future activity but for the most part doesn't produce a financial return.[90]

Rettberg was also worried about the impact that patents could have on iGEM, commenting that: 'There is not an actual research exemption. The courts struck that down. There is a de facto one, and will it work for iGEM in the future? I worry that some patent troll will send a letter to MIT and I will be shut down.'[91] Rettberg was referring to *Madey v Duke University*, in which it was held that only a 'very narrow and strictly limited' research exemption applies in the US.[92]

Thus key scientists in synthetic biology were concerned about the potentially damaging impact of the patent system on their research. That is, they were concerned that the patent system would not be effective in achieving its regulatory purpose, the stimulation of innovation, because of unintended negative consequences.[93] These synthetic biologists saw the need to go outside the existing regulatory mechanism of the patent system to avoid these negative impacts upon their research.

When the Registry was established, the founding scientists hoped it would facilitate a wide range of projects in biological engineering, and that other scientists would participate and contribute to the Registry.[94] The establishment of the Registry was part of broader activities in the parts agenda to move biology beyond work in single laboratories to collaboration between many research groups, which was seen as essential

---

[90] Ibid.

[91] Ibid., 38.

[92] *Madey v Duke University*, 307 3d 1351, 1362 (Fed Cir, 2002).

[93] Using the language of Brownsword and co-authors as set out in the Introduction, see Roger Brownsword and Han Somsen, 'Law, Innovation and Technology: Before We Fast Forward – A Forum for Debate' (2009) **1** *Law, Innovation and Technology* 1, 11, 24–5; Roger Brownsword, *Rights, Regulation and the Technological Revolution* (Oxford University Press, 2008), 134–45, 148–9; Roger Brownsword and Morag Goodwin, *Law and the Technologies of the Twenty-First Century: Text and Materials* (Cambridge University Press, 2012), 290–95, 301, 312–16.

[94] The Bio Fab Group: David Baker et al, 'Engineering Life: Building a Fab for Biology' (2006) June *Scientific American* 44, 46.

for the progress of biotechnology research.[95] The collective desire to promote these goals, and awareness of the potential limiting effects of intellectual property on biological research, are consistent with the decision to develop an open access system.

9.3 CHARACTERISING THE STANDARD BIOLOGICAL PARTS COMMONS

9.3.1 Open Source Technology

This section considers whether standard biological parts contributed under the BPA or accessed from the Registry are in fact 'open source' resources for synthetic biology researchers, as they are sometimes described.[96] In technology development, the term 'open source' was originally used in software programming to mean that the products or tools are 'free' to be used, modified and distributed; that these activities are not restricted by property rights.[97] Under the General Public License (GPL) used in many open source software projects, users of source code are free to copy, modify and distribute the code.[98] However, the GPL restricts the conditions on which users may distribute the accessed material. They must make the source code, or the modified version, available under the same free access conditions on which they originally accessed the source code. This is known as the 'viral' nature of such licenses.[99] Like viral particles, the open access terms are self-perpetuating. Users cannot impose any further conditions on the recipient. Users cannot incorporate the accessed code into products protected

---

[95] Ibid., 51.

[96] See, e.g., The Presidential Commission for the Study of Bioethical Issues, 'New Directions: The Ethics of Synthetic Biology and Emerging Technologies' (The Presidential Commission for the Study of Bioethical Issues, December 2010) <http://bioethics.gov/sites/default/files/PCSBI-Synthetic-Biology-Report-12.16.10_0.pdf>, 120.

[97] This definition is adapted from Free Software Foundation Inc, *The Free Software Definition* Free Software Foundation <http://www.gnu.org/philosophy/free-sw.html>.

[98] Free Software Foundation Inc., *The GNU General Public License (GPL) Version 2* <http://www.opensource.org/licenses/gpl-2.0.php>.

[99] Janet Hope, *Biobazaar: The Open Source Revolution and Biotechnology* (Harvard University Press, 2008), 11.

by intellectual property rights, or charge license fees for products that modify or incorporate the accessed code.[100]

Some features of the BPA model distinguish it from open source projects such as those that use the GPL. Under the BPA, users' future commercialisation and licensing strategies are not limited. Users are able to patent and commercialise any developments they make using BioBricks, as there is nothing in the BPA to prevent this. The BPA does not limit agreements the users might make with other parties regarding the contributed materials, stating that nothing in the BPA precludes users (or contributors) from voluntarily entering into a separate agreement with another party that may 'vary from or supersede' the BPA. This means that users are not limited in the conditions that they may impose on access to any intellectual property they develop using contributed material. If users wish to distribute their work based on contributed materials, they do not have to make this available on the same conditions under which they accessed the materials originally. There is nothing to prevent the charging of fees for licences.

A further difference is that users do not have to contribute any material or information to the Registry, such as any modifications or improvements they might make to parts accessed under the User Agreement. As Boyle has noted, in the absence of underlying property rights the strategic disclosure of synthetic biology information into the public domain does not provide a 'right to force subsequent innovators to release their code back into the commons'.[101]

The open source software model requires underlying intellectual property rights, because the copyright in the source code allows the original developer to enforce the conditions of the open access licence.[102] By contrast, sharing under the BPA is not based on an intellectual property holder's decision about how to implement their rights. Instead, the BioBricks Foundation and the iGEM Foundation have tried to be pre-emptive, to create 'sufficient prior art so basically patentability and enclosure would become impossible or at least very difficult'.[103] This is a deliberate and strategic disclosure of knowledge and data into the public

---

[100] Free Software Foundation Inc., *The GNU General Public License (GPL) Version 2* (n 98).

[101] Boyle, *The Public Domain* (n 4) 177.

[102] Ibid., 176.

[103] Richard A Johnson, 'Evidence to The United States National Academies, The Organisation for Economic Co-operation and Development, and The Royal Society, Opportunities and Challenges in the Emerging Field of Synthetic Biology' (10 July 2009) Washington DC, 59.

domain. The strategic disclosure approach was clear in the frequently asked questions for the draft BPA. In response to the question 'Why won't a patent "troll" file patents on all possible novel uses of BPA-contributed parts?', the FAQs stated that 'contributors and others can readily disclose at the time of contribution known and imaginable uses of so-contributed parts, making obvious many future uses'.[104]

Because the BioBricks model does not have the viral characteristics described above, it is quite different from open source projects. It is more accurately described as an 'open biotechnology' project, as discussed below. By contrast, iGEM teams and Registry users are requested to submit any new parts they create to the Registry. Thus the Registry does have a self-perpetuating characteristic, though not in the same way as open source software licences. This expectation to give back is not based on use of existing intellectual property rights in the contributed parts. It is a matter of culture and trust.[105] Trust and a sense of community are important when a group is developing self-regulation for collective resource management.[106]

The fact that the BioBricks model is not truly open source may limit its capacity to operate as a commons for synthetic biology. The model ensures that scientists will have free access to biological parts as originally contributed. If the researcher wants to access multiple parts, the BPA will assist them greatly as they will not have to arrange access with the developer of each part. This reduces the costs associated with material transfer agreements, a problem identified by Endy in 2005.[107] Thus the BPA serves the interests of synthetic biologists who wish to work on original standard biological parts, particularly those who wish to access many different biological parts. It allows them to patent useful applications developed from parts, as they may be required to do under funding arrangements.

However, if scientists wish to access improved or modified biological parts, devices composed from multiple parts, or systems that use parts to achieve particular outcomes with practical applications, the BPA would not serve this interest. The intended scope of the BPA's application is not

---

104 The BioBricks Foundation, *Frequently Asked Questions (BioBricks Public Agreement)* (2016) (n 19).

105 In 2009, Johnson suggested that the intellectual property in the synthetic biology community was based on trust, and stated that it would eventually need to move from trust to a contract model as the field developed: Johnson (n 103) 50.

106 Ostrom (n 42) 43.

107 Endy, 'Foundations for Engineering Biology' (n 6) 450.

made clear in the BPA or the accompanying information. The FAQs refer to the need for biotechnology to 'move towards a free-to-use "dictionary" of biological functions'.[108] This suggests the goal is to ensure free access to basic pre-competitive data and building blocks in synthetic biology – biological parts rather than applications using these parts. The scope of the BPA is sufficient to achieve this. It should avoid some of the problems encountered in other fields of biotechnology research, such as embryonic stem cell research and genetic disease research. In those areas basic research tools were patented, which impacted upon the ability of researchers to work in the field because access to the patented technology was required before any research could be conducted in that area. Thus this open, sharing approach is beneficial for the progress of research, and the promotion of socially desirable innovation in synthetic biology. As this is one of the key roles of regulation in relation to an emerging technology identified in the Introduction,[109] the sharing approach should be supported by government where possible.

9.3.2 Open Biotechnology

In recent years an ethos of 'openness' or 'sharing' can be observed in modern biological research communities more generally. Openness has arisen partly for practical reasons relating to the limitations of the intellectual property system. Due to the uncertainties and costs associated with patenting in biotechnology research, scientists have looked beyond patenting to other strategies such as contractual arrangements.[110] These 'access agreements' do not require underlying intellectual property rights.[111]

Current openness and sharing efforts in biological research are also underlain by a desire for improved outcomes of research efforts and better collaboration. This trend is particularly visible in medical research, in fields such as human genetic disease and cancer research.[112] The

---

[108] The BioBricks Foundation, *FAQ (BioBricks Public Agreement)* (2016) (n 19).

[109] See Brownsword and Somsen (n 93) 2.

[110] Joly (n 85) 418.

[111] Ibid.

[112] An example of a human genetics data sharing project is the SAGE Commons, see Sage Bionetworks, *The Sage Commons* <http://sagebase.org/commons/index.php>.

Presidential Commission noted this trend towards increased sharing of data and development of creative licensing strategies.[113]

These influences can also be seen at work in the standard biological parts commons. In this model, the ethos of sharing can be seen in the proactive disclosure of scientific knowledge and the commons building activities of the participants in the parts agenda. By building a commons which exists *outside* the patent system, the standard biological parts commons model avoids the conflict between the goal of promoting scientific progress through sharing and collaboration on the one hand and the secrecy of the intellectual property system on the other.[114] The practical problems and uncertainties of patenting in this field are avoided by designing a model of deliberate disclosure and non-assertion of intellectual property rights, based on contractual agreements.

Intellectual property and biotechnology expert Yann Joly has identified five defining characteristics of 'open biotechnology' projects. They must 'make use … of the internet and other information technologies', facilitate collaboration by scientists, 'include a strategy to ensure rapid public dissemination of information and research results', allow the use of results without restricting the 'freedom and integrity' of researchers, and 'not use intellectual property to limit access to the project' or its results.[115]

The standard biological parts commons has all of these characteristics. The Registry and the BioBricks Foundation both operate largely online, allowing members of the synthetic biology community to access and add information, contribute to discussions and participate in the development of technical standards. The open access nature of these resources facilitates collaboration within the community. Registered scientists can

---

[113] The Presidential Commission for the Study of Bioethical Issues (n 96) 120–21.

[114] Sociologist of science Robert K Merton argued that common ownership was 'an integral element of the scientific ethos'. This norm was said to be in conflict with the concept of intellectual property, with secrecy being the 'antithesis of this norm': see Robert K Merton, *The Sociology of Science: Theoretical and Empirical Investigations* (University of Chicago Press, 1973), 273–5. However, it should also be noted that Merton's norms of science have subsequently been criticised by sociologists of science. For example, Edmond and Mercer have argued that they are not borne out by ethnographic studies of scientists, and are more akin to ideals or norms that 'should exist if science's objectivity is to be guaranteed': Gary Edmond and David Mercer, 'Norms and Irony in the Biosciences: Ameliorating Critique in Synthetic Biology' (2009) **21**(3) *Law and Literature* 445, 465.

[115] Joly (n 85) 418.

contribute information, view information contributed by others, and use the online resources to identify those working on projects related to their own. The community building activities, including the Synthetic Biology X.0 conferences and iGEM, also facilitate collaboration. The BPA facilitates open dissemination and use of experimental results by providing a mechanism for the contribution of materials on the condition that any property rights identified will not be asserted against other members of the community. This means that intellectual property does not limit access to the materials contributed under the BPA. For these reasons, the model is more accurately described as 'open biotechnology' than 'open source'.

9.3.3 Democratising Innovation and Innovation Communities

In his book *Democratizing Innovation*, economist Eric von Hippel discussed the phenomenon of 'user innovation' – a trend in which users develop or improve products for their own use, rather than being limited to those products available from commercial manufacturers.[116] It has been suggested that synthetic biology is such a 'democratizing innovation effort'.[117] According to von Hippel, users innovate when they want customised products to fit their own specific requirements. He explains that users will innovate themselves when there are significant costs in hiring another party to design a product for them, and also because of sheer enjoyment of the innovation, problem-solving and learning processes. This theory also encompasses reasons why user innovators might 'freely reveal' their innovations rather than keep them secret or seek intellectual property rights.

Von Hippel's examples were drawn from innovation in a wide variety of areas including software and sports equipment. He did not apply this theory to the conduct of scientific research. However, a number of aspects of the theory of democratising innovation are echoed in synthetic biology's parts agenda. Democratising innovation means a democratisation of the 'opportunity to create'.[118] This can be seen in the parts agenda's aim to make biological engineering more predictable and easier to carry out by developing standardised methods and components. The founders of the parts agenda hope that this will open up the field to a

116 Eric von Hippel, *Democratizing Innovation* (The MIT Press, 2005).
117 Johnson (n 103) 50.
118 Von Hippel (n 116) 123–4.

wide range of researchers.[119] The agenda also extends the 'opportunity to create' to undergraduate students through iGEM. The distribution of sets of biological parts to iGEM teams and other Registry users mirrors the concept of manufacturers making 'tool kits' available for user innovation.[120] In the synthetic biology context, this is more accurately characterised as 'lead users' (those at the leading edge of their field, who would gain from solutions to problems within the field)[121] supplying other users with tool kits needed to innovate in the field.

User innovation can occur within 'innovation communities' when some users are willing to reveal their innovation, and other users find the information useful.[122] Innovation communities are:

> nodes consisting of individuals or firms interconnected by information transfer links which may involve face-to-face, electronic, or other communication. These can … exist within the boundaries of a membership group. They often … incorporate the qualities of communities for participants, where 'communities' is defined as meaning 'networks of interpersonal ties that provide sociability, support, information, a sense of belonging, and social identity'.[123]

The standard biological parts commons has many of the features of an 'innovation community'. Scientists participating in the commons are linked through both electronic communication and face-to-face interactions at the biannual Synthetic Biology X.0 conferences. Use of the Registry and the parts contributed under the BPA is limited to those who have registered, creating a membership group. The BioBricks Foundation and the Registry provide various technical support and training for registered users, such as workshops, online tutorials and scientific protocols.

The various community building and leadership activities of these organisations facilitate a sense of the emergence of a new community within biological research. Online forums and email discussion lists also provide tools to interact and share ideas within this community, as in von Hippel's concept of an innovation community. Innovation communities are often highly specialised and form 'repositories for information' for

---

[119] See The BioBricks Foundation, *FAQ (BioBricks Public Agreement)* (2016) (n 19).

[120] Von Hippel (n 116) 128.

[121] Ibid., 4.

[122] Ibid., 96.

[123] Ibid.

that particular area of innovation.[124] The Registry and BioBricks Foundation (through the BPA) function as repositories of information for the highly specialised field of research in synthetic biology's parts agenda. Making the information easily accessible, for example by collecting it in the Registry, increases its value to innovators.[125]

9.4 EVALUATING THE ACCESS ARRANGEMENTS

9.4.1 Patent Issues

Some commentary on the optimal form of an open source mechanism for synthetic biology has assumed that individual parts are patentable.[126] A search of the USPTO databases did not yield patents for inventions comprising BioBricks or standard biological parts themselves.[127] A number of patents have been granted for inventions which make use of BioBricks parts or 'BioBricks standard' DNA assembly techniques.[128] The parts and methods are elements of the patented methods or constructs. Some iGEM teams have also filed patent applications for constructs made with parts. Theoretically, an invention meeting the definition of a standard biological part or composition of such parts could be patented where the DNA has been sufficiently altered from a naturally occurring sequence.[129] Standard biological parts could also include stretches of DNA which are covered by now potentially invalid genetic sequence patents. Prior to the 2013 Supreme Court decision in *Myriad*, patent holders for those genetic sequences could have taken infringement action against synthetic biologists.[130] As the law now stands the potential for infringement of these patents on naturally occurring sequences has been ameliorated. The Registry does not provide information about the intellectual property status of registered parts. Intellectual property status

---

[124] Ibid.

[125] Ibid.

[126] See, e.g., Joachim Henkel and Stephen M Maurer, 'The Economics of Synthetic Biology' (2007) **3**(117) *Molecular Systems Biology* 1.

[127] Search of US Patent and Trademark Office Patents database 16 December 2016. Search terms: title field ('standard biological part', 'biological part', 'biobrick', 'standard biological'), all fields ('standard biological part', 'biobrick').

[128] Timothy Kuan-Ta Lu and Allen Cheng (2016) 'Massively Parallel Combinatorial Genetics', US Patent No 9,315,806.

[129] See Chapter 8.

[130] See *Association for Molecular Pathology, et al v Myriad Genetics Inc, et al*, 133 US (S Ct, 2013) 2107 and Chapter 8.

could include issues such as whether the part is patented and whether the contributors of that part have the authority to make the sequence of the part freely available or to allow the Registry to distribute the physical DNA comprising the part by registering it. The lack of information means that if a researcher wanted to be sure they were not infringing any patents; they would need to conduct patent searches for the sequences of each part they used and all experimental methods.

By contrast, under the BPA, contributors have to provide information about the intellectual property status of their contributed material 'to the best of their knowledge'. If they use BPA parts, users will know that the contributor is not going to pursue them for patent infringement, and will be aware of any patents held on that material by other parties. However, the utility of this mechanism depends on accurate disclosure of intellectual property status by contributors and will only apply prospectively. Prior to the BPA coming into effect, and even now if they use parts not covered by it, scientists will have to do some investigating if they wish to be sure that they are not infringing any patents when working with parts and devices. However, Rai stated in her evidence to the Commission that research shows academic life scientists are quite unconcerned about infringing patents in their work, and that the companies which own these patents tend not to sue them for infringement.[131] This suggests it would be valid to question how likely scientists working in synthetic biology are to be concerned about infringing patents when using material from the Registry. One of the goals of the BioBricks Foundation is to develop technical standards for use in parts-based synthetic biology research. If they use parts contributed under the BPA, scientists would be able to develop technical standards using material that is shared under the BPA. This would avoid the potential problem identified by Kumar and Rai of '"hold up" by a previously unknown patent holder who emerges only after others have invested large sums of money'.[132] This is particularly problematic if the patent relates to some standard that is then difficult to

---

[131] Arti K Rai, 'Evidence to The Presidential Commission for the Study of Bioethical Issues, Synthetic Biology' (13 September 2010) Philadelphia Penn. See John P Walsh et al, 'View from the Bench: Patents and Material Transfers' (2005) **309**(5743) *Science* 2002; see also Timothy Caulfield et al, 'Evidence and Anecdotes: An Analysis of Human Gene Patenting Controversies' (2006) **24**(9) *Nature Biotechnology* 1091; Naomi Hawkins, 'The Impact of Human Gene Patents on Genetic Testing in the United Kingdom' (2011) **13**(4) *Genetics in Medicine* 320.

[132] Kumar and Rai (n 4) 1758.

change.[133] In the synthetic biology context, this could involve, for example, a technical standard being developed for activities involving BioBricks. This might be used by many in the community as a basis for research, until a patent holder later appears and wants to enforce their patents against the researchers, causing difficulties.

9.4.2 Motivation to Contribute to the Commons

The Registry is available to iGEM teams and registered academic laboratories.[134] However, the synthetic biology 'community' envisaged by the iGEM Foundation also includes industry.[135] The BPA is expressly geared not only at academic researchers but also companies.[136] A question arises as to why scientists, particularly in commercial enterprises, will contribute parts or data to the Registry or under the BPA. This arises particularly in relation to the BPA, as users are not required to make any contribution in order to access the information contained there. If disclosing information means its value is potentially reduced, what would be the incentive? For example, under the BPA a user who develops a commercial application from contributed material can obtain their own patent over this. Their exclusive rights will be unencumbered by any obligations to the original contributor; there is no requirement that they give the original contributor any share in income generated by the patent. Commercial entities may be reluctant to contribute parts under these conditions.

In *Democratizing Innovation*, von Hippel identified reasons why user innovators might 'freely reveal' their innovations into a commons. Free revealing means that 'all intellectual property rights to that information are voluntarily given up by that innovator and all parties are given equal access to it', for example publishing information in a journal or publicly accessible website.[137] Open source software does not quite fit this definition because of the limitations imposed on use of the revealed code.[138] Similarly, revealing the sequence of a biological part under the BPA does not quite fit this definition as there are limitations on how the part can be used. However, von Hippel's theory is helpful in understanding why synthetic biologists may be willing to reveal.

---

133 Ibid.

134 iGEM Foundation, *Registry of Standard Biological Parts* (n 17).

135 iGEM Foundation, *iGEM Boilerplate* iGEM <http://igem.org/Press_Kit>.

136 The BioBricks Foundation, *The BioBrick™ Public Agreement* (n 39).

137 Von Hippel (n 116) 77.

138 Ibid.

Von Hippel explained that many innovators have a dim view of the value of obtaining a patent and that the time and cost involved in patenting can make this avenue difficult for individuals and small- or medium-sized firms.[139] So it might make practical sense for a user to freely release their innovation rather than protect it with intellectual property rights.[140] Further, 'innovators' active efforts to diffuse information about their innovations suggest that there are positive, private rewards to be obtained from free revealing'.[141] These include boosting the innovator's reputation amongst their peers, and benefits that flow from the fact that freely released innovations are more likely to be widely used by others, for example, improvement or 'debugging' of the innovation by other users.[142]

These influences can be seen at work in the standard biological parts commons. The formalised sharing and collaboration mechanisms under both the BPA and the Registry model enhance the benefits discussed above that scientists could expect from putting their work into the public domain individually. There may be limited incentives to patent biological parts, given that an open access resource already exists and could reduce the market for exploitation of the patent. The high costs of patenting are also problematic.[143]

Contributing a biological part to the commons could certainly be seen to enhance the reputation and credibility of the part's designers amongst their synthetic biologist peers by publicising the work, particularly if the part has been successfully used in experiments. Similar to benefits the designers would obtain from publishing a paper about their work, contributing a part to the commons allows the designers to be identified by interested peers, who might offer suggestions about using and improving the part or opportunities to perform collaborative research. All commons users are potentially able to resolve problems relating to a part or improve it. Users can add experience and information about using a part to the Registry, for the whole community's benefit. In this way the original designers can benefit from putting their part into the public domain. Currently the BPA does not have a mechanism for experience sharing, but it does allow identification of the designers and the potential for collaboration and discussion about using the part.

---

139 Ibid., 84.
140 Ibid., 85.
141 Ibid., 86.
142 Ibid., 86–9.
143 Drew Endy, 'Evidence to The Presidential Commission for the Study of Bioethical Issues, Synthetic Biology' (8 July 2010) Washington DC.

For these reasons, in the standard biological parts context revealing information could actually be seen to *increase* its value. Ostrom's theory of collective action is mirrored in the synthetic biology community in these ways, because the innovators gain more benefit by sharing and coordinating than they would by acting independently.

Further, it has been argued that *use* of a biological part will increase its value, because this generates knowledge about how the part actually worked, reducing the effort and cost involved in future use of that part.[144] This experimental experience and data can then be shared through the Registry, where all users can record experience using parts. Users would be more inclined to use parts that have already been successfully used by others.[145] The Registry's online catalogue provides information from users about whether a part actually works, and some parts are rated using Registry 'stars'. A part that is being successfully used and providing value to other Registry users could certainly boost the reputation of the original contributor.

Being involved in an open research community can also be a satisfying end in itself: 'creative programmers want to associate with each other: only their peers are able to truly appreciate their art ... it's also important that people want to share the beauty of what they have found. This sharing is another act that helps build community and friendship'.[146] In the highly specialised field of synthetic biology using standard biological parts, it can certainly be imagined that belonging to a community of researchers with similar passions and goals, and sharing one's achievements with them, would be rewarding activities in themselves. A sense of wanting to 'give back' has also been noted in open source software projects, where some programmers wish to contribute something back to those who have allowed them to access software source code.[147] As synthetic biology progresses and young scientists and those who have participated in iGEM as students become more senior in the field, this is likely to become an increasingly significant reason for making contributions.

9.4.3 Obligations to Funders

Some potential limitations of the standard biological parts commons arise from scientists' obligations to their research institutions and funders. In

144 Henkel and Maurer, 'The Economics of Synthetic Biology' (n 127) 1.
145 Ibid., 2.
146 Von Hippel (n 116) 124, quoting computer scientist Chris Hanson.
147 Ibid., 87.

the US, universities and other non-profit organisations or small businesses who receive federal funding for their research have obligations to the Federal Government under the Bayh Dole Act.[148] This may affect their ability to release data into the public domain. The key provision is 35 USC § 202. Under this provision, universities receiving federal funding can elect to keep the title to any inventions conceived in work carried out under the funding agreement. If the organisation does not elect to retain rights, the federal agency can receive the title or the inventor may request that they be granted the rights. If the funded organisation does retain the rights, they must agree to file a patent application within the statutory time limit of one year after disclosure.[149] If this is not done, the Federal Government may receive title to the inventions. The federal agency funding the research can require periodic reporting on the utilisation of the invention. Non-profit organisations cannot generally assign their rights to an invention without the approval of the funding agency.

This means that scientists whose research is federally funded could be in breach of their funding agreements if they do not patent their biological parts. Parts could still be disclosed to the commons *after* a patent application was filed, or a patent application could be filed up to one year after the first public disclosure of the part. But agreeing to the BPA may be problematic as this requires promising not to assert the future patent against users of the Registry. This will depend on the terms of the funding agreement. In any case, there would be little point in patenting a part that was going to be contributed under the BPA. It would have limited value because no revenue could be obtained from any users who obtained the part under the BPA. Similarly, if the organisation does not retain title to the invention, and title is granted to the scientist who designed the part, contributing the part to the commons may breach the terms of the grant of rights to the inventor.

The Bayh Dole Act only applies to *inventions*, so data that do not themselves comprise an invention could be released, such as experimental experience using parts.

---

148 See 35 USC (Office of the Law Revision Counsel, 2016) §§ 200–212. The inventions covered by the provisions of §§ 200–212 are referred to as 'subject inventions', defined in para. 201 (e) as 'any invention of the contractor conceived or first actually reduced to practice in the performance of work under a funding agreement'. Universities are included in the definition of 'non-profit organisation' in § 201 (i).

149 Ibid., §§ 202 (c)(3), see also time limits in §102 (b).

A university synthetic biologist's ability to participate in the open disclosure of parts also depends on the particular university's policies. The BioBricks Foundation could take a leadership role by liaising with university technology transfer staff to discuss how the BPA applies to scientists at their institution. Guidance could be developed on how the provisions of the Bayh Dole Act will interact with the duties and obligations under the BPA. Policy staff within government with responsibility for administering the Bayh Dole Act should also be involved in discussing its application to activities under the BPA.

In addition, many synthetic biologists are at least partly funded by commercial entities. Indeed, 'Distinguishing between academic, public and commercial research in synthetic biology is extremely difficult, as many researchers are active contributors in each domain.'[150] The factors identified above which encourage and reward disclosure of parts would have greater application to academic researchers than commercial entities, particularly those factors relating to reputation in the field. Commercial funders or collaborators may well impose conditions on disclosure of data and exercise of intellectual property rights such that the researcher would not have authority to disclose parts and other data or agree to the BPA.

9.4.4 Contributions and Usage

In practice, parts *are* being contributed to the Registry. In 2009 it had 3 500 parts, and has swelled to over 10 000 parts in 2016.[151] Rettberg has stated that much of the Registry's growth could be attributed to contributions by iGEM teams under the 'get and give' philosophy.[152]

Since the BPA was introduced, 22 separate contributions have been made under that agreement by six individuals including Endy.[153] Some of these contributions contain large sets of parts developed by the BIOFAB, a key institution which supports the parts agenda. According to the

---

150 The Presidential Commission for the Study of Bioethical Issues (n 96) 114.

151 See Rettberg (n 88) 36; The BioBricks Foundation, *FAQ (BioBricks Public Agreement)* (2016) (n 19).

152 Rettberg (n 88).

153 See The BioBricks Foundation, *Find a Part* (2016) The BioBricks Foundation <https://biobricks.org/bpa/find-a-part/>.

BioBricks Foundation website, Endy has contributed thousands of parts under the BPA.[154]

In relation to user uptake, the Registry is being widely used by all the teams in iGEM. Empirical research has suggested that 'the majority' of academic synthetic biologists use or contribute to the Registry.[155] It is not clear how many synthetic biologists have made use of the parts contained in the Registry or under the BPA. The BioBricks Foundation does not keep records of user agreements,[156] so the total number of users of the BPA is not currently known.

9.4.5 Practical and Enforcement Matters

There are some practical issues that may limit the effectiveness of the BPA which relate to the parties to the agreement and its enforcement. For biobanking facilities such as a stem cell bank, the deposit agreement would be made between the particular contributor and the banking facility and the access agreements would be made between the particular accessor and the banking facility. For example, the UK Stem Cell Bank has a Research Use Licence which sets out the terms for accessing and using stem cells, including intellectual property arrangements.[157] This agreement states that it is an agreement between the Secretary of State for Health (via the National Institute for Biological Standards and Control operating through the bank), the depositor of the stem cells and the person requesting access to the stem cells. This means that the bank could take a role in monitoring compliance and addressing any issues that arise.

In the BioBricks context however, once a contributor agrees to the contributor agreement, this forms a contract not with the BioBricks Foundation or any other institution, but with 'anyone who has signed or will later sign the User Agreement'.[158] Similarly, the user agreement is between the user and all the contributors.

---

[154] The BioBricks Foundation, *Free! Reliable! Parts! Use 'Em! Courtesy Drew Endy* (20 March 2013) The BioBricks Foundation <http://biobricks.org/2013/free-reliable-parts-use-em-courtesy-drew-endy/>.

[155] Linda J Kahl and Drew Endy, 'A Survey of Enabling Technologies in Synthetic Biology' (2013) **7**(13) *Journal of Biological Engineering* 1, 3.

[156] The BioBricks Foundation, *Users* (2013) The BioBricks Foundation <http://biobricks.org/bpa/users/agreement/>.

[157] UK Stem Cell Bank, *Research Use Licence* UK Stem Cell Bank <http://www.nibsc.org/asset.ashx?assetid=d54c90f7-95e8-4d90-b837-84c504404f74>.

[158] The BioBricks Foundation, *Contributors* (2013) The BioBricks Foundation <http://biobricks.org/bpa/contributors/>.

The BioBricks Foundation's disclaimer suggests that the Foundation envisages no role for itself in monitoring compliance or addressing an alleged breach of the BPA: 'The BioBricks Foundation is not a law firm and does not provide legal services ... The BioBricks Foundation makes no warranties regarding the information provided, and disclaims liability for damages resulting from its use.'[159]

The BioBricks Foundation could take no regulatory role in relation to the intellectual property or attribution requirements, as it is not a party to the BPA and no obligations are owed to it. By contrast, a biobank would keep records of all the agreements relating to access and may take a role in enforcing them in the situation of a breach. For example, anyone accessing resources from the UK Biobank enters into a material transfer agreement with the biobank, which can potentially intervene in the use of intellectual property in certain circumstances and would have records of all intellectual property agreements.[160]

Ostrom's design principles for collective management of common-pool resources emphasise the importance of monitoring compliance with the agreed rules. There must either be monitors who are accountable to those using the resource, or the monitoring must be done by the users themselves (mutual monitoring).[161] Ostrom proposed a further, related design principle: there must be graduated sanctions applicable to non-compliance.[162] Ostrom argued that these factors affect the likelihood that users will commit to the rules. Users want to be assured that others are also complying, and that they will not end up being exploited because they comply while others do not.[163] If they have this assurance, they are more likely to commit to compliance themselves.[164] The presence of

---

[159] Words preceding the User Agreement: The BioBricks Foundation, *The BioBrick™ User Agreement Version 1.0* (2011) (n 68).

[160] UK Biobank, 'Access Procedures: Application and Review Procedures for Access to the UK Biobank Resource' (UK Biobank, 2011) <http://www.ukbiobank.ac.uk/wp-content/uploads/2011/11/Access_Procedures_Nov_2011.pdf>, Annex II: Material Transfer Agreement for Data and/or Samples. See also discussion of access terms for biobanks in Dianne Nicol and Richard Gold, 'Standards for Biobank Access and Intellectual Property' in Rimmer and McLennan, *Intellectual Property and Emerging Technologies: The New Biology* (n 1).

[161] Ostrom (n 42) 90–100.

[162] Ibid.

[163] Ibid., 97, 186.

[164] Ibid., 186.

graduated sanctions deters non-compliance across the group and re-assures users.[165]

At present, the standard biological parts commons does not comply with Ostrom's design principles of monitoring and graduated sanctions. There is no entity with responsibility for monitoring compliance with the BPA, or addressing non-compliance by applying sanctions. Similarly, there is no visible mechanism for monitoring compliance with the rules for use of the Registry, or addressing non-compliance.

Further, there is no particular mechanism for resolving disputes in the standard biological parts commons. The need for low-cost, local-conflict resolution mechanisms forms another of Ostrom's design principles.[166] Within the BioBricks commons however, the BioBricks Foundation's discussion fora provide a place to discuss and resolve issues about the precise application or interpretation of the rules.[167]

There is potential for these issues to reduce the effectiveness of the standard biological parts commons over time. It is certainly possible that the iGEM Foundation, if it became aware of non-compliance, could cancel a user's registration and cease to supply physical parts to that entity or individual. Similarly, if non-compliance became apparent, the BioBricks Foundation could cancel a user's registration. However, as noted the BioBricks Foundation does not see itself taking a regulatory role. In either part system, users could informally ostracise a non-compliant user by no longer communicating and collaborating with them, but this could only occur if they were aware of the breach.

Instead, the goal of the BPA is to make the process of contributing and accessing parts as straightforward as possible. It is true that there may be less concern about breaches of the rules than in some other communities, because there is a norm of sharing. As Ostrom points out, 'in a setting in which there are strong norms against opportunistic behaviour, each appropriator will be less wary about the dangers of opportunism'.[168] However, the possibility needs to be considered that the BPA's terms may be breached, and hence it needs to be clear who could seek to enforce them or to resolve a conflict. There needs to be a central body with the function of collecting and maintaining records of all the agreements that are made. Otherwise, the obligations cannot be enforced as the relevant parties could not be determined. Users will have to resolve conflicts among themselves.

---

165 Ibid., 186–7.
166 Ibid., 90.
167 See ibid., 100–101.
168 Ibid., 36.

9.5 COMMENTS OF THE PRESIDENTIAL COMMISSION

The Commission's 2011 report discussed the open sharing of standard biological parts in the context of both intellectual property issues and public benefit. The Commission noted synthetic biology's dependence on standardisation: 'Because synthetic biology is in large part based on the application of engineering principles through the use of standardized modular parts, access to those standard components could be especially critical to the development of the field.'[169] It identified these potential problems but it did not assess potential solutions. This is part of the general lost opportunity to make specific recommendations discussed in Chapter 4.

Intellectual property and sharing were also considered within the discussion of the principle of 'public beneficence'. The Commission noted: 'Information sharing and reasonable access to discoveries and inventions have long fuelled the scientific enterprise. These activities enable scientists to leverage each other's work in order to more quickly advance new projects and translate basic research into products.'[170] The Commission expressed no particular view on the activities of the BioBricks Foundation. However, its statements could possibly be read as implicit support for the commons-building activities.

9.6 OTHER ACCESS MECHANISMS

9.6.1 Public Sector Licensing

Government agencies can take a role in promoting open access to standard biological parts and minimising potential patent problems in synthetic biology. The Commission's discussion of intellectual property emphasised the principle of public beneficence. It stated that: 'Information sharing is a critical mechanism for promoting scientific progress and innovation.'[171] It recommended that the government 'should consider best practices and other policy guidance, if needed, to ensure that access to basic research results and tasks is not unduly limited'.[172]

---

[169] The Presidential Commission for the Study of Bioethical Issues (n 96) 120–21.

[170] Ibid., 119.

[171] Ibid., 7.

[172] Ibid.

Because a great deal of public money is being put into synthetic biology, as discussed in Chapter 2, it is indeed important to promote public benefit from this research. The Commission rightly pointed out the importance of access to basic tools and information. Because of this, government policies should encourage sharing of basic tools such as BioBricks and support the open access mechanisms that have been developed. Sharing should be incentivised through the conditions imposed on public funding grants.[173]

Along with good-quality patenting, licensing behaviour is an important policy area to be considered in order to foster innovation in synthetic biology.[174] Licensing behaviour can make a considerable difference to access and impact of a patent. Public sector funding agencies can take a role in relation to licensing behaviour. If need be, the fact that federal funds are used for much research underpinning patent applications in synthetic biology can be leveraged to promote reasonable licensing behaviour and encourage non-troll behaviour. Where inventions have been created using government funds, the government retains certain rights. For example, where the grantee is a small business or non-profit organisation, the agency has a non-exclusive licence to use the patented technology and retains 'march-in rights'.[175] March-in rights allow funding agencies to require that licences be granted in certain circumstances. Further, these provisions allow the funding agency to require the patent holder to grant a licence on 'reasonable' terms. Many of the Venter Institute patent applications discussed in the previous chapter were developed with federal funding, as set out in Tables 8.1 and 8.2. Public sector funding agencies could also discourage research-blocking behaviour by encouraging or requiring non-exclusive licensing in their funding agreements.

In addition to march-in rights, the US Government could also use compulsory licensing to promote access to patented synthetic biology technology if this becomes necessary. Compulsory licensing refers to use of the patented invention without the agreement of the patent holder. There are several specific provisions for government use of patented inventions. First, where an invention has been created using federal funding, the Federal Government retains a royalty-free licence to practice the invention 'for or on behalf of the United States'.[176] It has been argued that this phrase extends the licence to government contractors, so that

---

173 Henkel and Maurer, 'Parts, Property and Sharing' (n 5).
174 Rai (n 132).
175 35 USC (n 149) §§ 202 (c) (4), 203.
176 Ibid § 202 (c) (4).

they are free to use inventions created with federal funding.[177] Secondly, if a patented invention is used or manufactured 'by or for the United States' without a licence, the patent holder can take action in the Court of Federal Claims to recover 'reasonable and entire compensation' for the use or manufacture.[178] This option is not dependent on the government having funded the development of the invention.[179]

Options such as march-in rights and compulsory licensing can have an impact on effect on patent access without actually being used, because the threat of use of the provisions encourages patent holders to negotiate and offer reasonable terms.[180] In the embryonic stem cell research context, the existence of march-in rights meant that the NIH had leverage to negotiate reasonable licence terms with the holder of the foundational patents.[181] If necessary, a similar approach could be taken in the synthetic biology context.

9.6.2 Patent Pools

The nature of synthetic biology as a complex technology suggests that there may be a role for patent pooling as a mechanism for accessing a group of technologies with reduced transaction costs.[182] For example, use or development of complex devices in synthetic biology could require access to multiple patented elements or parts. In a patent pool, each patent holder would license their technology to each other patent holder, and the whole set of licences would then be licensed to third parties

---

177 Terry K Tullis, 'Application of the Government License Defense to Federally Funded Nanotechnology Research: The Case for a Limited Patent Compulsory Licensing Regime' (2005) **53** *UCLA Law Review* 279, 306.

178 28 USC (Office of the Law Revision Counsel, 2016), § 1498 (a).

179 Jerome H Reichman, 'Compulsory Licensing of Patented Inventions: Comparing United States Law and Practice with Options under the TRIPS Agreement' (Duke University School of Law, 2006) <http://www.aals.org/documents/2006intprop/JeromeReichmanOutline.pdf>.

180 Mark A Lemley, 'Contracting Around Liability Rules' (2012) **100**(2) *California Law Review* 463, 476.

181 See Rebecca S Eisenberg and Arti K Rai, 'Proprietary Considerations' in R Lanza et al (ed), *Handbook of Stem Cells*, (Elsevier Academic Press, 2004) 793, 793–7.

182 Henkel and Maurer, 'Parts, Property and Sharing' (n 5) 1095.

outside the pool.[183] The license to third parties would be non-exclusive.[184] The benefit to users would be that they could obtain a 'one-stop licence' rather than having to negotiate with multiple patent holders.[185] This strategy could be useful where the patent environment is complex, for example in addressing patent thickets.[186] There are examples of patent pools and clearinghouses in biomedical science; though these collaborative models are not widely used in for-profit research.[187] However, a potential problem associated with patent pooling is difficulty in developing terms agreeable to all patent holders.[188]

In academic synthetic biology, the strong ethos of community and sharing could assist in developing terms for non-profit research access. If patent thickets begin to emerge, synthetic biologists should draw on the discussions and experience in other areas of complex technology.

Whether this will become necessary will depend on many factors, including the success of the standard biological parts commons and the trajectory of development of the scientific field.

9.6.3 The Defence of Experimental Use

Lastly, consideration could be given to law reform to enlarge the scope of the experimental use defence to patented infringement in the US. In the litigation over the *Myriad* patents, the Court of Appeals for the Federal Circuit stated that companies do not usually enforce their patents against researchers.[189] This was also noted in evidence to the Commission.[190] There is an impression among the synthetic biology community that current research is covered by an experimental use exemption, or that they will not be sued because it is not worthwhile for a company to do

---

183 Esther van Zimmeren, et al, 'Patent Pools and Clearinghouses in the Life Sciences' (2011) **29**(11) *Trends in Biotechnology* 569, 570.

184 Dianne Nicol, 'Patent Licensing in Medical Biotechnology in Australia: A Role for Collaborative Licensing Strategies?' (Centre for Law and Genetics Occasional Paper No 7, 2010) <http://www.lawgenecentre.org/occpaper_7.pdf>, 10.

185 Van Zimmeren et al (n 183) 570.

186 Ibid., 569; Nicol (n 184) 10.

187 Van Zimmeren et al (n 183) 573.

188 Henkel and Maurer, 'Parts, Property and Sharing' (n 5) 1095.

189 *Association for Molecular Pathology v United States Patent and Trademark Office*, 689 F 3d (CA Fed (NY), 2012) 1303, 1331 (Lourie J).

190 Ashley Stevens, 'Evidence to The Presidential Commission for the Study of Bioethical Issues, Synthetic Biology' (13 September 2010) Philadelphia.

so.[191] However, the legal scope of the experimental use defence to patent infringement is actually very narrow.[192] It may not provide a 'stable' legal basis for research in synthetic biology.[193] There has been some call for this to be widened for synthetic biology, as well as for other technologies.[194]

While a provision could be legislated to exempt non-commercial use from activities constituting patent infringement, this can be very difficult to define in the biotechnology research world, as discussed above. It is not yet clear whether this is necessary in synthetic biology, due to the relatively early stage of research and patenting in the field.

9.7 CONCLUSION

This chapter has highlighted that in the emerging field of synthetic biology, 'the proprietary model faces competition from open source alternatives'.[195] As yet it is unclear whether the diverging approaches of patenting and openness can co-exist and how the intellectual property environment will develop.[196]

Through the Registry, the BPA and other initiatives, leading synthetic biologists have created a commons in which their community can conduct its research. This commons exists outside the intellectual property system. Instead of pursuing private property rights, synthetic biologists are disclosing their work as a strategy to protect their field from future intellectual property problems.

Thus the standard biological parts commons provides an example of scientists creating a space for their own research, right from the establishment of a new field. They have created this commons due to

---

191 Arti K Rai, 'Unstandard Standardization: The Case of Biology' (2010) **53**(1) *Communications of the ACM* 37, 39; Drew Endy referred to an 'operational research exemption at universities' in his comments to the Presidential Commission: Endy, 'Evidence to The Presidential Commission for the Study of Bioethical Issues, Synthetic Biology' (n 143).

192 *Madey v Duke University* (n 92) 1361–2.

193 Van Zimmeren et al (n 183).

194 Arjun Bhutkar, 'Synthetic Biology: Navigating the Challenges Ahead' (2005) **8**(2) *The Journal of Biolaw and Business* 19, 25; Tullis (n 177) 300–303.

195 James Boyle, 'Monopolists of the Genetic Code?' (28 May, 2010) *The Public Domain* <http://www.thepublicdomain.org/2010/05/28/monopolists-of-the-genetic-code/>.

196 Jane Calvert, 'Ownership and Sharing in Synthetic Biology: A "Diverse Ecology" of the Open and the Proprietary' (2012) **7**(2) *BioSocietie* 169.

perceived limitations in the effectiveness of the patent system in promoting innovation in synthetic biology. Within this commons, access to information and innovation is not dependent on government intervening in the intellectual property market. Nor is access dependent on beneficent attitudes of key intellectual property holders. Instead, the people who are going to use the resource have taken steps to build their own commons and ensure access to what they need to do their research. The commons involves shared norms as well as intellectual property access mechanisms. The sense of community and the shared desire to see the field of synthetic biology move forward are likely to be significant contributors to the commons' success.

The BPA will facilitate open access to basic standard biological parts into the future. There are some concerns relating to the BPA, but overall it and the Registry are likely to operate so as to promote innovation in synthetic biology research, by avoiding some of the potential negative unintended consequences of the patent system. Because the standard biological parts community is focused on developing beneficial applications, the promotion of innovation through this commons is likely to contribute to desirable innovation and public benefit from synthetic biology. As this is one of the key roles for regulation in relation to emerging technologies, this commons should be supported through regulatory mechanisms where possible.

Further, due to the large investment of public funds in synthetic biology and the nature of standard biological parts as fundamental building blocks, public sector agencies should take a role in supporting sharing efforts and promoting reasonable licensing behaviour. Adoption of patent pools and expansion of experimental use may provide other strategies to address access problems if they arise.

10. Conclusion and future directions

10.1 OVERVIEW

> The continued advance of technological development will bring many blessings, but human values must be preserved as we evaluate the deployment and use of powerful new technologies. Some advances warrant caution and careful oversight: the proliferation of nanomaterials, synthetic life-forms, and surveillance drones are examples of new technologies rife with promise and potential, but in need of review and safeguards.
>
> Al Gore[1]

This quote from Al Gore's *The Future* captures the issues examined in this book: the power, potential and danger of emerging technologies, the need to evaluate their impact, and the need for careful regulation in some areas. The key benefit of analysing these issues in the context of synthetic biology is an understanding of the specific, distinctive regulatory challenges posed by this emerging field. The regulatory challenges are shaped by the distinctive features of the synthetic biology's social and scientific context: the science itself and its underlying engineering ethos, the communities within synthetic biology, the novel risks and high uncertainty, and debates about access, ownership and control.

This book has investigated whether law reform measures or new regulatory approaches are required to address the regulatory challenges posed by synthetic biology in the intellectual property, biosafety, environmental protection and biosecurity. The analysis has taken the overarching goal of regulation in relation to emerging technology to be to improve the 'basic conditions of human social existence'.[2] That is, both to facilitate desirable innovation and beneficial technology development

---

[1] Al Gore, *The Future* (Random House, 2013), 373.

[2] Adopting the language of Brownsword and Somsen in Roger Brownsword and Han Somsen, 'Law, Innovation and Technology: Before We Fast Forward – A Forum for Debate' (2009) **1** *Law, Innovation and Technology* 1, 2–3.

and its deployment, and to minimise potential harms or negative effects from the emerging technology by managing risks.[3]

Achieving these goals requires a new, bespoke, connected regulatory regime for synthetic biology. This chapter brings together the key features of synthetic biology's distinctive social and scientific context, and the key regulatory needs arising from it. It explores the possibility of creating a new Synthetic Biology Agency to carry out enhanced regulation. This chapter also highlights the need for greater discussion of synthetic biology regulation at the international level, and for investigation of issues at the intersections of the realms of law discussed throughout this book.

10.2 SYNTHETIC BIOLOGY'S DISTINCTIVE SOCIAL AND SCIENTIFIC CONTEXT

10.2.1 The Revolutionary Science of Synthetic Biology

This research has illuminated the ways in which the science of synthetic biology itself is distinctive from previous biotechnologies such as recombinant DNA engineering. The endeavours of synthetic biologists represent a significant departure from previous life sciences research in several respects. First, molecular biologists have previously been focusing on seeking knowledge of living systems by breaking organisms and cells down into smaller and smaller entities, and understanding how these entities operate. This can be described as a reductionist approach to understanding biology.[4] Some research in synthetic biology involves beginning to *build* life, taking these tiny parts and building new living systems from the bottom up.

Secondly, synthetic biology represents a shift from engineering genes to 'genome engineering'.[5] This is the age of synthetic genomics – the 'large-scale re-making of a genome' in order to use microbial cells as

---

3 Ibid.

4 By contrast, Cho et al argue that minimal genome research and creating synthetic life *are* reductionist because they seek to understand life in terms of DNA: Mildred K Cho et al, 'Ethical Considerations in Synthesizing a Minimal Genome' (1999) **286**(5447) *Science* 2087.

5 This term is used by Church and Regis in George Church and Ed Regis, *Regenesis: How Synthetic Biology will Reinvent Nature and Ourselves* (Basic Books, 2013), see e.g., 59.

manufacturing centres.[6] Synthetic biology aims to construct new 'biological entities' including novel genetic circuits and cells which do not exist in nature, and to substantially redesign natural biological systems.[7] This means that there is a greater difference between organisms occurring in nature and those produced with synthetic biology than was the case in relation to the earlier recombinant DNA technology. Over time, genome engineering will become 'more ambitious and radical' as we get better at designing and assembling genomes for our purposes.[8] The difference between the naturally occurring world and the systems and tools of synthetic biology will only become greater as this field progresses. This will be particularly so as we move towards construction of truly 'synthetic' cells.

Thirdly, synthetic biology represents a departure from previous scientific work because it involves the direct integration of the principles of engineering into biology.[9] The engineering ethos also creates some of the unique aspects of its social and scientific context, and the regulatory challenges posed. For example, the use of engineering concepts such as standards impacts upon the intellectual property issues raised by the technology.[10] The emphasis on de-skilling of biological engineering to create systems that can be assembled from standard parts to work in predictable ways generates particular risks in relation to the emergence of do-it-yourself biology and biosecurity hazards.[11]

Fourthly, the enabling technologies of synthetic biology, particularly DNA synthesis and assembly techniques, have profound implications for capability in biological engineering. We have moved from being able to determine the sequence of DNA and converting this sequence into information that can be stored in a computer, to being able to move from

---

[6] Ibid., 4–5.

[7] See Engineering Biology Research Consortium, *What is Synthetic Biology?* (2017) Engineering Biology Research Consortium <https://www.ebrc.org/what-is-synbio>.

[8] Church and Regis (n 5) 7.

[9] For a clear explanation of these principles and what they mean in the biology context see ibid., 160–62.

[10] Arti K Rai, 'Evidence to The Presidential Commission for the Study of Bioethical Issues, Synthetic Biology' (13 September 2010) Philadelphia Penn; Arti K Rai, 'Unstandard Standardization: The Case of Biology' (2010) **53**(1) *Communications of the ACM* 37.

[11] Gautam Mukunda et al, 'What Rough Beast? Synthetic Biology, Uncertainty, and the Future of Biosecurity' (2009) **28**(2) *Politics and the Life Sciences* 2; M Schmidt, 'Diffusion of Synthetic Biology: A Challenge to Biosafety' (2008) **2**(1–2) *Systems and Synthetic Biology* 1.

the digital information or 'software' back to physical DNA that works in cells.[12] That is, we can now design new DNA sequences and build these to our specifications. These developments, along with work on minimal cells, mean that 'at the microbial level, nature has been redesigned and recoded in significant ways'.[13]

In addition to its distinctiveness from previous biotechnologies, synthetic biology is a transformative technology of 'unprecedented power'.[14] It is a field driven by developing beneficial applications in response to real-world problems. Synthetic biology is transformative because it 'has the potential to transform our fundamental understanding of biological systems as well as our ability to manipulate them for human benefit'.[15] Its applications in areas such as biofuels and bioremediation have the potential to improve economic and environmental sustainability.[16] Synthetic biology also has potential to transform the way in which scientists respond to emerging infectious disease threats, and other diseases such as cancer.[17] Governments and industry have recognised this transformative potential, and vast amounts of money, both public and private, are being devoted to synthetic biology research.[18] This makes the selection of the right regulatory approach all the more important.

A final important aspect of the science of synthetic biology is the very rapid pace of developments in this field. This pace can be 'bewildering'

---

12 J Craig Venter, 'Evidence to The Presidential Commission for the Study of Bioethical Issues, Synthetic Biology' (8 July 2010) Washington DC.

13 Church and Regis (n 5) 7.

14 Ibid., 17; Adam Rutherford, *Creation: The Origin of Life; The Future of Life* (Penguin Books, 2013), The Future of Life 47, 91.

15 Geoff Baldwin et al, *Synthetic Biology – A Primer* (Imperial College Press, 2012), xiii.

16 Markus Schmidt, 'Executive Summary' in Markus Schmidt (ed.), *Synthetic Biology: Industrial and Environmental Applications* (Wiley-Blackwell, 2012) xxi, xxi.

17 Church and Regis (n 5) 115–17; Venter, 'Evidence to The Presidential Commission for the Study of Bioethical Issues, Synthetic Biology' (n 12); J Craig Venter, 'Evidence to United States House of Representatives Committee on Energy and Commerce, Effects of Developments in Synthetic Genomics and Implications for Energy and Health' (27 May 2010) Washington DC.

18 As an example from the private sphere, BP has invested $500 million in synthetic biology biofuels research at the University of California Berkeley: Church and Regis (n 5) 173. As an example of public investment, the Department of Energy allocated $30 million to investigate biological design under its Biological and Environmental Research Program: 'National Bioeconomy Blueprint' (The White House, 2012) <http://www.whitehouse.gov/sites/default/files/microsites/ostp/national_bioeconomy_blueprint_april_2012.pdf>, 22.

even to scientists themselves.[19] The fast pace of progress impacts upon the regulatory challenges posed. It increases uncertainty about the specific likelihood of various risks. It also creates difficulty in predicting when certain developments may eventuate. These factors make the task of creating legitimate, effective regulation – which remains connected to the state of scientific development in synthetic biology – all the more complex. They increase the need for monitoring of the state of scientific development and adaptation of regulation.

10.2.2 Risk and Uncertainty

Issues of risk and uncertainty are pervasive in the debate surrounding synthetic biology. The emergence of the field creates potential risks in the areas of environmental harm, biosafety and biosecurity. The science of synthetic biology is novel, but also relatively immature, and its impacts are uncertain.[20] This high degree of uncertainty surrounding the risks of synthetic biology is a distinctive feature of its social and scientific context. Because the difference between naturally occurring organisms and those produced with synthetic biology is greater than for previous biotechnologies, uncertainty about environmental risk is greater. It is more difficult to predict the environmental effects as there is significant uncertainty about both the nature and likelihood of potential environmental harm.

Synthetic biology has raised novel risks as the ability to experiment with biotechnology becomes distributed beyond the traditional, regulated venues. The challenges of biosafety in relation to do-it-yourself biology are particularly acute due to uncertainty about precisely what research is occurring in this area and how dangerous this may (or may not) be.

Synthetic biology has also raised novel risks in the area of biosecurity. The risks in this area may be greater than previous security threats from chemical and nuclear weapons.[21] There is also great uncertainty about who would wish to misuse such technology, the ease with which they could do so and the scale of potential harm that could be caused. The distributed nature of synthetic biology activities exacerbates these significant uncertainties.

---

[19] Rutherford (n 14) 90–91.

[20] Roger Brownsword and Morag Goodwin, *Law and the Technologies of the Twenty-First Century: Text and Materials* (Cambridge University Press, 2012) 129; Brownsword and Somsen (n 2) 27; Gregory N Mandel, 'Regulating Emerging Technologies' (2009) **1** *Law, Innovation & Technology* 75, 78.

[21] Church and Regis (n 5) 231.

Controversy about the balance of risks and benefits offered by synthetic biology creates difficulty in achieving legitimate regulation. Democratic deliberation, as facilitated by the Commission, can go some way to addressing this.[22] Uncertainty about the risks also poses particular challenges for effective regulation. These factors create a need for adaptive regulatory mechanisms to assist in ensuring regulatory connection.

10.2.3 Novel Communities

The various communities of researchers in synthetic biology contribute to the distinctiveness of its social and scientific context, and its regulatory challenges.

The standard biological parts community has constructed its own commons for synthetic biology research.[23] The sharing arrangements created by the BioBrick Public Agreement have characteristics of an 'open biotechnology' project, an 'innovation community' and a 'common pool resource' managed through collective action. Sharing through the Registry of Standard Biological Parts is more akin to an open source arrangement, though it is not underpinned by intellectual property rights in the shared material. The researchers are highly engaged in regulatory discussions through their leadership by key figures such as Drew Endy. The commons is supported by shared goals, both scientific and ethical, and a shared culture. While communities for sharing are being established in other areas of life sciences also, the proactive, community-led approach to intellectual property and the establishment of a commons supported by contractual mechanisms so early in the field's development make the BioBricks community distinctive.

The standard biological parts commons advances the regulatory goal of promoting desirable innovation in synthetic biology in two ways. First, the legal mechanisms which have been adopted for the sharing of BioBricks reduce the intellectual property barriers to these tools for researchers. Secondly, the community fosters norms including positive contribution and safety. However, the ethical and safety requirements within the BioBrick Public Agreement (BPA) are not enforceable.

---

[22] Brownsword and Goodwin (n 20) 58–9.

[23] In relation to the open revealing strategy, see James Boyle, *The Public Domain: Enclosing the Commons of the Mind* (Yale University Press, 2008), 178.

The technology of synthetic biology is becoming distributed, which is a departure from previous biotechnology research.[24] Developments in the technology of synthetic biology also enable greater amateur experimentation beyond the realms of professional, institutional and industrial science. The do-it-yourself biology community has emerged as another distinctive community within synthetic biology, with a particular culture and different motivations. The do-it-yourself biology community consists of a network of individuals and groups who are enthusiastically exploring biology outside the usual venues of companies and universities. This is not as cohesive a group as the BioBricks community, but DIYbio.org has provided a focal point for discussion, sharing of information and development of biosafety principles and best practices within the community. There are strong links in culture, people and motivations between this community and the general synthetic biology community.

To the extent that general motivations can be determined, the key matter valued by the do-it-yourselfers is making biotechnology experimentation accessible to a wider range of people. They are also focused on designing cheaper equipment, the ability to explore neglected topics which may not be feasible to study within institutions, and the ability to look into one's own genes and disease risks. For some, working around intellectual property constraints is also a factor. The do-it-yourself biologists value openness and safety and are engaged in attempts to develop best practices and ethical principles. While the founders of this group are engaging in discussions about safety, the regulatory problems in relation to biosafety are exacerbated by the fact that not all regulatees are identifiable and amenable to discussions about regulation.

10.2.4 Open vs Closed, Free vs Constrained Science

Ownership and control of synthetic biology, and access to the technology, have been key themes in this book. Persistent questions have emerged as to who is able to 'do' synthetic biology and what constraints are placed on them; who is able to access and use synthetic biology tools and information, and who is able to influence its development.

There is a recurring tension between freedom to do research in synthetic biology and various forms of control, or restriction, on the ability to do this research. In the BioBricks context, the language of freedom refers to the ability to freely access the basic research tools

---

[24] Baldwin et al (n 15) 131, though these authors note that distribution of technology is not unique to synthetic biology.

represented by BioBricks and share these with others. It also refers to freedom from the restrictions of the monopolies granted under the patent system.

Issues relating to the democratisation of science are also relevant within this theme of 'open' vs 'closed' synthetic biology. The technology of synthetic biology is becoming distributed, which is a departure from previous biotechnology research.[25] The rhetoric of freedom and liberty is prevalent in the do-it-yourself biology context. There it refers to freedom to do research in biology, regardless of formal qualifications and outside the traditional venues of universities and companies. It also refers to the freedom to 'play and create'[26] through doing biological research, without controls imposed by government.

Similarly, a tension was evident in the Commission's report and deliberations between the value of intellectual freedom in scientific research and 'letting science rip' on the one hand and the need to address and manage risks posed by synthetic biology on the other.[27]

Issues of open and closed science also arise in relation to the questions of who should decide how synthetic biology is regulated and who should be able to influence the trajectory of its development. Some scientists seem to prefer a 'closed' model where it is mainly within their own expert communities that issues of governance and research priorities are determined. There is a contrasting perspective which holds that the activities of scientists should be open in that they are monitored and scientists are accountable to those *outside* their field.[28] As an extension of

---

25 Ibid.

26 Adopting the language used by Lange in relation to the concept of a 'commons': David Lange, 'Reimagining the Public Domain' (2003) **66**(1–2) *Law and Contemporary Problems* 463, 483.

27 The Presidential Commission for the Study of Bioethical Issues, 'New Directions: The Ethics of Synthetic Biology and Emerging Technologies' (The Presidential Commission for the Study of Bioethical Issues, December 2010) <http://bioethics.gov/sites/default/files/PCSBI-Synthetic-Biology-Report-12.16.10_0.pdf>, 124, 141–3; John Arras 'Comments in Commission Deliberations, The Presidential Commission for the Study of Bioethical Issues, Synthetic Biology' 16 November 2010.

28 See discussion of external accountability in Joy Y Zhang et al, 'The Transnational Governance of Synthetic Biology: Scientific Uncertainty, Cross-borderness and the "Art" of Governance' (BIOS Working Paper No: 4, BIOS (Centre for the Study of Bioscience, Biomedicine, Biotechnology and Society), May 2011) <http://www2.lse.ac.uk/BIOS/publications/Working%20Papers/WP4TransnationalGovernance.pdf>, 4.

this issue, there is also tension within synthetic biology about the extent to which sociological and ethical perspectives should be part of research in synthetic biology.[29]

The rhetoric of open and closed science is a prominent feature of intellectual property debates in synthetic biology. While research occurring in the standard biological parts commons is seen as open, patent monopolies such as those being sought by the Venter Institute could be seen to close off access to technology. The standard biological parts commons is underpinned by the express goal of putting information in the public domain so that it cannot be closed off with intellectual property rights. In this commons, innovation is occurring in a distributed and cooperative way. The very nature and existence of intellectual property monopolies have also spurred some of the innovation in the do-it-yourself community.

There is tension between the value of openness and sharing in some areas of synthetic biology and the regulatory goals of managing biosecurity and biosafety risks. There is disagreement about whether safety and security are maximised by 'opening' or 'closing' synthetic biology research. On the one hand, some argue that the more widely the information is distributed and the more brains working on it, the safer we are.[30] On the other hand, some see control of scientific information and limitations on who can access and experiment with biotechnology as more promising avenues for risk management.

These features of the science of synthetic biology, the risks and uncertainties it raises, the unique communities that have emerged and the tensions between 'open' and 'closed' synthetic biology form the elements of its distinctive social and scientific context. This unique context presents distinctive regulatory challenges in managing the novel communities that have emerged, the new scientific capabilities, the new risks and the uncertainty about these, as well as facilitating socially beneficial innovation in a new intellectual property environment characterised by both proactive sharing and proactive patenting. These challenges require a distinctive regulatory approach for synthetic biology.

---

[29] See Comments of conference attendees during debate on paper: Paul Freemont, 'Synthetic Biology – Do We Need New Regulatory Systems?' (Paper presented at the 21st Century Borders/Synthetic Biology: Focus on Responsibility and Governance, 2013) <http://www.scienceforglobalpolicy.org/LinkClick.aspx?fileticket=VqAc-vBzvMk%3d&tabid=133>, 21–22.

[30] Rob Carlson, *Biology is Technology: The Promise, Peril, and New Business of Engineering Life* (Harvard University Press, 2010), 235–6.

10.3 WHAT DO WE NEED NOW?

10.3.1 Current Regulatory Needs for Synthetic Biology

The Commission's inquiry facilitated democratic deliberation in relation to the issues of risk, benefit, ownership, access and control raised by synthetic biology. It has improved the procedural legitimacy of future government action.[31] However, the Commission did not fully seize its opportunity to be proactive, and was minimalist in its recommendations, which largely pointed out areas where more information is required. The Commission could have gone further in its substantive recommendations, particularly in the areas of intellectual property and biosecurity. It focused on similarities between synthetic biology and earlier genetic technologies, to the detriment of consideration of the distinctiveness of the regulatory challenges posed by synthetic biology.

Enhanced regulatory action should be taken to address synthetic biology's distinctive challenges, and harness the opportunities it creates. There are various regulatory needs which must be addressed if regulation is to perform this dual role.

First, regulation of synthetic biology should be directed towards encouraging sharing and open access to basic research and tools and promoting patent quality. In synthetic biology, there are already many patent applications. These include aspects of gene synthesis, synthetic cell construction and large-scale metabolic engineering, as well as numerous applications for biofuels. The USPTO has already granted a number of synthetic biology patents. The key processes underpinning synthetic biology will meet the tests for patentable subject matter set out in *Bilski*; *Mayo*; and *Myriad*.[32] The patentability of synthetic DNA is important as this will be an increasingly used tool in various aspects of

---

[31] The Presidential Commission itself pointed out that 'public participation frequently fosters the perceived political legitimacy of the policies and practices that are ultimately chosen': The Presidential Commission for the Study of Bioethical Issues (n 27) 152.

[32] *Bilski v Kappos*, 130 US (S Ct, 2010) 3218; *Mayo Collaborative Services v Prometheus Laboratories Inc.*, 132 US (S Ct, 2012) 1289; *Association for Molecular Pathology v United States Patent and Trademark Office*, 653 F 3d (CA Fed (NY), 2011) 1329; *Association for Molecular Pathology v United States Patent and Trademark Office*, 689 F 3d (CA Fed (NY), 2012) 1303.

synthetic biology. A stretch of DNA which is not identical in sequence to naturally occurring DNA is likely to meet the subject matter tests in the case law.[33]

The potential for BioBricks to infringe existing DNA patents has been negated under the Supreme Court's decision in *Myriad*.[34] Where they are different in sequence from natural DNA, BioBricks would be likely to be considered patentable subject matter themselves. This would also apply to composite parts and devices.

Some features of synthetic biology research make it vulnerable to patent thicket and hold-up problems, in particular its reliance on standards and the potential for patents on BioBricks or other standard parts.[35] I have argued that while the academic synthetic biology community often believes its work is covered by an experimental use exemption, this exemption is actually very narrow and does not cover work with BioBricks. The potential problem of patent holders seeking to enforce the patents involved in a standard should be taken into account in the standard-setting work occurring in synthetic biology. If problems in accessing groups of parts arise, the mechanism of patent pooling could be considered.

The standard biological parts commons promotes access to foundational building blocks in synthetic biology research, serving the regulatory goal of facilitating desirable innovation. Given this, and the large expenditure of public funds in supporting synthetic biology research, government agencies should take a role in promoting sharing and supporting the standard biological parts commons where possible. This may include threat of the use of march-in rights if necessary.

The norms of the BioBricks community are demonstrated through its websites, and the BPA includes ethical and safety requirements. This shows positive engagement with the broader issues around synthetic biology, but these terms cannot be enforced. The commons is different from a traditional public biobank in this regard, creating room for a regulatory agency to oversee these aspects of synthetic biology research.

The second regulatory need is for precautionary regulation in relation to environmental risk. I have criticised the adoption of the principle of prudent vigilance by the Commission on the grounds that it overlooks the

---

33 *Association for Molecular Pathology, et al v Myriad Genetics Inc., et al*, 133 US (S Ct, 2013) 2107.

34 Ibid.

35 See Chapter 8 and Rai, 'Unstandard Standardization: The Case of Biology' (n 10); Rai, 'Evidence to The Presidential Commission for the Study of Bioethical Issues, Synthetic Biology' (n 10).

utility and implementation of the precautionary principle and was based on a caricatured understanding of precaution. The principle is more nuanced than the Commission acknowledges, and need not be seen as out of step with existing approaches to environmental and health risk regulation in the US. It has been interpreted and applied over several decades, unlike the un-tested concept of prudent vigilance. Its application does not require the blocking of scientific progress.

A precautionary approach to the environmental risks of synthetic biology should be preferred to prudent vigilance due to the differences between synthetic biology and earlier genetic technologies. The environmental risks should be approached by adopting a precautionary stance and avoiding reliance on technical safety measures built into synthetic organisms.

Further, the adoption of prudent vigilance is not conducive to international leadership on the regulatory issues raised by synthetic biology. International discussion and coordination on these issues is vital for future regulation of the field.

Thirdly, there is a need for greater regulation of do-it-yourself biology in order to reduce biosafety risks. There are presently identifiable biosafety risks associated with the community's activities. The Commission's analysis of biosafety issues largely neglected the issue of safety at the research and experimentation stage, focusing on safety at the stage of environmental release of synthetic biology products. While universities and companies using federal funding must comply with NIH biosafety procedures and have institutional biosafety frameworks, these do not apply to do-it-yourselfers. These individuals or groups are also not subject to the regulatory requirements which apply to commercial research and development of transgenic organisms and state-based occupational health and safety standards. This means that there are distinctive, novel risks associated with this community within synthetic biology.

The activities of the visible do-it-yourself biologists associated with DIYbio show a high level of engagement with safety issues and law enforcement agencies. Therefore, among this visible group there is currently a minimal risk of harm from intentional misuse of biology. The main issues arise from the potential for accidental problems to occur, due to lack of capacity to implement safety procedures and low levels of technical ability and safety understanding among do-it-yourself biologists, and the lack of institutional oversight of these activities.

The Commission's approach, which emphasised fostering responsibility *within* the community of do-it-yourselfers rather than creating oversight, is insufficiently cautious. While the principle of intellectual freedom was at the centre of the Commission's reasoning, do-it-yourselfers should accept that a degree of responsibility comes with this freedom. Experimentation with biotechnology should not be completely democratised and decentralised. While there may be difficulty in implementing criminal provisions specific to this group, civil measures could include extending the current NIH guidelines for biosafety to the do-it-yourself community in a form that is accessible. Further, a system of registration for do-it-yourself activities, perhaps at the level of community labs, should be considered. It is important for the effectiveness of these measures and the activities of the do-it-yourself community to be reviewed so that the measures can adapt over time in order to remain connected to the state of do-it-yourself biology.

Fourthly, there is a need for enforceable oversight in the biosecurity area. Screening of gene synthesis requests by the synthesis industry should be made compulsory. While arguments have been made that voluntary mechanisms which leave the problem to industry are preferable, such approaches are not cautious enough and may not be effective. There is a risk that compliance will not occur. Because of the high degree of uncertainty about the biosecurity risks of synthetic biology and the rapid pace of change in scientific development, a cautious approach should be taken including enforceable oversight based on a rule of law approach. Self-regulation will lack external accountability and procedural legitimacy.

Fifthly, there is a need for an adaptive mechanism to review the performance of policies, obtain new information about scientific development and modify the regulatory regime as needed. The concept of planned adaptation has utility for managing the risks raised by synthetic biology by facilitating the gathering of new knowledge about the science and the effectiveness of regulatory mechanisms. This would increase the regulatory connection of measures adopted to regulate the fast-moving science. While a process of adaptive regulation may not neutralize all matters of uncertainty, it can be used in combination with a precautionary approach to update regulatory settings over time.

Sixthly, there is a need for wide stakeholder engagement including with industry, academic science, do-it-yourself biologists, members of the public and non-government organisations.

10.3.2 A Synthetic Biology Agency

One way to address these regulatory needs would be to establish a unique institution for bespoke regulation of synthetic biology. This institution could be known as a Synthetic Biology Agency.

A Synthetic Biology Agency would provide enhanced regulation by extending the current regulatory environment in several ways. Allowing synthetic biology to proceed in a way that is most able to create societal benefit requires regulation to be more cautious regarding the risks posed. Therefore, the Synthetic Biology Agency would implement a regime of risk assessment for environmental harm which is underpinned by the precautionary principle, in recognition of the high degree of uncertainty surrounding synthetic biology's potential impacts. The precautionary approach would be implemented such that the burden would fall on entities wishing to release organisms to provide information regarding their risk profile.

A Synthetic Biology Agency would engage with the do-it-yourself and synthetic biology communities to develop appropriate oversight measures for the do-it-yourself world. It would liaise with industry and research stakeholders in developing compulsory regulatory provisions for oversight of DNA synthesis activities and other regulatory instruments. It would enforce these regulatory mechanisms.

A Synthetic Biology Agency would take responsibility for regulatory policy making to encourage sharing of tools such as BioBricks, to promote innovation in synthetic biology. It would monitor the impact of patents in the field of synthetic biology and liaise with the USPTO where necessary to promote patent quality. It would also encourage and incentivise non-troll licensing behaviour. It would liaise with funding agencies to do this through funding agreements.

Engagement with do-it-yourself and BioBricks communities, industry and academic scientists in synthetic biology would also form a critical aspect of a Synthetic Biology Agency's work. As there are concerns within the do-it-yourself community regarding the impact of regulation on their activities, this consultation and communication will be very important in developing regulation that has 'buy in' from this community.

This will also apply in relation to biosecurity. The gene synthesis industry has current knowledge about technological development and industry trends.[36] This should be utilised in regulatory development. Industry buy-in should be maximised through consultation regarding any mandatory oversight measures. The DNA synthesis industry's existing

---

[36] Amy Smithson, 'Renovating Governance Strategies for Synthetic Biology and Other "Dual-Use" Technologies' (Paper presented at the 21st Century Borders/Synthetic Biology: Focus on Responsibility and Governance, 2013) <http://www.scienceforglobalpolicy.org/LinkClick.aspx?fileticket=VqAc-vBzvMk%3d&tabid=133>, 28.

efforts to agree on principles for screening should be capitalised upon in this process.

This model would also promote procedural legitimacy and external accountability.[37] These important features of a regulatory approach are lacking in the self-regulation approaches critiqued in Chapters 3, 6, and 7. There needs to be public trust that perspectives outside the world of scientists working in synthetic biology are being taken into account.[38] The Synthetic Biology Agency should therefore undertake public engagement activities.

Promoting the best balance between risk and benefit also requires adaptive governance in order to promote regulatory connection. The Synthetic Biology Agency would be tasked with important adaptive activities. It would review the stage of technological development and commission updated studies about the environmental, health and biosecurity risks. Identifying the gaps in risk research would be a key role.[39] It would review biosafety procedures and requirements in relation to synthetic biology generally and in their particular application to groups such as do-it-yourselfers. It would review environmental risk regulation and implementation. Keeping abreast of the state of technological development will also be critically important in relation to biosecurity, as technology for DNA synthesis improves in the industry develops and spreads.

Adaptation is a strategy that is not limited to managing uncertainty about the science. The Synthetic Biology Agency would engage in review of both technological development and the *impact* and effectiveness of regulatory mechanisms. The Synthetic Biology Agency could also review the operation of the patent system in respect of synthetic biology.

More consideration of the specific optimal mechanism of adaptive regulation is needed. For example, the question of whether the functions of acquiring new information and recommending policy adjustments should occur within the same agency, or be separate, would need to be explored.[40] The administrative law issues that can impact on adaptive

---

[37] See discussion of external accountability in Zhang et al (n 28) 4.

[38] In relation to the importance of public trust and confidence in regulation of emerging technologies generally see Mandel (n 20).

[39] As proposed for regulation of emerging technologies generally in ibid., 83.

[40] For example, in their study of planned adaptation McCray et al found that the roles of 'learning', that is acquiring new knowledge, and 'changing', that is adapting the regulatory regime in response to the new knowledge, were separate in all the examples identified. Some of the relevant knowledge-generating institutions, such as the National Academies of Science, were independent from

approaches would also require greater consideration, particularly the way in which decision-making might be adaptive while preserving proper public participation. The particular types of decisions likely to be made in relation to synthetic biology that are best suited to adaptive decision-making also need further consideration, as does the integration of precautionary regulation with adaptation.

There are advantages of combining these enhanced regulatory activities within one institution. First, this model would help to address the current 'patchwork'[41] array of regulatory measures in place for synthetic biology and the problem of coordination between agencies. Improvements in coordination are likely to make the system more efficient than the status quo.[42] If it is well resourced, this agency will be able to assess and approve synthetic biology research projects in a timely manner.

Secondly, by combining these regulatory functions, institutional expertise in the unique science and technology of synthetic biology and the legal issues it raises can be generated. Consideration would need to be given to how the responsibilities of this agency would be accommodated within the Coordinated Biotechnology Framework.

Thirdly, cross-fertilisation in policy making can be achieved where the different regulatory challenges discussed in this book interact with each other. In this way, the linkages and interactions between the different regulatory challenges raised by synthetic biology can be elucidated. The model could be reviewed several years after implementation to determine whether combining the different functions has worked effectively, or whether some functions (e.g., those relating to intellectual property) are better dealt with by separate agencies.

Fourthly, this model would also foster the continuation of work that has already begun in bringing together professionals within the different scientific fields which make up synthetic biology and professionals in social sciences such as law, sociology, ethics, policy and public awareness of science.

In this model, the precautionary principle would be applied in a proportionate way rather than as a means to halt the development of a controversial technology. This is intended to be an initial approach to be taken at this stage of synthetic biology's development, when uncertainty

---

government: Lawrence E McCray et al, 'Planned Adaptation in Risk Regulation: An Initial Survey of US Environmental, Health, and Safety Regulation' (2010) **77** *Technological Forecasting & Social Change* 951, 958.

[41] The Presidential Commission for the Study of Bioethical Issues (n 27) 102.

[42] See discussion of agency coordination generally in Mandel (n 20) 87–8.

is pervasive. As the technology develops and more information about its likely impacts on the environment becomes available, the understanding of the risk profile may change such that a modification of this approach is appropriate. Understanding of the 'risk profile' of synthetic biology in relation to security and safety threats will also change over time, though it is doubtful that a relaxing of the enforceable approach to regulation in the biosecurity area will be appropriate.

This model of enhanced regulation would certainly require expenditure, in setting up and resourcing the new agency. However, given the large amounts of public funding already being expended in synthetic biology, the potentially transformative benefits that could be achieved, along with the pervasive uncertainty, these are valuable activities in enhanced regulation to promote maximum benefit to the public and minimal harm.

If in future it is found that this regulatory model works well, extension of the idea to other types of emerging technology could be considered.

10.4 FUTURE DIRECTIONS: TRANSNATIONAL REGULATION, INTERSECTIONAL ISSUES AND BENEFIT SHARING

This section highlights some further issues that require greater investigation and discussion as the field of synthetic biology moves forward. These include developing a transnational approach, investigating how the regulatory challenges and goals intersect, and using regulation as a tool for benefit sharing.

10.4.1 International Discussion of the Regulation of Synthetic Biology

As noted at the outset, the development of transnational regulation is a challenge in relation to emerging technologies.[43] Across jurisdictions, there are likely to be differing attitudes to technological development and risk, and different values and ethical perspectives.[44]

---

[43] Brownsword and Somsen (n 2) 11.

[44] For discussion of this challenge of 'regulatory cosmopolitanism' see Brownsword and Somsen (n 2) 11, 31–2 and Roger Brownsword, *Rights, Regulation and the Technological Revolution* (Oxford University Press, 2008), 185–211.

The enhanced regulation I have proposed thus far would operate at the level of the domestic regulatory environment in the US. Domestic law reform processes such as that conducted by the Commission may help to address diverging views regarding the risks and benefits of synthetic biology within the US,[45] but there is also a need to consider diverging perspectives on the risks and benefits, and potential ethical differences between countries.

The scientific field of synthetic biology and the regulatory issues it raises are international in nature. In order to see the truly international reach of synthetic biology, one only has to look at the rapid expansion of the iGEM competition to include science students from all over the world. Similarly, interest in DIYbio has spread internationally. This international nature of the field poses regulatory challenges,[46] adding to the complexity of regulating this field. In particular, the international nature of the gene synthesis industry and the international dissemination of research information about synthetic biology indicate that a multilateral approach to biosecurity regulation is required.

There needs to be a focused international discussion of the regulatory issues raised by synthetic biology, with a view to developing an international approach to these issues and potentially a multi-lateral agreement. A potential forum for a general discussion could be the UN Educational, Scientific and Cultural Organization. This was the forum for the adoption of the Universal Declaration on the Human Genome and Human Rights in 1997.[47] This is an example of an international discussion of a novel and transformative technological development.

International fora in which specific regulatory issues raised by synthetic biology could be further debated include the World Health Organization, the World Intellectual Property Organization and the Convention on Biological Diversity. Indeed, discussion of synthetic biology's impacts is already occurring within the framework of the Convention on Biological Diversity.

As the field moves forward, regulatory approaches need to be studied comparatively across jurisdictions. In developing regulation at the domestic level, developments in other jurisdictions and regulation at the international level need to be considered. While this book has focused

---

45 Brownsword and Goodwin (n 19) 59.

46 The transnational nature of synthetic biology is pointed out as a challenge for governance in Zhang et al (n 28).

47 UN Educational, Scientific and Cultural Organization, 'Universal Declaration on the Human Genome and Human Rights' (29th General Conference, 11 November 1997).

largely on biosecurity regulation within the US, it would also be prudent to investigate the policy discussions and regulatory developments occurring in other jurisdictions, particularly in Europe, as major scientific and social sciences research in synthetic biology is occurring in that jurisdiction. This would allow comparison between the regulatory environment in the US and other jurisdictions, yielding insight into how regulatory approaches in the US might be improved.

Greater understanding of developments in other jurisdictions will also assist in assessing the possibilities for the development of international regulation.

10.4.2 Intersectional Issues

This book has been an effort to map the regulatory challenges posed by synthetic biology, and thus has not explored in depth the interactions *between* the various regulatory issues raised. For example, issues arise at the intersection between intellectual property law and policy and environmental or security law and policy. These interactions require further consideration, and a selection are highlighted below.

(a) Enhancing innovation in renewable energy, environmental remediation and climate change mitigation

The way in which intellectual property rights are used in synthetic biology and how the tension between the approaches of patenting and openness plays out will have broad significance beyond the issue of how best to foster innovation in science. The interaction between intellectual property rights and synthetic biology will impact on wider policy debates in areas such as access to medicines and synthetic vaccines.[48] It will also impact upon issues of accessibility and prospects for technology transfer of clean technologies such as advanced biofuels.[49]

Further consideration is needed of the way in which intellectual property will affect the environmental and climate change related applications of synthetic biology, because this will impact upon their deployment. Synthetic biology and energy companies are investing heavily in research into using synthetic biology to produce cleaner technologies, such as biofuels. These applications raise policy questions such as how to

---

[48] Henk van den Belt, 'Synthetic Biology, Patenting, Health and Global Justice' (2013) *Systems and Synthetic Biology* 1.

[49] Ibid., 5–8; on the interaction between intellectual property, climate change and clean technology see Matthew Rimmer, *Intellectual Property and Climate Change: Inventing Clean Technologies* (Edward Elgar Publishing, 2011).

provide incentives for development of clean technologies. There are also applications in pollutant detection and environmental remediation. More consideration is needed of the best model for encouraging synthetic biology research (both private and public innovation) and its deployment.

The role of intellectual property in the development and diffusion of green technologies has been debated internationally.[50] For synthetic biology, key issues are how to stimulate innovation in energy, clean technology and environmental remediation, and how to ensure these technologies are accessible to developing countries. Indeed, 'with synthetic biology playing an increasingly prominent role in creation of second and third generation biofuels, commons and property issues are likely to fuse with debates over climate change and development'.[51] Issues arise as to what role synthetic biology patents will have in promoting development of clean technologies, and how patenting of synthetic biology will affect technology transfer. There are also questions as to how international trade agreements impact upon the patentability and diffusion of synthetic biology research products.

(b) Interactions between intellectual property and biosecurity

The intersection between intellectual property and biosecurity regulation warrants further investigation. For example, there could be investigation into the potential role for the intellectual property system in regulating or monitoring dual-use research. As noted in the introduction, the bargain underpinning the patent system is that in order to receive the legal monopoly, the inventor must publicly disclose the invention. This means that patent applications contain a wealth of detailed, publicly available information that may not be published anywhere else.[52] Thus the patent system may provide a useful way of monitoring developments in synthetic biology. It should be considered whether it is likely to form an effective avenue for regulation, or whether there are other more effective ways to manage biosecurity issues.

---

[50] See Rimmer, ibid., 39–82.

[51] Kenneth A Oye and Rachel Wellhausen, 'The Intellectual Commons and Property in Synthetic Biology' in Markus Schmidt et al (eds), *Synthetic Biology: The Technoscience and its Societal Consequences* (Springer Academic Publishing, 2009) 121, 138.

[52] Hector MacQueen et al, *Contemporary Intellectual Property: Law and Policy* (Oxford University Press, 2008), 365.

10.4.3 Benefit Sharing

The work in this book has been aimed at promoting the beneficial development of the technology at the research stage by looking into the issues of intellectual property ownership and sharing of information. The principled *distribution* of benefits of the technology was beyond the scope of this work. Going forward, there is a need for consideration of benefit sharing in relation to synthetic biology's products. This should consider the potential of synthetic biology products to provide benefits in the developing world. Some scientists have strongly advocated for the potential of synthetic biology to assist in addressing problems facing developing nations.[53] Deployment may involve use of the mechanisms of intellectual property law, such as compulsory licensing, and international trade law relating to intellectual property.

It has also been suggested that legal mechanisms can be used to stimulate innovation in areas that would be of particular benefit to developing countries.[54] This idea should be explored in relation to synthetic biology.

10.5 CONCLUSION

There is a revolution happening, and it is in our hands.
Adam Rutherford.[55]

It is certainly worthwhile considering the general, recurring challenges posed by emerging technologies, to avoid reinventing the 'regulatory wheel'.[56] There is value in developing general ideas and principles to apply to the development of regulation of emerging technologies where

---

[53] For example, Maria Mercedes Roca, 'The Challenges of Deploying Synthetic Biology Technologies in Developing Countries' (Paper presented at the 21st Century Borders/Synthetic Biology: Focus on Responsibility and Governance, 2013) <http://www.scienceforglobalpolicy.org/LinkClick.aspx?fileticket=VqAc-vBzvMk%3d&tabid=133>.

[54] Brownsword and Somsen (n 2) 53.

[55] Rutherford (n 15) 91.

[56] Roger Brownsword, 'So What Does the World Need Now? Reflections on Regulating Technologies' in Roger Brownsword and Karen Yeung (eds), *Regulating Technologies: Legal Futures, Regulatory Frames, and Technological Fixes* (2008) 23, 30–32.

possible.[57] However, it is *also* critical to look at what is distinctive about a particular technology and the regulatory challenges it raises.[58] This analysis should consider distinctiveness both in terms of the science itself, and the social context of the technology. This will promote regulatory legitimacy, effectiveness and connection, supporting the dual regulatory goals of maximising benefit and minimising harm.

It is indeed the case that there are certain regulatory challenges observed across emerging technologies. The key recurring themes in this book fall within a wider scholarly discussion and debate about how regulation, law and governments should respond to the challenges posed by emerging science and technologies. This discussion is currently playing out in the synthetic biology context, but the challenges of legitimacy, effectiveness, connection, risk management and cosmopolitanism will continue to be raised as new technologies emerge in all areas of scientific research.[59] The particular nature of the challenges posed and the features of the regulatory context will vary from technology to technology, but there are higher-order issues as to how to minimise the way in which law lags behind scientific development[60] and how to make regulation more able to respond to these developments, how to respond to risk and uncertainty using regulatory instruments, how to engage the public and scientific communities, and how to promote innovation for the purposes of societal benefit. Though tensions between open and closed visions of science are particularly acute in synthetic biology, they are also present in wider debates about whether access to certain scientific information should be open or closed, both in terms of intellectual property and risk issues.

This book has taken a regulatory perspective, but is part of a broader scholarly effort to understand all of these challenges. This effort is occurring not only in the field of law, but also in science, science policy, bioethics, political science and sociology. This scholarly discourse will

---

57 As has been done, e.g., by Mandel, (n 20). Gutmann has also noted that there are some 'parallels between the ethical issues raised by synthetic biology and those of emerging technologies generally': Amy Gutmann, 'The Ethics of Synthetic Biology: Guiding Principles for Emerging Technologies' (2011) **41**(4) *Hastings Center Report* 17, 18.

58 See e.g., Brownsword, *Rights, Regulation and the Technological Revolution* (n 44) 27.

59 See discussion of these challenges in general terms in Brownsword, 'So What Does the World Need Now? Reflections on Regulating Technologies' (n 56); Brownsword and Somsen (n 2); Brownsword (n 44); Brownsword and Goodwin (n 20).

60 Brownsword, 'So What Does the World Need Now? Reflections on Regulating Technologies' (n 56) 26–7.

impact upon how modern technologies are developed, constrained and deployed, and how they will benefit society.

Synthetic biology is a technology with transformative power and the potential to have a profound impact in many areas of human life. It is a field focused on developing positive applications in response to the world's most pressing problems. Given the transformative potential benefits of synthetic biology, the particular risks and uncertainties it raises and the huge amounts of public and private funding being devoted to the field, this is a high-stakes discussion. Understanding and responding to the regulatory issues posed by synthetic biology is a critical part of addressing its risks and obtaining the most societal benefit from its applications. In the synthetic biology environment, this discussion has occurred in tandem with scientific developments from an early stage. This is one of synthetic biology's distinctive attributes, and a trend that should continue. As the debates surrounding synthetic biology progress and domestic regulatory adaptations are implemented, discussion will need to move beyond the domestic sphere. The key challenge will become international cooperation and the development of transnational regulation for this burgeoning field. Further, strategies will be needed to get the most out of synthetic biology's exciting potential by promoting the development of its most transformative applications and the sharing of these benefits throughout the world.

Bibliography and sources

Alberts B and A Klug (2000), 'The Human Genome Itself Must Be Freely Available to All Humankind', **404**(6776) *Nature* 325.

Amyris, Artemesinin (2013) Amyris <http://www.amyris.com/Products/176/Artemisinin>.

Amyris, *Breakthrough Science* (2013) Amyris <http://198.55.195.109/Innovation/155/BreakthroughScience>.

Amyris, *Fuels* (2013) Amyris <http://198.55.195.109/Products/173/Fuels>.

Annaluru N et al (2014), 'Total Synthesis of a Functional Designer Eukaryotic Chromosome' **344**(6179) *Science* 55.

Antunes M S et al (2011), 'Programmable Ligand Detection System in Plants through a Synthetic Signal Transduction Pathway', **6**(1) *PLoS One* e16292.

Arkin A (2008), 'Setting the Standard in Synthetic Biology', **26**(7) *Nature Biotechnology* 771.

Arkin A P and D Endy (1999), 'A Standard Parts List for Biological Circuitry' (7 October 1999) *DARPA White Paper* <http://dspace.mit.edu/bitstream/handle/1721.1/29794/Arkin.Endy.DARPA.pdf?sequence=1>.

Arras J, *'Comments in Commission Deliberations, the Presidential Commission for the Study of Bioethical Issue, Synthetic Biology'*, (16 November 2010) Atlanta Ga.

Association for Molecular Pathology et al, (2012), 'Brief for Petitioners' in Association for Molecular Pathology v United States Patent and Trademark Office, *appeal docketed No. 12-398* (US S Ct, heard 15 April 2013).

Ayres I and J Braithwaite (1995), *Responsive Regulation: Transcending the Deregulation Debate*, Oxford University Press.

Baker D et al (2006), 'Engineering Life: Building a Fab for Biology', **294**(6) *Scientific American* 44.

Baldwin G et al (2012), *Synthetic Biology – A Primer*, Imperial College Press.

Baldwin R and J Black (2008), 'Really Responsive Regulation', **71**(1) *The Modern Law Review* 59.

Balmer A and P Martin (2008), 'Synthetic Biology: Social and Ethical Challenges, An Independent Review Commissioned by the Biotechnology and Biological Sciences Research Council' (Biotechnology and Biological Sciences Research Council, 2008) <http://www.bbsrc.ac.uk/organisation/policies/reviews/scientific_areas/0806_synthetic_biology.pdf>.

Bassler B, *'Evidence to The Presidential Commission for the Study of Bioethical Issues, Synthetic Biology'* (8 July 2010), Washington DC.

Beck U (English Edition, 2009) *World at Risk*, Polity Press.

Bedau M et al (2010), 'Life After the Synthetic Cell', **465**(7297) *Nature* 422.

Belshaw P J et al (2007), *US Patent No. 8,008,005.*

Benders G A et al (2016), *US Patent Application No. 12/783489.*

Benders G A et al (2016), *US Patent Application No. 12/718911.*

Benkler Y (1999), 'Free as the Air to Common Use: First Amendment Constraints on Enclosure of the Public Domain', **74** *NYUL Rev* 354.

Bennet G and P Rabinow, *Synthetic Biology 2007: From Manifestos to Ramifying Research Programs Connexions* <http://cnx.org/content/m18812/1.2/>.

Berg P et al (1974), 'Potential Biohazards of Recombinant DNA Molecules', **185**(4148) *Science* 303.

Berg P et al (1975), 'Summary Statement of The Asilomar Conference on Recombinant DNA Molecules', **72**(6) *Proceedings of the National Academy of Sciences* 1981.

Bhutkar A, (2005) 'Synthetic Biology: Navigating the Challenges Ahead', **8**(2) *The Journal of Biolaw and Business* 19.

Biegel S (2003), *Beyond Our Control?: Confronting the Limits of Our Legal System in the Age of Cyberspace*, MIT Press.

Bill and Melinda Gates Foundation (2011), 'Apply Synthetic Biology to Global Health Challenges', Grand Challenges in Global Health, <http://www.grandchallenges.org/Explorations/Topics/Pages/SyntheticBiologyRound7.aspx>.

Bimber B and D H Guston (1995), 'Politics by the Same Means: Government and Science in the United States' in Sheila Jasanoff et al (eds), *Handbook of Science and Technology Studies*, Sage Publications.

Biotechnology and Biological Sciences Research Council, 'Business Secretary Announces £40M for UK Synthetic Biology', (2015) Biotechnology and Biological Sciences Research Council <http://www.bbsrc.ac.uk/news/policy/2015/150129-pr-business-secretary-40m-for-synbio/>.

Biotechnology and Biological Sciences Research Council, 'Joint Synthetic Biology Initiative', (2011) <http://www.bbsrc.ac.uk/jointsyntheticbiology>.

Biotechnology and Biological Sciences Research Council, 'Synthetic Biology for Growth Programme', (2016) Biotechnology and Biological Sciences Research Council <http://www.bbsrc.ac.uk/research/programmes-networks/synthetic-biology-growth-programme/>.

Biotechnology and Biosciences Research Council, 'Networks in Synthetic Biology Biotechnology and Biosciences Research Council', <http://www.bbsrc.ac.uk/funding/opportunities/2007/synthetic-biology.aspx>.

Bobe J, *'DIYbio Meetup in Boston on Thursday, October 16'* (11 October 2008) DIYbio Blog http://diybio.org/2008/10/.

Bobe J, *'Don't Phage Me, Bro'* (24 May 2008) DIYbio Blog http://diybio.org/2008/05/.

Bobe J, *'Evidence to The Presidential Commission for the Study of Bioethical Issues, Synthetic Biology'*, (14 September 2010), Philadelphia Penn.

Boldt J and O Müller (2008), 'Newtons of the Leaves of Grass: Certain Ethical Implications of Synthetic Biology Research Go Beyond Those of Genetic Engineering', **26**(4) *Nature Biotechnology* 387.

Bourzac K (2011), 'Synthetic Cells', **114**(3) *Technology Review* 56.

Boustead G, 'The Biohacking Hobbyist', (11 December 2008), *Seed Magazine.* <http://seedmagazine.com/content/article/the_biohacking_hobbyist/>.

Boyle J (2003), 'Foreword: The Opposite of Property?', **66**(1–2) *Law and Contemporary Problems* 1.

Boyle J (2003), 'The Second Enclosure Movement and the Construction of the Public Domain', **66**(1–2) *Law and Contemporary Problems* 33.

Boyle J (2008), *The Public Domain: Enclosing the Commons of the Mind*, Yale University Press.

Boyle J (2010), 'Monopolist of the Genetic Code', The Public Domain, <http://www.thepublicdomain.org/2010/05/28/monopolists-of-the-genetic-code/>.

Brownsword R (2008) 'So What Does the World Need Now? Reflections on Regulating Technologies' in Roger Brownsword and Karen Yeung (eds), *Regulating Technologies: Legal Futures, Regulatory Frames, and Technological Fixes*, Hart Publishing.

Brownsword R (2008), *Rights, Regulation, and the Technological Revolution*, Oxford University Press.

Brownsword R (2011), 'Responsible Regulation: Prudence, Precaution and Stewardship', **62**(5) *Northern Ireland Legal Quarterly* 573.

Brownsword R and M Goodwin (2012), *Law and the Technologies of the Twenty-first Century: Text and Materials*, Cambridge University Press.

Brownsword R and H Somsen (2009), 'Law, Innovation and Technology: Before We Fast Forward—A Forum for Debate', **1**(1) *Law, Innovation and Technology* 1.

Bubela T et al (2012), 'Synthetic Biology Confronts Publics and Policy Makers: Challenges for Communication, Regulation and Commercialization', **30**(3) *Trends in Biotechnology* 132.

Bügl H et al (2007), 'DNA Synthesis and Biological Security', **25**(6) *Nature Biotechnology* 627.

Butler D (2000), 'US/UK Statement on Genome Data Prompts Debate on "Free Access"', **404**(6776) *Nature* 324.

Callaway E (2016), '"Minimal" Cell Raises Stakes in Race to Harness Synthetic Life', **531**(7596) *Nature Reviews Microbiology* 557.

Callaway E and A Coghlan (2010), 'The Meaning of Artificial Life', **206**(2762) *New Scientist* 6.

Callura J M et al (2010), 'Tracking, Tuning, And Terminating Microbial Physiology Using Synthetic Riboregulator', **107**(36) *Proceedings of the National Academy of Sciences* 15898.

Calvert J (2012), 'Ownership and Sharing in Synthetic Biology: A "Diverse Ecology" of the Open and the Proprietary?', **7**(2) *BioSocieties* 169.

Cameron D E et al (2014), 'A Brief History of Synthetic Biology', **12**(5) *Nature Reviews Microbiology* 381.

Cameron J (1999), 'The Precautionary Principle: Core Meaning, Constitutional Framework and Procedures for Implementation' in Ronnie Harding and Elizabeth Fisher (eds), *Perspectives on the Precautionary Principle*, The Federation Press.

Cantley M and M Lex (2011), 'Genetically Modified Foods and Crops' in Jonathan B Wiener et al (eds), *The Reality of Precaution: Comparing Risk Regulation in the United States and Europe*, Routledge.

Canton B, A Labno and D Endy (2008), 'Refinement and Standardization of Synthetic Biological Parts and Devices', **26**(7) *Nature Biotechnology* 787.

Capron A M (2002), 'Governmental Bioethics Commissions: The Nature of the Beast', **9**(3) *Medical Ethics* 4.

Carlson R (2003), 'The Pace and Proliferation of Biological Technologies' **1** *Biosecurity and Bioterrorism: Biodefense Strategy, Practice, and Science* 203.

Carlson R (2009), 'The Changing Economics of DNA Synthesis' **27**(12) *Nature Biotechnology* 1091.

Carlson R (2010), *Biology is Technology: The Promise, Peril and New Business of Engineering Life*, Harvard University Press.

Carlson R, *'Evidence to The Presidential Commission for the Study of Bioethical Issues, Synthetic Biology'*, (8 July 2010), Washington DC.

Carlson R, 'Garage Biology in Silicon Valley', (8 March 2010), Synthesis <http://www.synthesis.cc/2010/03/garage-biology-in-silicon-valley.html>.

Carter S R and R M Friedman, DNA Synthesis and Biosecurity: Lessons Learned and Options for the Future, (J Craig Venter Institute, 2015) <http://www.jcvi.org/cms/fileadmin/site/research/projects/dna-synthesis-biosecurity-report/report-complete.pdf>.

Carter S R et al, Synthetic Biology and the US Biotechnology Regulatory System: Challenges and Options, (J Craig Venter Institute, 2014) <http://www.jcvi.org/cms/fileadmin/site/research/projects/synthetic-biology-and-the-us-regulatory-system/full-report.pdf>.

Caruso D (2008), 'Synthetic Biology: An Overview and Recommendations for Anticipating and Addressing Emerging Risks', *Science Progress, Center for American Progress*, Washington, DC. <http://www.scienceprogress.org/wp-content/uploads/2008/11/syntheticbiology.pdf>

Caudill D S (2009), 'Synthetic Science: A Response to Rabinow', **21**(3) *Law and Literature* 431.

Caulfield T et al (2006), 'Evidence and Anecdotes: An Analysis of Human Gene Patenting Controversies', **24**(9) *Nature Biotechnology* 1091.

Cello J et al (2002), 'Chemical Synthesis of Poliovirus cDNA: Generation of Infectious Virus in the Absence of Natural Template', **297**(5583) *Science* 1016.

Centre for Synthetic Biology and Innovation, 'About CSynBI', (2016) Kings College London <http://www.kcl.ac.uk/sspp/departments/sshm/research/CSynBI/Index.aspx>.

Check E (2005), 'Synthetic Biologists Face Up to Security Issues', **436**(7053) *Nature* 895.

Cho M K et al (1999), 'Ethical Considerations in Synthesizing a Minimal Genome', **286**(5447) *Science* 2087.

Church C, *'Evidence to The Presidential Commission for the Study of Bioethical Issues, Synthetic Biology'*, (8 July 2010), Washington DC.

Church C, 'George M Church's Tech Transfer, Advisory Roles, and Funding Sources', (2012), Harvard Molecular Technology Group and Lipper Center for Computational Genetics <http://arep.med.harvard.edu/gmc/tech.html>.

Church G (2005), 'Let Us Go Forth and Safely Multiply', *Nature*, **438**(7067), 423.

Church G M and E Regis (2014), *Regenesis: How Synthetic Biology Will Reinvent Nature and Ourselves*, Basic Books.

Church G M et al (2014), 'Realizing the Potential of Synthetic Biology', **15**(4) *Nature Reviews Molecular Cell Biology* 289.

Cohen S N and A C Chang (1973), 'Recircularisation and Autonomous Replication of a Sheared R-factor DNA Segment in Escherichia Coli

Transformants', **70**(5) *Proceedings of the National Academy of Sciences* 1293.

Cook-Deegan R (1994) '*The Gene Wars: Science, Politics and the Human Genome*', Norton.

Cowell M, 'iGEM Closes Doors to Amateurs', (10 April 2009), *DIYbio Blog* http://diybio.org/2009/04/10/igem-closes-doors-to-amateurs/.

Cowell M, 'Low-Throughput Lab Automation', (Speech delivered at Synbio Slam, Stanford University, Ca, 14 June 2011) <http://www.youtube.com/watch?v=WoE2DnIDikw>.

Craig R K and J B Ruhl (2014), 'Designing Administrative Law for Adaptive Management', **67**(1) *Vanderbilt Law Review* 1.

Dana G V et al (2012), 'Synthetic Biology: Four Steps to Avoid a Synthetic-Biology Disaster', **483**(7387) *Nature* 29.

Dayton L, 'Geneticist Creates First Artificial Life', *The Australian*, (21 May 2010).

de Lorenzo V (2010), 'Environmental Biosafety in the Age of Synthetic Biology: Do We Really Need a Radical New Approach?', **32**(11) *BioEssays* 926.

Defense Advanced Research Projects Agency, 'Living Foundries Industry Day', (3 June 2011) FedBizOpps.gov <https://www.fbo.gov/index?s=opportunity&mode=form&id=77b10102cba5c98c6be6b5c3d7a11387&tab=core&_cview=0>.

Demissie H T (2010), 'Is Beneficent Regulation the New Better Regulation? Nano-Regulation in the Wake of the "New Better Regulation" Movement', **2**(1) *Law, Innovation and Technology* 115.

Demissie H T (2011), 'Justice or Beneficence: What Regulatory Virtue for Nano-Governance?', **2**(3) *European Journal of Law and Technology* 1.

Department of Agriculture Forestry Service, *FAQs on 2012 Planning Rule* The Forest Planning Rule <https://www.fs.usda.gov/detail/planningrule/faqs#1>.

Department of Energy, 'Obama Administration Announces New Investments to Advance Biofuels Industry and Enhance America's Energy Security', (Press Release, 2 July 2012).

Department of Health and Human Services et al, 'Biosafety in Microbiological and Biomedical Laboratories 5th Edition' (US Department of Health and Human Services, Public Health Service, Centers for Disease Control and Prevention, National Institutes of Health, 2009) <http://www.cdc.gov/biosafety/publications/bmbl5/BMBL.pdf>.

Department of Health and Human Services, 'Charter: Recombinant DNA Advisory Committee', (2011), Office of Science Policy – National Institutes of Health <http://oba.od.nih.gov/oba/RAC/Signed_RAC_Charter_2011.pdf>.

Department of Health and Human Services, 'Screening Framework Guidance for Providers of Synthetic Double Stranded DNA' (19 November 2010) Public Health Emergency <http://www.phe.gov/Preparedness/legal/guidance/syndna/Pages/factsheet.aspx>.

Department of Health and Human Services, *Frequently Asked Questions: Screening Framework Guidance for Providers of Synthetic Double-Stranded DNA* (19 November 2010) Public Health Emergency <http://www.phe.gov/Preparedness/legal/guidance/syndna/Pages/faqs.aspx>.

DIYbio, An Institution for the Do-It-Yourself Biologist DIYbio <http://diybio.org/>.

Dutfield G (2012), 'Patent Law, Biotechnologies and Language in Subject Matter Expansionism' in M Rimmer and A McLennan (eds), *Intellectual Property and Emerging Technologies: The New Biology*, Edward Elgar.

Editorial (2001), 'Slippery Slopes?', **19**(1) *Nature Biotechnology* 1.

Editorial (2007), 'What is Biosecurity?' **63**(2) *Bulletin of the Atomic Scientists* 18.

Edmond G and D Mercer (2009), 'Norms and Irony in the Biosciences: Ameliorating Critique in Synthetic Biology' **21**(3) *Law and Literature* 445.

Eisenberg R S and A K Rai (2004), 'Proprietary Considerations' in R Lanza et al (eds), *Handbook of Stem Cells*, (Elsevier Academic Press).

Eisenstein M (2010), 'Synthetic DNA Firms Embrace Hazardous Agents Guidance but Remain Wary of Automated "Best-Match"', **28**(12) *Nature Biotechnology* 1225.

Endy D (2010) Prepared Statement, *'Evidence to United States House of Representatives Committee on Energy and Commerce, Effects of Developments in Synthetic Genomics and Implications for Energy and Health'*, (27 May), Washington DC.

Endy D, 'Evidence to The Presidential Commission for the Study of Bioethical Issues, Synthetic Biology' (8 July 2010) Washington DC.

Endy D, 'Synthetic Biology' (2003) <http://dspace.mit.edu/bitstream/handle/1721.1/38455/SyntheticBiologyStudy.pdf>.

Endy D (2003), 'Strategy for Biological Risk and Security' <http://dspace.mit.edu/bitstream/handle/1721.1/30595/BioRisk.v2.pdf?sequence=1>.

Endy D (2005), 'Foundations for Engineering Biology', **438**(7067) *Nature* 449.

Endy D (2008), 'Synthetic Biology: Can We Make Biology Easy to Engineer?', **4**(4) *Industrial Biotechnology* 340.

Endy D, 'Evidence to The United States National Academies, The Organisation for Economic Co-operation and Development, and The

Royal Society, Opportunities and Challenges in the Emerging Field of Synthetic Biology', (9 July 2009), Washington DC.

Enemark C (2010), 'Law in the Time of Anthrax: Biosecurity Lessons from the United States', **17**(5) *Journal of Law and Medicine* 748.

Engineering Biology Research Consortium, 'Core Activities', (2017) Engineering Biology Research Consortium <https://www.ebrc.org/core-activities>.

Engineering Biology Research Consortium, 'Frequently Asked Questions', (2017) Engineering Biology Research Consortium <https://www.ebrc.org/faq>.

Engineering Biology Research Consortium, 'What is Synthetic Biology?' Engineering Biology Research Consortium <https://www.ebrc.org/what-is-synbio>.

Environmental Protection Agency et al, 'Modernizing the Regulatory System for Biotechnology Products: Final Version of the 2017 Update to the Coordinated Framework for the Regulation of Biotechnology' (2017) <https://www.epa.gov/sites/production/files/2017-01/documents/2017_coordinated_framework_update.pdf>.

Environmental Protection Agency, 'Vocabulary Catalog', Environmental Protection Agency <https://ofmpub.epa.gov/sor_internet/registry/termreg/searchandretrieve/glossariesandkeywordlists/search.do?details=&glossaryName=Exposure%20Factors%20Glossary>.

Epstein G L (2008), 'The Challenges of Developing Synthetic Pathogens' (19 May) *Bulletin of the Atomic Scientists* <http://www.thebulletin.org/web-edition/features/the-challenges-of-developing-synthetic-pathogens>.

Epstein R A and F S Kieff (2011), 'Questioning the Frequency and Wisdom of Compulsory Licensing for Pharmaceutical Patents', **78** *The University of Chicago Law Review* 71.

ERASynBio, 'ERASynBio 1st Joint Call for Transnational Research Projects: Building Synthetic Biology Capacity through Innovative Transnational Projects' (ERASynBio, 2013) <http://www.erasynbio.eu/lw_resource/datapool/_items/item_40/erasynbio1_call_text_130422_final.pdf>.

ERASynBio, 'ERASynBio' ERASynBio <http://www.erasynbio.eu/project>.

Establishing the Presidential Commission for the Study of Bioethical Issues, Executive Order 13521 of 24 November 2009, Federal Register, 74(228), (30 November 2009).

ETC Group, *ETC Group: A Brief History* ETC Group <http://www.etcgroup.org/en/about/History_of_etcgroup_page>.

ETC Group et al, 'The Principles for the Oversight of Synthetic Biology', (ETC Group, 13 March 2012) <http://libcloud.s3.

amazonaws.com/93/ae/9/2287/1/Principles_for_the_oversight_of_synthetic_biology.pdf>.

ETC Group, 'As Extreme Genetic Engineers Gather in Hong Kong, Critics Warn of Corporate Grab on Plant Life: SynBio 4.0 = SynBio-4-profit' (News Release, 9 October 2008).

ETC Group, *Backgrounder: J Craig Venter Institute's Patent Application on World's First Human-Made Species* (2007) ETC Group <http://www.etcgroup.org/sites/www.etcgroup.org/files/publication/pdf_file2/etc_synthbckgrnder_final2.pdf>.

ETC Group, 'Extreme Genetic Engineering: An Introduction to Synthetic Biology' (ETC Group, January 2007) <http://www.etcgroup.org/upload/publication/602/01/synbioreportweb.pdf>.

ETC Group, 'Extreme Monopoly: Venter's Team Makes Vast Patent Grab on Synthetic Genomes', (8 December 2007) <http://www.etcgroup.org/en/node/663>.

ETC Group, 'Global Coalition Sounds the Alarm on Synthetic Biology, Demands Oversight and Societal Debate' (News Release, 19 May 2006).

ETC Group, 'NGOs Blast US Presidential Commission on Bioethics' (16 December 2010) <http://www.etcgroup.org/en/node/5244>.

ETC Group, 'Patenting Pandora's Bug: Goodbye, Dolly…Hello, Synthia! J Craig Venter Institute Seeks Monopoly Patents on the World's First-Ever Human-Made Life Form', (7 June 2007) <http://www.etcgroup.org/en/node/631>.

ETC Group, 'Syns of Omission: Civil Society Organizations Respond to Report on Synthetic Biology Governance from the J Craig Venter Institute and Alfred P Sloan Foundation', (News Release, 17 October 2007) <http://www.etcgroup.org/en/node/654>.

ETC Group, 'Synthia is Alive … and Breeding: Panacea or Pandora's Box?', (News Release, 20 May 2010).

ETC Group, 'The G(e)nomes of Zurich: Civil Society Calls for Urgent Controls on Synthetic Life', (News Release, 25 June 2007).

ETC Group, 'The New Biomassters: Synthetic Biology and the Next Assault on Biodiversity and Livelihoods', (ETC Group, 2010) <http://www.etcgroup.org/upload/publication/pdf_file/biomassters_27feb2011.pdf>.

ETC Group, 'Venter Institute Builds Longest Sequence of Synthetic DNA (that Doesn't Work)', (News Release, 24 January 2008).

European Commission (2000), 'Communication from the Commission on the Precautionary Principle', Commission of the European Communities <http://eur-lex.europa.eu/LexUriServ/LexUriServ.do?uri=COM:2000:0001:FIN:EN:PDF>.

European Commission, 'Synthetic Biology', (European Commission Directorate General for Research, 2007) <http://ec.europa.eu/research/fp6/nest/pdf/nest_synth_bio.pdf>.

European Commission (2016), 'What are the Rules of Gene Regulation?', Horizon 2020 <https://ec.europa.eu/programmes/horizon2020/en/news/what-are-rules-gene-regulation>.

European Group on Ethics in Science and New Technologies to the European Commission, 'Ethical Aspects of Synthetic Biology: Proceedings of the Roundtable Debate', (European Commission, 19 May 2009) http://ec.europa.eu/bepa/european-group-ethics/docs/publications/round_table_ethical_aspects_of_synthetic_biology.pdf.

ExxonMobil, 'Algae Biofuels', <http://www.exxonmobil.com/Corporate/energy_climate_con_vehicle_algae.aspx>.

Farahany N (2010) 'Comments in Commission Deliberations, The Presidential Commission for the Study of Bioethical Issues, Synthetic Biology' 16 November 2010.

Fatehi L and J Kuzma (2013) 'Policy Innovation in Synthetic Biology Governance', (Paper presented at the 21st Century Borders/Synthetic Biology: Focus on Responsibility and Governance, 2013) <http://www.scienceforglobalpolicy.org/LinkClick.aspx?fileticket=VqAc-vBzvMk=&tabid=133>.

Fidler DP and L O Gostin (2008) *Biosecurity in the Global Age: Biological Weapons, Public Health and the Rule of Law*, Stanford University Press.

Fischer M and S M Maurer (2010), 'Harmonizing Biosecurity Oversight for Gene Synthesis', **28**(1) *Nature Biotechnology*, 20.

Fisher E C et al (eds), (2006), *Implementing the Precautionary Principle: Perspectives and Prospects*, Edward Elgar Publishing.

Flitner D Jr (1986), *The Politics of Presidential Commissions*, Transnational Publishers Inc.

Free Software Foundation Inc., The GNU General Public License (GPL) Version 2 <http://www.opensource.org/licenses/gpl-2.0.php>.

Freemont P '*Synthetic Biology – Do We Need New Regulatory Systems?*' (Paper presented at the 21st Century Borders/Synthetic Biology: Focus on Responsibility and Governance, 2013) <http://www.scienceforglobalpolicy.org/LinkClick.aspx?fileticket=VqAc-vBzvMk%3d&tabid=133>.

Friedman R, *'Governance of Synthetic Biology'* (Paper presented at the 21st Century Borders/Synthetic Biology: Focus on Responsibility and Governance, 2013) <http://www.scienceforglobalpolicy.org/LinkClick.aspx?fileticket=VqAc-vBzvMk%3d&tabid=133>.

Gaines S E (1990), 'Science, Politics, and the Management of Toxic Risks Through Law' **30**(3) *Jurimetrics* 271.

Gaisser S and T Reiss (2009), 'Shaping the Science–Industry–Policy Interface in Synthetic Biology', **3**(1–4) *Systems and Synthetic Biology* 109.

Galligan D J (2007), 'Citizen's Rights and Participation in the Regulation of Biotechnology' in Francesco Francioni (ed.), *Biotechnologies and International Human Rights*, Hart Publishing Ltd.

Garfinkel M S et al, 'Synthetic Genomics: Options for Governance' (The J Craig Venter Institute, Massachusetts Institute of Technology, Center for Strategic and International Studies, October 2007) <http://www.jcvi.org/cms/fileadmin/site/research/projects/synthetic-genomics-report/synthetic-genomics-report.pdf>.

Garza A, 'Comments in Commission Deliberations, the Presidential Commission for the Study of Bioethical Issues, Synthetic Biology' (16 November 2010).

Geist M (2003), 'Cyberlaw 2.0', **44** *Boston College Law Review* 323.

Gen9, 'Powering the Synthetic Biology Revolution', (2013) Gen9 <http://gen9bio.com/>.

Gibson D G (2008), *US Patent Application No. 12/247126.*

Gibson D G et al (2009), *US Patent Application No. 12/371543.*

Gibson D G et al (2008), 'Complete Chemical Synthesis, Assembly, and Cloning of a Mycoplasma Genitalium Genome', **319**(5867) *Science* 1215.

Gibson D G et al (2008), 'One-Step Assembly in Yeast of 25 Overlapping DNA Fragments to Form a Complete Synthetic Mycoplasma Genitalium Genome', **105**(51) *Proceedings of the National Academy of Sciences* 20404.

Gibson D G et al (2009), 'Enzymatic Assembly of DNA Molecules up to Several Hundred Kilobases', **6**(5) *Nature Methods* 343.

Gibson D G et al (2010), 'Creation of a Bacterial Cell Controlled by a Chemically Synthesized Genome', **329**(5987) *Science* 52.

Gibson D G and H O Smith (2010), *US Patent No. 7,776,532.*

Gibson D G et al (2011), 'Methods For In Vitro Joining and Combinatorial Assembly of Nucleic Acid Molecules [Kit]', *United States Patent Application No. 13/035,699.*

Glass J I et al (2006), *US Patent Application No. 11/546364.*

Glass J I et al (2006), *US Patent Application No. 11/644,713.*

Glass J I et al (2006), 'Essential Genes of a Minimal Bacterium', **103**(2) *Proceedings of the National Academy of Sciences of the United States of America* 425.

Glass J I et al (2010), *US Patent Application No. 12/598414.*

Glass J I et al (2016), *US Patent No. 9,434,974.*

Glass J I et al (2016), *US Patent No. 9,481,880.*

Gold E R et al (2007), 'Gene Patents—More Evidence Needed, But Policymakers Must Act', **25**(4) *Nature Biotechnology* 388.

Goldstein B D and R S Carruth (2003), 'Implications Of The Precautionary Principle For Environmental Regulation In The United States: Examples From The Control Of Hazardous Air Pollutants In The 1990 Clean Air Act Amendments', **66**(4) *Law and Contemporary Problems* 247.

Gore A (2013), *The Future*, Random House.

Gostin L O and D P Fidler, (2007) 'Biosecurity Under the Rule of Law', **38** *Case Western Journal of International Law* 438.

Gray G et al (2011), 'Beef, Hormones and Mad Cows', in Jonathan B Wiener et al (eds), *The Reality of Precaution: Risk Regulation in the United States and Europe*, Routledge.

Gunningham N (2007), 'Regulating Biotechnology: Lessons from Environmental Policy', in Han Somsen (ed.), *The Regulatory Challenge of Biotechnology: Human Genetics, Food and Patents*, Edward Elgar Publishing.

Gunningham N, P Grabosky and D Sinclair (1998), *Smart Regulation: Designing Environmental Policy*, Clarendon Press.

Gutmann A (2011), 'The Ethics of Synthetic Biology: Guiding Principles For Emerging Technologies', **41**(4) *Hastings Center Report* 17.

Gutmann A, *'Comments in Commission Deliberations, The Presidential Commission for the Study of Bioethical Issues, Synthetic Biology'* 16 November 2010.

Gutmann A and D Thompson (1997), 'Deliberating About Bioethics', **27**(3) *Hastings Center Report* 38.

Hardin G (1968), 'The Tragedy of the Commons', **162**(3859) *Science* 1243.

Hauser S, 'Comments in Commission Deliberations, the Presidential Commission for the Study of Bioethical Issues, Synthetic Biology', (16 November 2010).

Hawkins N (2011), 'The Impact of Human Gene Patents on Genetic Testing in the United Kingdom', **13**(4) *Genetics in Medicine* 320.

Heller M A and R S Eisenberg (1998), 'Can Patents Deter Innovation? The Anticommons In Biomedical Research', **280**(5364) *Science* 698.

Henkel J and S M Maurer (2007), 'The Economics of Synthetic Biology', **3**(1) *Molecular Systems Biology* 117.

Henkel J and S M Maurer (2009), 'Parts, Property and Sharing', **27**(12) *Nature Biotechnology* 1095.

Herman E, 'The Challenge to Meet Global Need for Protein Sources for Animal Production', (Paper presented at the 21st Century Borders/ Synthetic Biology: Focus on Responsibility and Governance, 2013)

<http://www.scienceforglobalpolicy.org/LinkClick.aspx?fileticket=VqAc-vBzvMk%3d&tabid=133>.

Hillson N J (2016), *US Patent No. 9,361,427.*

Hillson N J, 'Tools and Opportunities to Enhance Risk Analysis', National Academy of Sciences <http://nas-sites.org/biotech/files/2016/04/Tools_Risk_Analysis_Hillson.pdf>.

Hope J (2009), *Biobazaar: The Open Source Revolution and Biotechnology*, Harvard University Press.

Hutchison III C A et al (2016), 'Design and Synthesis of a Minimal Bacterial Genome', **351**(6280) *Science* 6253.

iGEM Foundation, *Our Philosophy* Registry of Standard Biological Parts <http://parts.igem.org/Help:Philosophy>.

iGEM Foundation, *Registry of Standard Biological Parts* <http://parts.igem.org/Main_Page>.

iGEM Foundation, 'iGEM: Synthetic Biology Based on Standard Parts', iGEM <http://ung.igem.org/Main_Page>.

Institute on Science for Global Policy, '21st Century Borders/Synthetic Biology: Focus on Responsibility and Governance' (2013) <http://www.scienceforglobalpolicy.org/LinkClick.aspx?fileticket=VqAc-vBzvMk%3d&tabid=133>.

International Association for Synthetic Biology, 'The IASB Code of Conduct for Best Practices in Gene Synthesis', International Association for Synthetic Biology, <http://www.ia-sb.eu/tasks/sites/synthetic-biology/assets/File/pdf/iasb_code_of_conduct_final.pdf>.

International Gene Synthesis Consortium, 'Harmonized Screening Protocol: Gene Sequence and Customer Screening to Promote Biosecurity', International Gene Synthesis Consortium <http://www.genesynthesisconsortium.org/wp-content/uploads/2012/02/IGSC-Harmonized-Screening-Protocol1.pdf>.

International Research Panel of the Presidential Commission for the Study of Bioethical Issues, 'Research Across Borders: International Research Panel of the Presidential Commission for the Study of Bioethical Issues' (September 2011) <http://bioethics.gov/sites/default/files/IRP-%20Research%20Across%20Borders.pdf>.

J Craig Venter Institute, 'Venter Institute Scientists Create First Synthetic Bacterial Genome' (Press Release, 24 January 2008).

J Craig Venter Institute, 'First Minimal Synthetic Bacterial Cell Designed and Constructed by Scientists at Venter Institute and Synthetic Genomics, Inc.' (Press Release, 24 March 2016).

J Craig Venter Institute, 'First Self Replicating Synthetic Bacterial Cell: Frequently Asked Questions', <http://www.jcvi.org/cms/research/projects/first-self-replicating-synthetic-bacterial-cell/faq/>.

J Craig Venter Institute, 'Venter Institute Scientists Create First Synthetic Bacterial Genome' (Press Release, 24 January 2008).

Jackson A, 'What Do We Need From Synthetic Biology?' (Paper presented at the Fifth International Conference on Synthetic Biology, Stanford University, 2011).

Jaffer Z, 'Eri Gentry's Garage Biotech Revolution', (15 July 2010) Wired UK <http://www.wired.co.uk/news/archive/2010-07/15/eri-gentry-garage-biotech-revolution>.

Jasanoff S (2009), *The Fifth Branch: Science Advisers as Policymakers*, Harvard University Press.

Jensen M K and J D Keasling (2015), 'Recent Applications of Synthetic Biology Tools for Yeast Metabolic Engineering', **15**(1) *FEMS Yeast Research* 1.

Johnson N, *'Steal This Genome!'*, (30 March 2005), East Bay Express http://www.eastbayexpress.com/ebx/steal-this-genome/Content?oid=1077148.

Johnson R A 'Evidence to The United States National Academies, The Organisation for Economic Co-operation and Development, and The Royal Society, Opportunities and Challenges in the Emerging Field of Synthetic Biology' (10 July 2009) Washington DC.

Joly Y (2010), 'Open Biotechnology: Licenses Needed', **28**(5) *Nature Biotechnology* 417.

Jones J (2007), 'Regulatory Design for Scientific Uncertainty: Acknowledging the Diversity of Approaches in Environmental Regulation and Public Administration', **19**(3) *Journal of Environmental Law* 347.

Jones J (2008), 'Certainty as Illusion: The Nature and Purpose of Uncertainty in the Law' in Gabriele Bammer and Michael Smithson (eds), *Uncertainty and Risk: Multidisciplinary Perspectives*, Earthscan.

Jorgensen E, 'Isn't That Cute? Little Johnny's First New Life Forms', (Speech delivered at Synthetic Biology Slam, Stanford University, Ca, 14 June 2011) <http://www.youtube.com/watch?v=LqHdK_yYbiM&list=UUFHpWUDcjnjNAXrhGSb9cqw&index=7&feature=plcp>.

Kahl L J and D Endy (2013), 'A Survey of Enabling Technologies in Synthetic Biology', **7**(13) *Journal of Biological Engineering* 1.

Kahn L H 'DIY Biology', (18 June 2012), *Bulletin of the Atomic Scientists*, <http://www.thebulletin.org/web-edition/columnists/laura-h-kahn/diy-biology>.

Kaiser J (2007), 'Attempt to Patent Artificial Organism Draws a Protest', **316**(5831) *Science* 1557.

Keasling J D (2008), 'Synthetic Biology for Synthetic Chemistry', **3**(1) *ACS Chemical Biology* 64.

Keasling J D, 'Evidence to United States House of Representatives Committee on Energy and Commerce, Developments in Synthetic Genomics and Implications for Health and Energy', (27 May 2010).

Keasling Lab, 'Metabolic Engineering', Keasling Lab, Lawrence Berkeley National Laboratory <http://keaslinglab.lbl.gov/research_areas/index.html>.

Khalil A S and J J Collins (2010), 'Synthetic Biology: Applications Come of Age', **11**(5) *Nature Reviews Genetics*, 367.

Kuiken T, 'Do-It-Yourself Biology: Reality and the Path toward Innovation' (Paper presented at the 21st Century Borders/Synthetic Biology: Focus on Responsibility and Governance, 2013) <http://www.scienceforglobalpolicy.org/LinkClick.aspx?fileticket=VqAc-vBzvMk%3d&tabid=133>.

Kumar S and A Rai (2006), 'Synthetic Biology: The Intellectual Property Puzzle', **85** *Texas Law Review* 1745.

Kuzma J and T Tanji (2010), 'Unpackaging Synthetic Biology: Identification of Oversight Policy Problems and Options', **4**(1) *Regulation & Governance* 92.

Lange D (2003), 'Reimagining the Public Domain', **66**(1–2), *Law & Contemporary Problems* 463.

Lartigue C et al (2007), 'Genome Transplantation in Bacteria: Changing One Species to Another', **317**(5838) *Science* 632.

Lartigue C et al (2009), 'Creating Bacterial Strains from Genomes that Have Been Cloned and Engineered in Yeast', **325**(5948) *Science* 1693.

Ledford H (2010), 'Gene-synthesis Rules Favour Convenience', **467**(7318) *Nature* 898.

Ledford H (2015), 'CRISPR, the Disruptor', **522**(7554) *Nature* 20.

Lemley M A (2012), 'Contracting Around Liability Rules', **100**(2) *California Law Review* 463.

Lessig L (2001), 'The Architecture of Innovation', **51** *Duke LJ* 1783.

Lewens T (2007), 'Introduction: Risk and Philosophy' in Tim Lewens (ed.), *Risk: Philosophical Perspectives*, Routledge.

Litman J (1990), 'The Public Domain', **39** *Emory L J* 965.

Lou C et al (2012), *US Patent Application No. 13/489,205.*

LS9, 'Technology Overview', (2011) LS9 <http://ls9.com/technology/technology-overview>.

Lu T K T and A Cheng (2016), *US Patent No. 9,315,806.*

MacQueen H et al (2008), *Contemporary Intellectual Property: Law and Policy*, Oxford University Press.

Mandel G N (2009), 'Regulating Emerging Technologies', **1**(1) *Law, Innovation and Technology* 75.

Marris C and N Rose (2012), 'Let's Get Real on Synthetic Biology', **214**(2868) *New Scientist* 28.

Martin K V et al (2012), *US Patent No. 8,309,323*.

Maurer S M, 'Before it's too Late: Why Synthetic Biologists Need an Open-Parts Collaboration – and How to Build One', (2009), **10**(8) *EMBO Reports* 806.

Maurer S M et al, 'Working Paper – Making Commercial Biology Safer: What the Gene Synthesis Industry Has Learned About Screening Customers and Orders' (Goldman School of Public Policy and Boalt Law School, University of California, 2009) <http://gspp.berkeley.edu/assets/uploads/page/Maurer_IASB_Screening.pdf>.

Maurer S M and L Zoloth (2007), 'Synthesizing Biosecurity', **63**(6) *Bulletin of the Atomic Scientists* 16.

McCray L E et al (2010), 'Planned Adaptation in Risk Regulation: an Initial Survey of US Environmental, Health, and Safety Regulation', **77**(6) *Technological Forecasting and Social Change* 951.

McFadden M et al (2008), 'Uncertainty, Terrorism and Law Enforcement' in Gabriele Bammer and Michael Smithson (eds), *Uncertainty and Risk: Multidisciplinary Perspectives* (Earthscan).

McKenna, P (2009) 'Rise of the Garage Genome Hackers' **2689** *New Scientist* 20.

McLennan A (2012) 'Building with Biobricks: Constructing a Commons for Synthetic Biology Research' in M Rimmer and A McLennan (eds), *Intellectual Property and Emerging Technologies: The New Biology*, Edward Elgar, 176.

McLennan A (2017) 'Patent Law and the Emerging Science of Synthetic Biology: An Examination of Principle and Practice', **36**(2) *Biotechnology Law Report* 59.

McLennan A (2017) 'Patent Law and the Emerging Science of Synthetic Biology: An Examination of Principle and Practice, Part 2', **36**(3) *Biotechnology Law Report* 105.

McLennan A and M Rimmer (2012), 'Inventing Life: Patent Law and Synthetic Biology', The Conversation <http://theconversation.com/inventing-life-patent-law-and-synthetic-biology-5178>.

Merton R K (1973), *The Sociology of Science: Theoretical and Empirical Investigations*, University of Chicago Press.

Miller S and M J Selgelid (2007), 'Ethical and Philosophical Consideration of the Dual-use Dilemma in the Biological Sciences', **13**(4) *Science and Engineering Ethics* 523.

Minshull J and R Wagner (2009), 'Preventing the Misuse of Gene Synthesis', **27**(9) *Nature Biotechnology* 800.

Moe-Behrens G H et al (2013), 'Preparing Synthetic Biology for the World', **4** *Frontiers in Microbiology* 5.

Mohan-Ram V S and J M Waxman (2008), 'Synthetic Biology Patent Applications Expected to Present New Challenges', **2**(12) *Life Sciences Law & Industry Report* 1.

Mooallem J (2010), 'Do-It-Yourself Genetic Engineering', *The New York Times Online* (Online), 10 February <http://www.nytimes.com/2010/02/14/magazine/14Biology-t.html?_r=1&pagewanted=1>.

Mukunda G et al (2009), 'What Rough Beast? Synthetic Biology, Uncertainty, and the Future of Biosecurity', **28**(2) *Politics and the Life Sciences* 2.

Nair P (2009), 'Straight talk with ... Mac Cowell and Jason Bobe', **15**(3) *Nature Medicine* 230.

National Forest System Land Management Planning, 36 CFR Part 219, Federal Register 77(68) 21162, 9 April 2012.

National Science Advisory Board for Biosecurity, 'Addressing Biosecurity Concerns Relating to Synthetic Biology', (2010) <http://oba.od.nih.gov/biosecurity/pdf/NSABB%20SynBio%20DRAFT%20Report-FINAL%20(2)_6-7-10.pdf>.

National Science Advisory Board for Biosecurity, 'Addressing Biosecurity Concerns Related to the Synthesis of Select Agents' (National Institutes of Health, Department of Health and Human Services, 2006) <http://oba.od.nih.gov/biosecurity/pdf/Final_NSABB_Report_on_Synthetic_Genomics.pdf>.

National Science Advisory Board for Biosecurity, 'Strategies to Educate Amateur Biologists and Scientists in Non-life Science Disciplines about Dual Use Research in the Life Sciences' (National Science Advisory Board for Biosecurity, June 2011) <http://oba.od.nih.gov/biosecurity/pdf/FinalNSABBReport-AmateurBiologist-NonlifeScientists_June-2011.pdf>.

National Science Foundation, University of California Berkeley (2013) Gen-3 Engineering Research Centers (ERC): (ERC) Active Awards <http://www.nsf.gov/awards/award_visualization.jsp?org=NSF&pims_id=5502&ProgEleCode=1480&RestrictActive=on&BooleanElement=true&BooleanRef=true&from=fund#region=US-CA&instId=0013128000>.

Neave M (2005), 'Institutional Law Reform In Australia: The Past and the Future', *Windsor Year Book of Access to Justice*, **23**, 343.

Nicholas A, J B Phillips and M D Pedersen (2009), 'Creating Genetic Devices', *United States Patent Application No. 12/552829.*

Nicol D and R Gold (2012), 'Standards for Biobank Access and Intellectual Property' in Matthew Rimmer and Alison McLennan (eds), *Intellectual Property and Emerging Technologies: The New Biology*, Edward Elgar.

Nicol D and J L Nielsen (2003), 'Patents and Medical Biotechnology: An Empirical Analysis of Issues Facing the Australian Industry'. (Centre for Law and Genetics Occasional Paper No. 6, Centre for Law and Genetics, 2013) <http://www.ipria.org/publications/reports/BiotechReportFinal.pdf>.

Nicol D (2010), 'Patent Licensing in Medical Biotechnology in Australia: A Role for Collaborative Licensing Strategies?' (Centre for Law and Genetics Occasional Paper No. 7, 2010) <http://www.lawgenecentre.org/occpaper_7.pdf>.

NIH Guidelines for Research Involving Recombinant or Synthetic Nucleic Acid Molecules, Federal Register 81 (No 73) 15 April 2016.

Obama B, 'Letter from President Barack Obama to Dr Amy Gutmann' (20 May 2010).

Office of Biotechnology Activities, 'About Recombinant DNA Advisory Committee (RAC)', Office of Science Policy – National Institutes of Health <http://oba.od.nih.gov/rdna_rac/rac_about.html>.

O'Malley M et al (2008), 'Knowledge-making Distinctions in Synthetic Biology', **30**(1) *BioEssays* 57.

Ostrom E (1990), *Governing the Commons: The Evolution of Institutions for Collective Action*, Cambridge University Press.

Oye K A (2009), 'Embracing Uncertainty', **26**(1) *Issues in Science and Technology* 91.

Oye K A and R Wellhausen, (2009) 'The Intellectual Commons and Property in Synthetic Biology' in Markus Schmidt et al (eds), *Synthetic Biology: the Technoscience and its Societal Consequences*, Springer Academic Publishing.

Paddon C J et al (2013), 'High-Level Semi-Synthetic Production of the Potent Antimalarial Artemisinin', **496**(7446) *Nature* 528.

Palmer A, 'Craig Venter's Research is Scary, But Not in the Way You Think', *Telegraph* (UK) online (22 May 2010).

Patterson M (2010), 'A Biopunk Manifesto', LiveJournal <http://maradydd.livejournal.com/496085.html>.

Paxson R E and J F Hicklin (2010), *US Patent No. 7,769,576*.

Picataggio S et al (2015), *US Patent No. 8,114,641*.

Pleiss J (2006), 'The Promise of Synthetic Biology', **73**(4) *Applied Microbiology and Biotechnology* 735.

Posner R A (2004), *Catastrophe: Risk and Response*, Oxford University Press.

Pottage A (2009), 'Protocell Patents: Property Between Modularity and Emergence' in M A Bedau and E C Parke (eds), *The Ethics of Protocells: Moral and Social Implications of Creating Life in the Laboratory*, MIT Press.

Prather L J K, 'Evidence to The Presidential Commission for the Study of Bioethical Issues, Synthetic Biology', (8 July 2010), Washington DC.

Prepared Statement of Venter J C, 'Evidence to United States House of Representatives Committee on Energy and Commerce, Effects of Developments in Synthetic Genomics and Implications for Energy and Health', (27 May 2010), Washington DC.

Purnick P E and R Weiss (2009), 'The Second Wave of Synthetic Biology: From Modules to Systems', **10**(6) *Nature Reviews Molecular Cell Biology* 410.

Rai A (2010), 'Law and Technology Unstandard Standardization: The Case of Biology', **53**(1) *Communications of the ACM*, 37.

Rai A K (2003), 'Engaging Facts and Policy: A Multi-Institutional Approach to Patent System Reform' **103**(5) *Columbia Law Review* 1035.

Rai A K 'Evidence to the Presidential Commission for the Study of Bioethical Issues, Synthetic Biology', (13 September 2010), Philadelphia Penn.

Rai A K (1999) 'Regulating Scientific Research: Intellectual Property Rights and the Norms of Science', **94**(1) *Northwestern University Law Review* 77.

Rai A and J Boyle (2007), 'Synthetic Biology: Caught Between Property Rights, the Public Domain, and the Commons', **5**(3) *PLoS biology* e58.

Rai A K and R S Eisenberg (2003), 'Bayh-Dole Reform and the Progress of Biomedicine', **66**(1–2) *Law and Contemporary Problems* 289.

Rai A, S J Graham and M E Doms 'Patent Reform: Unleashing Innovation, Promoting Economic Growth and Producing High-Paying Jobs – A White Paper from the U.S. Department of Commerce' (United States Department of Commerce, 13 April 2010) <https://www.esa.doc.gov/sites/default/files/reports/documents/patentreform_0.pdf>.

Redford K H et al (2013), 'Synthetic Biology and Conservation of Nature: Wicked Problems and Wicked Solutions' **11**(4) *PLoS Biol* e1001530.

Registry of Standard Biological Parts, 'Help: About the Registry' <http://partsregistry.org/Help:About_the_Registry>.

Reichman J H (2006), *Compulsory Licensing of Patented Inventions: Comparing United States Law and Practice with Options under The TRIPS Agreement*, in American Association of Law Schools Mid-Year Workshop on Intellectual Property, <http://www.aals.org/documents/2006intprop/JeromeReichmanOutline.pdf>.

Reinhoff H, 'Comments in Panel Discussion', (Speech delivered at Outlaw Biology Symposium, Los Angeles, 29–31 January 2010) <http://artsci.ucla.edu/videos/outlaw_biology/OB9_Panel_Discussion1/>.

Relman D, 'Evidence to The Presidential Commission for the Study of Bioethical Issues, Synthetic Biology', (14 September 2010) Philadelphia Penn.

Renninger N S and D J McPhee (2008), *US Patent No. 7,846,222.*

Renninger N S et al (2010), *US Patent No. 7,659,097.*

Rettberg R, 'Evidence to The United States National Academies, The Organisation for Economic Co-operation and Development, and The Royal Society, Opportunities and Challenges in the Emerging Field of Synthetic Biology', (10 July 2009), Washington DC.

Riley M F and R A Merrill (2005), 'Regulating Reproductive Genetics: A Review of American Bioethics Commissions and Comparison to the British Human Fertilisation and Embryology Authority', **6** *Columbia Science & Technology Law Review* 1.

Rimmer M and A McLennan (eds) (2012), *Intellectual Property and Emerging Technologies: The New Biology*, Edward Elgar Publishing.

Rimmer M (2008), *Intellectual Property and Biotechnology: Biological Inventions*, Edward Elgar Publishing.

Rimmer M (2009), 'The Sorcerer II Expedition: Intellectual Property and Biodiscovery', **6** *Macquarie Journal of Intellectual Property* 147.

Rimmer M (2011), *Intellectual Property and Climate Change: Inventing Clean Technologies*, Edward Elgar Publishing.

Roca M M (2013), 'The Challenges of Deploying Synthetic Biology Technologies in Developing Countries', (Paper presented at the 21st Century Borders/Synthetic Biology: Focus on Responsibility and Governance); <http://www.scienceforglobalpolicy.org/LinkClick.aspx?fileticket=VqAc-vBzvMk%3d&tabid=133>.

Roessler P G et al (2012), *US Patent No. 8,124,400.*

Rutz B (2009), 'Synthetic Biology and Patents', **10**(Supplementary Issue) *EMBO Reports*, S14.

Samuelson P (2006), 'Enriching Discourse on Public Domains', **55**(4) *Duke Law Journal* 783.

Sandin P (2007), 'Common-sense Precaution and Varieties of the Precautionary Principle' in Tim Lewens (ed.), *Risk: Philosophical Perspectives*, Routledge.

Sandin P (2009), 'A New Virtue Based Understanding of the Precautionary Principle' in M A Bedau and E C Parke (eds), *The Ethics of Protocells: Moral and Social Implications of Creating Life in the Laboratory*, MIT Press.

Schmidt M (2008), 'Diffusion of Synthetic Biology: A Challenge to Biosafety', **2**(1–2) *Systems and Synthetic Biology* 1.

Schmidt M (2009) 'Do I Understand What I Can Create? Biosafety Issues in Synthetic Biology' in Markus Schmidt et al (eds), *Synthetic*

Biology: the Technoscience and its Societal Consequences, Springer Academic Publishing.
Schmidt M, 'Safeguarding the Genetic Firewall with Xenobiology' (Paper presented at the 21st Century Borders/Synthetic Biology: Focus on Responsibility and Governance, 2013) <http://www.scienceforglobalpolicy.org/LinkClick.aspx?fileticket=VqAc-vBzvMk%3d&tabid=133>.
Schmidt M et al (2009), 'A Priority Paper for the Societal and Ethical Aspects of Synthetic Biology', **3**(1) *Systems and Synthetic Biology* 3.
Schmidt M (2012), 'Executive Summary' in M Schmidt (ed.), *Synthetic Biology: Industrial and Environmental Applications*, Wiley-Blackwell.
Screening Framework Guidance for Providers of Synthetic Double Stranded DNA, Federal Register 75 (197) 62820, (13 October 2010).
Second International Meeting on Synthetic Biology, 'Declaration of the Second International Meeting on Synthetic Biology', (2006) syntheticbiology.org <http://syntheticbiology.org/SB2Declaration.html>.
Service R F (2016), 'Synthetic Biology. Synthetic Microbe Has Fewest Genes, But Many Mysteries', **351**(6280) *Science (New York, NY)* 1380.
Silver P, 'Evidence to The United States National Academies, The Organisation for Economic Co-operation and Development, and The Royal Society, Opportunities and Challenges in the Emerging Field of Synthetic Biology', (10 July 2009), Washington DC.
Simirenko L et al, 'Bliss: The Black List Sequence Screening Pipeline', Synthetic Biology Engineering Evolution Design Conference <http://synbioconference.org/2016/proceeding/paper/bliss-black-list-sequence-screening-pipeline>.
Simon B M and C T Scott (2011), 'Unsettled Expectations: How Recent Patent Decisions Affect Biotech' **29**(3) *Nature Biotechnology* 229.
Smith H O et al (2003), 'Generating a Synthetic Genome by Whole Genome Assembly: ϕX174 Bacteriophage from Synthetic Oligonucleotides', **100**(26) *Proceedings of the National Academy of Sciences* 15440.
Smithson A, 'Renovating Governance Strategies for Synthetic Biology and Other "Dual-Use" Technologies', (Paper presented at the 21st Century Borders/Synthetic Biology: Focus on Responsibility and Governance, 2013) <http://www.scienceforglobalpolicy.org/LinkClick.aspx?fileticket=VqAc-vBzvMk%3d&tabid=133>.
Smolke C, 'Evidence to The United States National Academies, The Organisation for Economic Co-operation and Development, and The Royal Society, Opportunities and Challenges in the Emerging Field of Synthetic Biology', (9 July 2009), Washington DC.
Snow A, 'Evidence to the Presidential Commission for the Study of Bioethical Issues, Synthetic Biology' (8 July 2010) Washington DC.

Snowball W D (1993), 'Rhetorical Constraints on Government Commissions' in C M Miller and B C McKinney (eds), *Government Commission Communication*, Praeger.

Staff Writer (2010), 'And Man Made Life: Artificial Life, the Stuff of Dreams and Nightmares, Has Arrived', *The Economist*, 22 May, 11.

Staff Writer (2010), 'Genesis Redux', *The Economist*, 22 May, 81.

Stanford University, 'Biosafety Training Requirements Stanford University: Environmental Health and Safety', <https://ehs.stanford.edu/reference/biosafety-training-requirements>.

Stevens A, 'Evidence to The Presidential Commission for the Study of Bioethical Issues, Synthetic Biology', (13 September 2010), Philadelphia.

Sunstein C R (2002), *Risk and Reason: Safety, Law, and the Environment*, Cambridge University Press.

Sunstein C R (2005), *Laws of Fear: Beyond the Precautionary Principle*, Cambridge University Press.

SynBioComm, 'SynBioComm: Towards a European Synthetic Biology Community' SYNBIOSAFE <http://www.synbiosafe.eu/index.php?page=synbiocomm>.

SYNBIOSAFE, 'Safety and Ethical Aspects of Synthetic Biology', SYNBIOSAFE <http://www.synbiosafe.eu/>.

Synthetic Biology Engineering Research Centre, 'About SynBERC', <http://www.synberc.org/about>.

Synthetic Biology Engineering Research Centre, 'Devices', (2013) SynBERC <http://synberc.org/content/devices>.

Synthetic Biology Engineering Research Centre, 'Parts', (2013) SynBERC <http://synberc.org/content/parts>.

Synthetic Biology Engineering Research Centre, *Practices* (2013) Synthetic Biology Engineering Research Centre <http://www.synberc.org/practices>.

Synthetic Biology Engineering Research Centre, 'Research', (2013) Synthetic Biology Engineering Research Centre <http://synberc.org/content/research>.

Synthetic Biology Research Community, 'FAQ – What is Synthetic Biology?' <http://syntheticbiology.org/FAQ.html>.

Synthetic Genomics Inc., 'Next Generation Fuels & Chemicals', (2012) Synthetic Genomics <http://www.syntheticgenomics.com/what/renewablefuels.html>.

Synthetic Genomics Inc., 'Novel Applications', (2015) Synthetic Genomics <http://www.syntheticgenomics.com/novelapplications.html>.

Synthetic Genomics Inc., 'Synthetic Genomics Inc and J Craig Venter Institute Form New Company, Synthetic Genomics Vaccines Inc

(SGVI), to Develop Next Generation Vaccines', (Press Release, 7 October 2010).

Synthetic Genomics Inc., 'What We Do', (2012) Synthetic Genomics Inc. <http://www.syntheticgenomics.com/what/>.

Synthetic Genomics Inc., 'Flu Vaccine' (2016) Synthetic Genomics, <https://www.syntheticgenomics.com/flu-vaccine>.

Tait J and D Castle (2012), 'Artificial Microbes: Balanced Regulation of Synthetic Biology', **484**(7392) *Nature* 37.

The Bio Fab Group: Baker D et al (2006), 'Engineering Life: Building a Fab for Biology', *Scientific American*, June, 44.

The BioBricks Foundation, *About* (2012) <http://biobricks.org/about-foundation/>.

The BioBricks Foundation, FAQ <http://www.biobricks.org/FAQ.php>.

The BioBricks Foundation, *Education Program* (2012) <http://biobricks.org/programs/education-program/>.

The BioBricks Foundation, 'SBx.0 Conference Series', (2013) The BioBricks Foundation <http://biobricks.org/programs/sbx-0-conference-series/>.

The BioBricks Foundation, *The BioBrick™ Public Agreement* (2013) The BioBricks Foundation <http://biobricks.org/bpa/>.

The BioBricks Foundation, 'Technical Program', (2016) The BioBricks Foundation <http://biobricks.org/programs/technical-program/>.

The BioBricks Foundation, 'The BioBricks Foundation', (2013) <http://www.biobricks.org/index.php>.

The BIOFAB, *About the BIOFAB* <http://biofab.org/about>.

The BIOFAB, 'Biofab Launches World's First Biological Design-Build Facility', (2010) Biofab <http://biofab.org/datasheet>.

The Editors (2006), 'How to Kill Synthetic Biology', **294**(6) *Scientific American* 4.

The Presidential Commission for the Study of Bioethical Issues, 'New Directions: The Ethics of Synthetic Biology and Emerging Technologies' (The Presidential Commission for the Study of Bioethical Issues, December 2010) <http://www.bioethics.gov/documents/synthetic-biology/PCSBI-Synthetic-Biology-Report-12.16.10.pdf>.

The Presidential Commission for the Study of Bioethical Issues, 'History of Bioethics Commissions', <http://bioethics.gov/cms/history>.

The President's Council on Bioethics, 'Reproduction and Responsibility: The Regulation of New Biotechnologies' (March 2004) <http://bioethics.georgetown.edu/pcbe/reports/reproductionandresponsibility/index.html>.

The Royal Academy of Engineering, 'Synthetic Biology: Public Dialogue on Synthetic Biology' (The Royal Academy of Engineering, 2009) <http://www.raeng.org.uk/news/publications/list/reports/syn_bio_dialogue_report.pdf>.

The Royal Academy of Engineering, 'Synthetic Biology: Scope, Applications and Implications' (The Royal Academy of Engineering, 2009) <http://www.raeng.org.uk/news/publications/list/reports/Synthetic_biology.pdf>.

The Royal Society and Organisation for Economic Co-operation and Development, 'Symposium on Opportunities and Challenges in the Emerging Field of Synthetic Biology: Synthesis Report' (The Royal Society, Organisation for Economic Co-operation and Development, 2010) <http://www.oecd.org/science/biotech/45144066.pdf>.

The Woodrow Wilson International Center for Scholars, 'US Trends in Synthetic Biology Research Funding', (September 2015), <http://www.synbioproject.org/site/assets/files/1386/final_web_print_sept2015.pdf?>.

The Woodrow Wilson International Center for Scholars: Science Technology and Innovation Program, *Comprehensive Environmental Assessment and Synthetic Biology Applications Workshop: Background Materials* (2011) <http://www.synbioproject.org/process/assets/files/6609/_draft/background_packet_for_cea_and_syn_bio_workshop_jul_28_2011.pdf>.

The Woodrow Wilson International Center for Scholars: Science Technology and Innovation Program, *Comprehensive Environmental Assessment and Synthetic Biology Applications Workshop: Conference Notes Version 1.0* (2011) <http://www.synbioproject.org/process/assets/files/6609/_draft/final_cea_and_syn_bio_workshop_notes_wilson_center.pdf>.

Thomas J, 'Evidence to The Presidential Commission for the Study of Bioethical Issues, Synthetic Biology', (8 July 2010), Washington DC.

Torrance A W (2010), 'Synthesizing Law for Synthetic Biology', **11**(2) *Minnesota Journal of Law, Science and Technology* 629.

Torrance A W and L J Kahl, 'Synthetic Biology Standards and Intellectual Property' (The National Academies, 2012) <http://synberc.org/sites/default/files/Torrance%20&%20Kahl_Syn%20Bio%20Standards%20and%20IP_NAS%20Report%202012.pdf>.

Trouwborst A (2006), *Precautionary Rights and Duties of States*, Martinus Nijhoff Publishers.

Tucker J B (2010), 'Double-Edged DNA: Preventing the Misuse of Gene Synthesis', **26**(3) *Issues in Science and Technology* 23.

Tucker J B and R A Zilinskas (2006), 'The Promise and Perils of Synthetic Biology', Spring *The New Atlantis* 25.

Tullis T K (2005), 'Application of the Government License Defense to Federally Funded Nanotechnology Research: The Case for a Limited Patent Compulsory Licensing Regime', **53** *UCLA L. Rev*. 279.

Tumpey T M et al (2005), 'Characterization of the Reconstructed 1918 Spanish Influenza Pandemic Virus', **310**(5745) *Science* 77.

UK Biobank, 'UK Biobank Ethics and Governance Framework', (UK Biobank, 2007), <http://www.ukbiobank.ac.uk/wp-content/uploads/2011/05/EGF20082.pdf?phpMyAdmin=trmKQlYdjjnQIgJ%2CfAzikMhEnx6>.

UK Stem Cell Bank, 'Code of Practice for the Use of Human Stem Cell Lines', (UK Stem Cell Bank, 2010), <http://www.nibsc.org/asset.ashx?assetid=36019c52-bba2-497b-903f-df33e5c39dbc>.

University of California Berkeley (2012) 'Jon Christopher Anderson', Berkeley Research, <http://vcresearch.berkeley.edu/jon-christopher-anderson>.

Upbin B, 'The Bioweathermap Sees the Germs Around Us', (13 July 2011), *Forbes* http://www.forbes.com/sites/bruceupbin/2011/07/13/the-bioweathermap-sees-the-germs-around-us/.

Van den Belt H (2013), 'Synthetic Biology, Patenting, Health and Global Justice', **7**(3) *Systems and Synthetic Biology* 87.

Van Zimmeren E et al (2011), 'Patent Pools and Clearinghouses in the Life Sciences', **29**(11) *Trends in Biotechnology* 569.

Venter J C (2007), *A Life Decoded: My Genome, My Life*, Viking.

Venter J C et al (2006), *US Patent Application No*. 11/635355.

Venter J C, 'Designing and Synthesising Life', (Speech delivered at Ted Talks, 2008) <http://www.youtube.com/watch?v=nKZ-GjSaqgo>.

Venter J C, *'Evidence to The Presidential Commission for the Study of Bioethical Issues, Synthetic Biology'*, (8 July 2010), Washington DC.

Venter J C 'First Self-Replicating Synthetic Bacterial Cell', (Speech delivered at J Craig Venter Institute Press Conference 20 May 2010) <http://www.jcvi.org/cms/research/projects/first-self-replicating-synthetic-bacterial-cell/video>.

Venter J C 'Synthetic Life', (Speech delivered at Ted Med 2009, San Diego, October 2009) <http://www.youtube.com/watch?v=dvBV2qnSZwo>.

Venter J C et al (2010), 'Synthetic Genomics: Where Next?' **2762** *New Scientist* 3.

Vergano D, 'Scientists Create 1st Bacteria Strain from Man-made DNA', *USA Today* (20 May 2010).

Von Hippel E (2005), *Democratizing Innovation*, MIT Press.

Von Schomberg R (2006), *'The Precautionary Principle and its Normative Challenges'* in E Fisher, J Jones and R von Schomberg (eds),

Implementing the Precautionary Principle: Perspectives and Prospects, Edward Elgar Publishing.

Wagner R, 'Evidence to The Presidential Commission for the Study of Bioethical Issues, Synthetic Biology', (14 September 2010), Philadelphia, Penn.

Walsh J P et al (2005), 'View from the Bench: Patents and Material Transfers', **309**(5743) *Science* 2002.

Way J C et al (2014), 'Integrating Biological Redesign: Where Synthetic Biology Came from and Where It Needs to Go', **157**(1) *Cell* 151.

Weinberger S (2010), 'Pentagon Turns to "Softer" Sciences', **464**(7291) *Nature* 970.

Welch M and C Gustafsson (2008), *US Patent Application No. 12/184,233*.

Wellhausen R and Mukunda G (2009), 'Aspects of the Political Economy of Development and Synthetic Biology', **3**(1–4) *Systems and Synthetic Biology* 115.

White House, 'National Bioeconomy Blueprint' (The White House, 2012), <http://www.whitehouse.gov/sites/default/files/microsites/ostp/national_bioeconomy_blueprint_april_2012.pdf>.

White House 'National Strategy for Biosurveillance' (The White House, 2012) <http://www.whitehouse.gov/sites/default/files/National_Strategy_for_Biosurveillance_July_2012.pdf>.

Wiener J B (2011), 'The Rhetoric of Precaution' in Jonathan B Wiener et al (eds), *The Reality of Precaution: Comparing Risk Regulation in the United States and Europe*, Routledge.

Wilson M P and M R Schwarzman (2009), 'Toward a New US Chemicals Policy: Rebuilding the Foundation to Advance New Science, Green Chemistry, and Environmental Health', **117**(8) *Environmental Health Perspectives* 1202.

Wimmer E (2006), 'The Test-Tube Synthesis of a Chemical Called Poliovirus: The Simple Synthesis of a Virus Has Far-Reaching Societal Implications', **7**(Special Issue) *EMBO Reports* S3.

Wohlsen M (2011), *Biopunk: DIY Scientists Hack the Software of Life*, Current.

Wolinsky H (2009), 'Kitchen Biology', **10**(7) *EMBO Reports* 683.

World Health Organization, 'Biorisk Management: Laboratory Biosecurity Guidance' (World Health Organization, 2006) <http://www.who.int/csr/resources/publications/biosafety/WHO_CDS_EPR_2006_6.pdf>.

World Health Organization, 'Laboratory Biosafety Manual 3rd Edition' (World Health Organization, 2004) <http://www.who.int/csr/resources/publications/biosafety/Biosafety7.pdf>.

Wynne B (1992), 'Uncertainty and Environmental Learning: Reconceiving Science and Policy in the Preventive Paradigm', **2**(2) *Global Environmental Change* 111.

Wyss Institute for Biologically Inspired Engineering, 'About Us', (2013) Wyss Institute for Biologically Inspired Engineering <http://wyss.harvard.edu/viewpage/about-us/about-us>.

You E H, 'Evidence to the Presidential Commission for the Study of Bioethical Issues, Synthetic Biology', (9 July 2010), Washington DC.

Young L (2010), *US Patent No. 7,704,690.*

Young L et al (2010), *US Patent No. 7,723,077.*

Zegart A B (2004), 'Blue Ribbons, Black Boxes: Toward A Better Understanding of Presidential Commissions', **34**(2) *Presidential Studies Quarterly* 366.

Zhang J et al (2011), 'The Transnational Governance of Synthetic Biology: Scientific Uncertainty, Cross-Borderness and the "art" of Governance', (BIOS Working Paper No: 4, BIOS (Centre for the Study of Bioscience, Biomedicine, Biotechnology and Society), May 2011) <http://www2.lse.ac.uk/BIOS/publications/Working%20Papers/WP4TransnationalGovernance.pdf>.

CASES

Alice Corporation Pty Ltd. v CLS Bank International et al, 134 (US Supreme Court, 2014) 2347

Alpharma Inc v Council of the European Union (T-70/99) [2002] ECR II-03495

American Fruit Growers v Brogdex Co., 283 US (S Ct, 1931) 1

Appellate Body Report, *Measures Affecting the Importation of Apples (Japan v United States)*, DS 245 (26 November 2003)

Ariad Pharmaceuticals Inc. v Eli Lilly and Company, 598 F 3d (CA Fed (Mass), 2010) 1336

Association for Molecular Pathology et al v Myriad Genetics Inc. et al, 133 US (S Ct, 2013) 2107

Association for Molecular Pathology v United States Patent and Trademark Office, 702 F Supp 2d (SDNY, 2010) 181

Association for Molecular Pathology v United States Patent and Trademark Office, 653 F 3d (CA Fed (NY), 2011) 1329

Association for Molecular Pathology v United States Patent and Trademark Office, 689 F 3d (CA Fed (NY), 2012) 1303

Bilski v Kappos, 130 US (S Ct, 2010) 3218

Diamond v Chakrabarty, 447 US (S Ct, 1980) 303

Funk Bros Seed Co v Kalo Inoculant Co, 333 US (S Ct, 1948) 127

Gottschalk v Benson, 409 US (S Ct, 1972) 63
Graham v John Deere Co of Kansas City, 383 US (S Ct, 1966) 1
Hartranft v Wiegmann, 121 US (S Ct, 1887) 609
In re Aller, 220 F 2d 454 (CCPA 1955)
In re Bilski, 545 F 3d (Fed Cir, 2008) 943
In re Dane K Fisher v Lagudi, 421 F 3d (Fed Cir, 2005) 1365
In re Marek Z Kubin, 561 F 3d (Fed Cir, 2009) 1351
In re O'Farrell, 853 F 2d (CA Fed, 1988) 894
In re Wands, 858 F 2d (CA Fed, 1988) 731
International Tribunal for the Law of the Sea, *Responsibilities and Obligations of States Sponsoring Persons and Entities with Respect to Activities in the Area (Advisory Opinion)*, Case No 17 (1 February 2011)
KSR Intern Co v Teleflex Inc., 550 US (SCt, 2007) 398
Madey v Duke University, 307 3d 1351, 1362 (Fed Cir, 2002)
Mayo Collaborative Services Inc. v Prometheus Laboratories Inc., 132 US (S Ct, 2012) 1289
Panel Report, *European Communities – Measures Affecting the Approval and Marketing of Biotech Product (United States v European Communities)*, DS291 (29 September 2006)
Pulp Mills on the River Uruguay (Argentina v Uruguay) (Judgment) [2010] ICJ Rep 14
Reserve Mining Company v Environmental Protection Agency, 514 F 2d (Court of Appeals, Federal Circuit, 1975) 492
Telstra Corporation Ltd v Hornsby Shire Council (2006) 146 LGERA 10
Vas-Cath Inc. v Mahurkar, 935 F 2d (Fed Cir, 1991) 1555

LEGISLATION, BILLS, INTERNATIONAL AGREEMENTS AND INSTRUMENTS

5 USC (Office of the Law Revision Counsel, 2017)
15 USC (Office of the Law Revision Counsel, 2017)
16 USC (Office of the Law Revision Counsel, 2017)
18 USC (Office of the Law Revision Counsel, 2017)
28 USC (Office of the Law Revision Counsel, 2016)
33 USC (Office of the Law Revision Counsel, 2017)
35 USC (Office of the Law Revision Counsel, 2017)
42 USC (Office of the Law Revision Counsel, 2016)
36 CFR (US Government Publishing Office – e-CFR, 2017)
40 CFR (US Government Publishing Office – e-CFR, 2017)
42 CFR (US Government Publishing Office – e-CFR, 2016)

Cartagena Protocol on Biosafety to the Convention on Biological Diversity, opened for signature 29 January 2000 2226 UNTS 208, (entered into force 11 September 2003).

Constitution of the United States

Convention on Biological Diversity, opened for signature 5 June 1992 1760 UNTS 79 (entered into force 29 December 1993).

Convention on the Prohibition of the Development, Production and Stockpiling of Bacteriological (Biological) and Toxin Weapons and on their Destruction opened for signature 10 April 1972 (entered into force 26 March 1975).

Decisions Adopted by the Conference of the Parties to the Convention on Biological Diversity at its Eleventh Meeting, Annex I UN Doc UNEP/CBD/COP/11/35 (8–19 October 2012).

Marrakesh Agreement Establishing the World Trade Organization: Annex 1A Agreement on the Application of Sanitary and Phytosanitary Measures, opened for signature 15 April 1994 (entered into force 1 January 1995).

Rio Declaration on Environment and Development Annex 1 UN Doc A/CONF.1.51/26 (Vol. I) (12 August 1992).

United Nations Framework Convention on Climate Change opened for signature 9 May 1992 1771 UNTS 107 (entered into force 21 March 1994).

Universal Declaration on the Human Genome and Human Rights (United Nations Educational, Scientific and Cultural Organization, 29th General Conference, 11 November 1997).

Weapons of Mass Destruction Prevention and Preparedness Act 2012, HR 2356, 112th Congress (2012) (Bill).

Index